THE MILITARY MEMOIRS OF LIEUT.-GENERAL SIR JOSEPH THACKWELL

G.C.B., K.H.; COLONEL 16TH LANCERS

Yours most faithfully
Jos: Thackwell

THE MILITARY MEMOIRS OF LIEUT.-GENERAL SIR JOSEPH THACKWELL

G.C.B., K.H.; COLONEL 16TH LANCERS

ARRANGED FROM DIARIES
AND CORRESPONDENCE

BY COLONEL H. C. WYLLY, C.B.

AUTHOR OF "THE CAMPAIGN OF MAGENTA AND SOLFERINO"

PREFACE

It is probable that these Memoirs would never have seen the light, but for the fact that in some books, comparatively recently published, which deal with the history of the campaigns of the Sutlej and of the Indus, it appeared that on at least one occasion blame was imputed to Sir Joseph Thackwell, either owing to the writers not being in possession of a complete knowledge of all the attendant circumstances, or because they were not aware of the terms of the actual orders from Lord Gough by which the actions of Sir Joseph Thackwell were guided or controlled. General Thackwell was not only the most loyal of subordinates, but he had ever carefully refrained, even in his private correspondence and journals, from anything approaching comment or criticism of the actions or commands of his superiors. In all the pages of his diaries—bridging the lapse of years between Corunna and Gujerat—there is no word of adverse, much less of hostile, criticism. He had, too, a horror of anything like literary or journalistic controversy in regard to military operations; he strongly disapproved of officers writing for the Press—in a letter written before his son Edward came to India he warns him, as he values his own prospects and his father's good opinion, "never to write in the newspapers." The "Narrative of the Second Seikh War" was written, as the author admits, without Sir Joseph's knowledge or countenance, and it is unquestionable that its publication gave him no little annoyance. Nearly half a century has, however, passed away since the General's death, and it is thought that in justice to his memory some attempt may now well be made to show how circumstances shaped the events in which he bore a leading part, and to what extent such of his actions as have been questioned were fettered by his loyal observance of the orders of those who were responsible for the carrying out of the general scheme of operations.

Major-General W. de W. R. Thackwell, C.B., V.D., J.P., of Wynstone Place, Gloucestershire, the eldest surviving son of Sir Joseph Thackwell, has for some time past wished to publish his father's memoirs, and has unreservedly placed at my disposal for this purpose, not only all the late Sir Joseph Thackwell's journals and the manuscript record he compiled of the services of the 15th Hussars in Spain, but a mass of correspondence—private letters and copies of public documents—connected with the events of a long and particularly varied military career.

I am greatly indebted to my brother, Major Wylly, of the staff of the Royal United Service Institution, for much help in the way of reference to the literature of the wars wherein Sir Joseph Thackwell was engaged, and to Mr. William Foster, of the India Office, for assistance in regard to maps.

In the matter of the spelling of Indian place-names, I have thought it less confusing to ignore the different manner in which these are spelt by the various writers quoted, and to spell each one throughout in the way to which the ordinary reader is best accustomed.

H. C. W.

March, 1908.

LIST OF WORKS CONSULTED

Life of Sir David Baird	Hook.
Diary of Sir John Moore	Maurice.
Records of the 15th Hussars	Cannon.
Records of the 10th Hussars	Liddell.
Narrative of the Campaign in the Peninsula .	Londonderry.
Life of Lord Seaton	Moore Smith.
Passages in the Early Life of Major-General Sir G. Napier	Napier.
History of the War in the Peninsula . . .	Napier.
Campaign of the Army in Spain by Sir J. Moore.	Moore.
Life of Lord Lynedoch	Delavoye.
History of the German Legion	Beamish.
Memoirs of Viscount Combermere . . .	Knollys.
Wellington's Operations in the Peninsula . .	Butler.
How England saved Europe	Fitchett.
Wellington's Men	Fitchett.
Life of Lord Castlereagh	Alison.
Annals of the Peninsula Campaigns.	
Miscellaneous pamphlets.	
Wellington's Despatches	Gurwood.
Waterloo Letters	Siborne.
Journal of Lieutenant Woodberry.	
Cavalry in the Waterloo Campaign . . .	Wood.
Memoirs of the 18th Hussars	Malet.
The War in Afghanistan	Havelock.
The Afghan Campaign	Helsham Jones.
The Campaign in Afghanistan	Hough.
The First Afghan War	Durand.
Memorials of Afghanistan	Stocqueler.
The War in Afghanistan	Kaye.
Military Services in the Far East . . .	Mackinnon.
History of India	Keene.
The Indian Administration of Lord Ellenborough	Colchester.
Life of Hugh Viscount Gough	Rait.
The Sikhs and the Sikh Wars	Gough and Innes.

Title	Author
Despatches of Hardinge and Gough.	
Hardinge (Rulers of India)	Hardinge.
History of the 3rd Light Dragoons	Kauntze.
Life of Lord Clyde	Shadwell.
The Second Seikh War	Thackwell.
Autobiography of Sir H. Smith	Moore Smith.
Records of the 16th Lancers	Cannon.
The 9th Queen's Royal Lancers	Reynard.
The Punjab Campaign, 1848-49	Lawrence-Archer
Dalhousie (Rulers of India)	Hunter.
Barracks and Battle-fields	Caine.
Rambling Reminiscences of the Punjaub Campaign	Macpherson.
Life of Sir Hope Grant	Knollys.
Battles of Chillianwallah and Gujerat	Sir Charles Gough.
Records of the 14th Hussars	Hamilton.
The Punjab in Peace and War	Thorburn.
Decisive Battles of India	Malleson.
Parliamentary Papers.	
Life of Sir Henry Durand	Durand.
History of the Bengal Artillery	Stubbs.
Life of Sir Charles Napier	Napier.
A Year on the Indian Frontier	Edwardes.
The Calcutta Review, vol. xv.	Durand.
History of the British Empire in India	Trotter.
Essays, Military and Political	Lawrence.
History of the Sikhs	Cunningham.
Calcutta Review, vol. vi.	Edwardes.
The Sikhs	Gordon.
Memoirs of Colonel Mountain	Mountain.

CONTENTS

CHAPTER I

CHAPTER II

CHAPTER III

CHAPTER IV

CHAPTER V

CHAPTER VI

CHAPTER VII

CHAPTER VIII

CHAPTER IX

CHAPTER X

CHAPTER XI

CHAPTER XII

CHAPTER XIII

CHAPTER XIV

CHAPTER XV

CHAPTER XVI

CHAPTER XVII

CHAPTER XVIII

CHAPTER XIX

CHAPTER XX

LIST OF MAPS AND ILLUSTRATIONS

PLAN SHOWING POSITION & MOVEMENTS OF 15TH HUSSARS

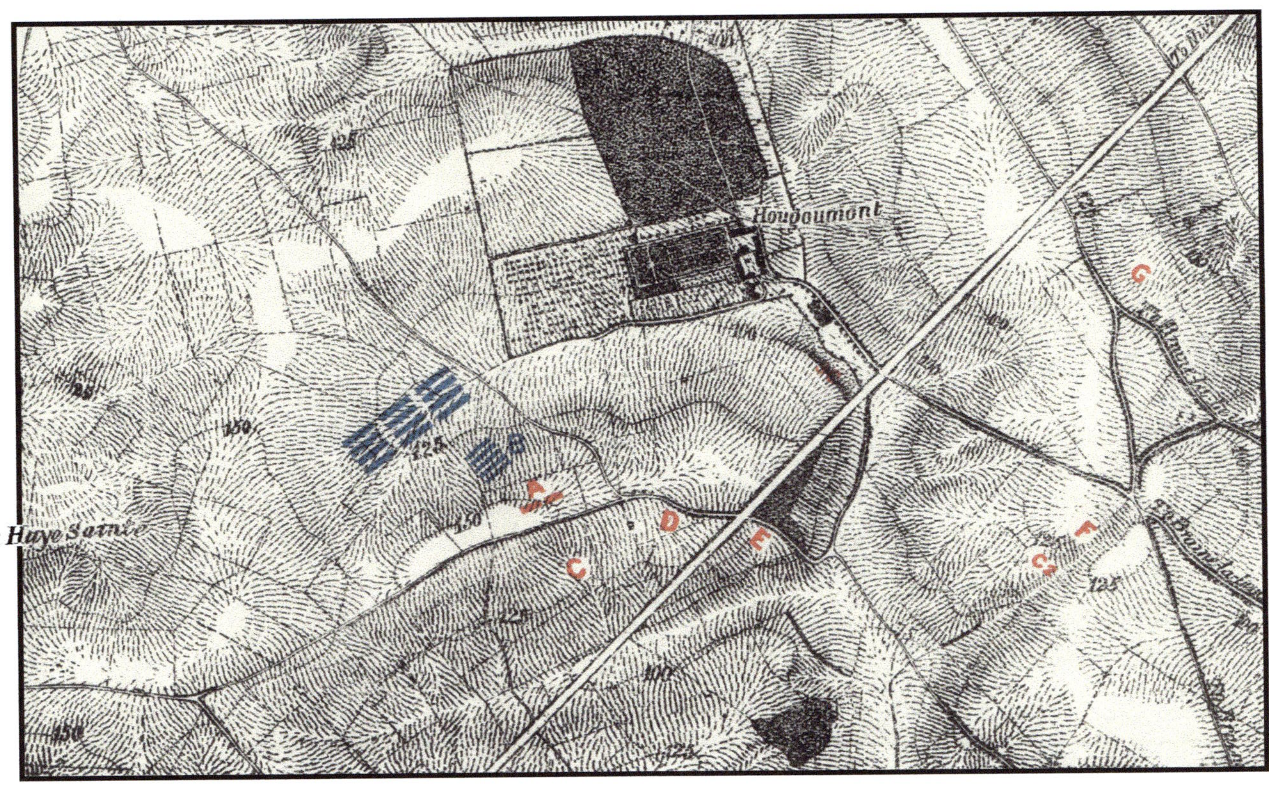

PLAN SHOWING POSITIONS & MOVEMENTS OF GRANT'S CAVALRY BRIGADE

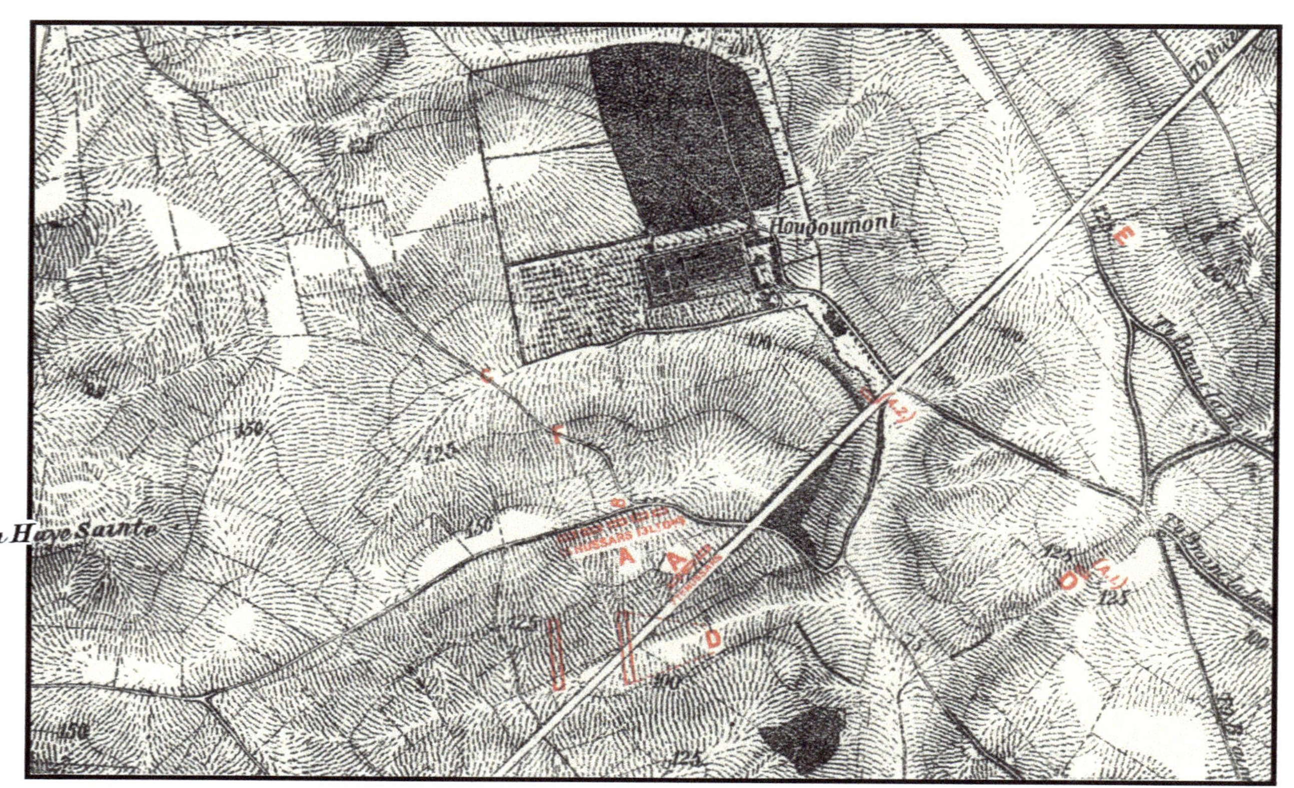

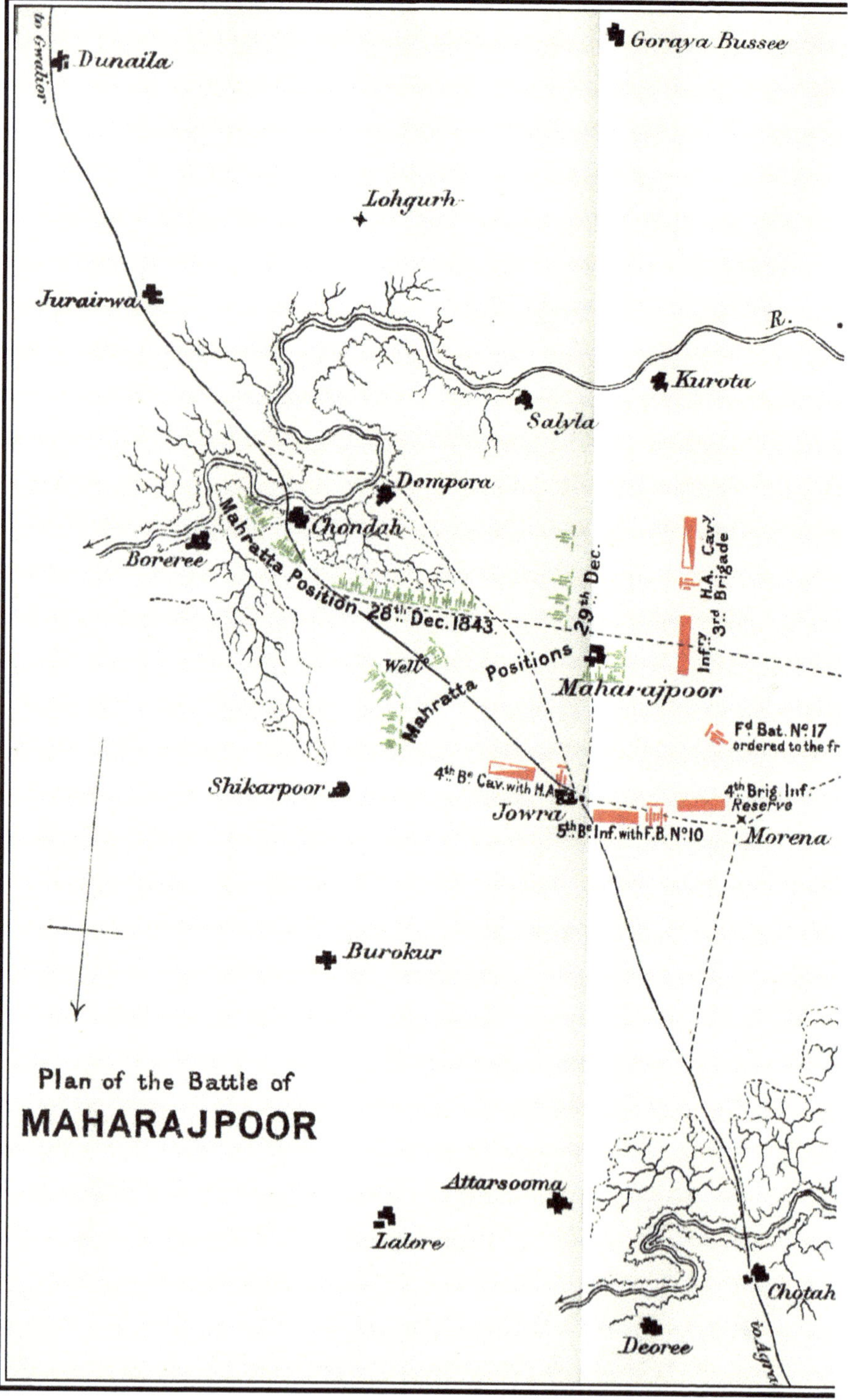

to Gwalior
Dunaila
Goraya Bussee
Lohgurh
Jurairwa
R.
Kurota
Salyla
Dompora
Chondah
Boreree
Mahratta Position 28th Dec. 1843
29th Dec.
3rd Brigade
H.A. Cavy
Infry
Well
Mahratta Positions
Maharajpoor
Fd Bat. No 17
ordered to the fr
4th Be Cav. with H.A.
Shikarpoor
Jowra
4th Brig. Inf.
Reserve
5th Be Inf. with F.B. No 10
Morena
Burokur
Plan of the Battle of
MAHARAJPOOR
Attarsooma
Lalore
Chotah
Deoree
to Agra

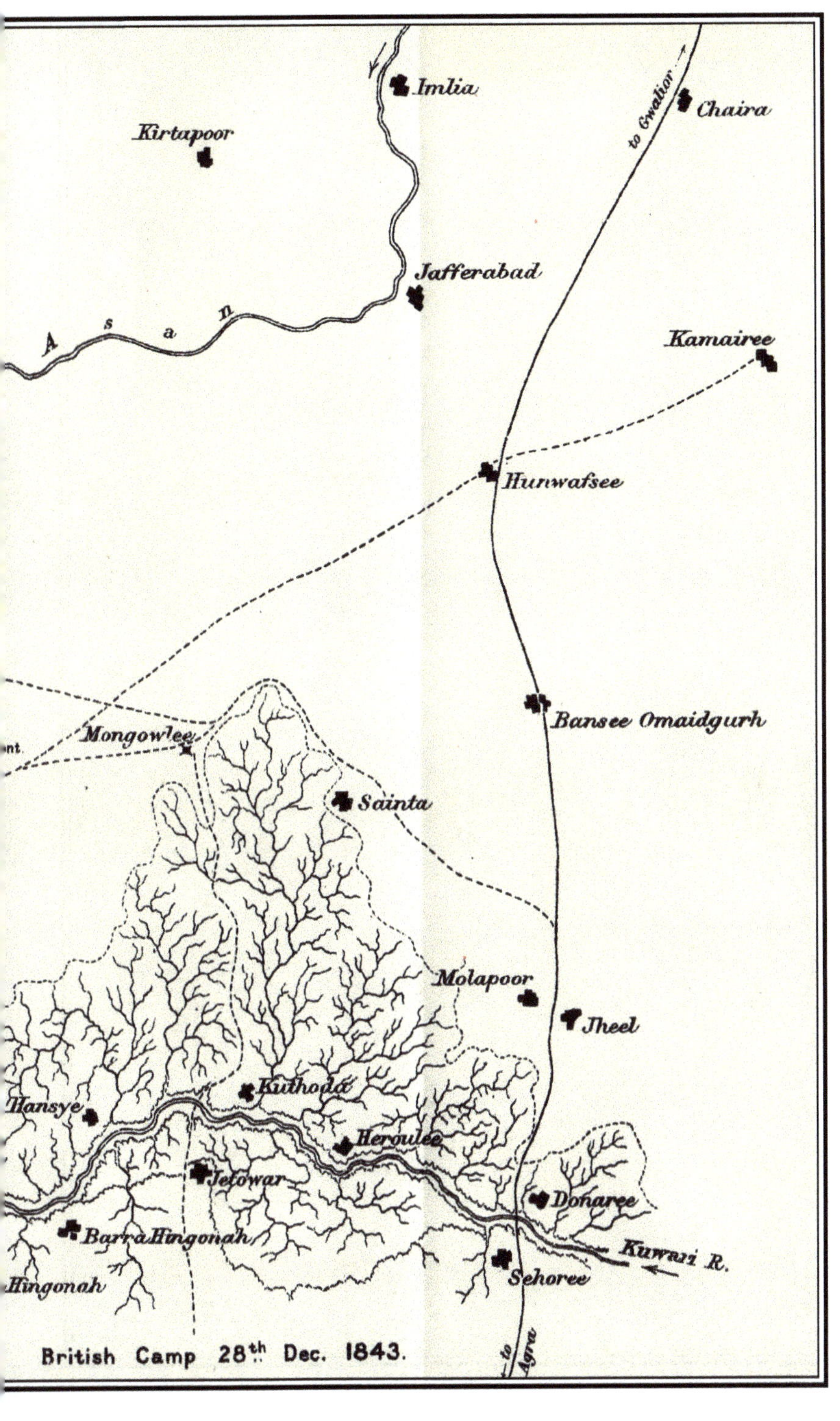

Imlia
Chaira
to Gwalior
Kirtapoor
Jafferabad
A s a n
Kamairee
Hunwafsee
Bansee Omaidgurh
Mongowlee
Sainta
Molapoor
Jheel
Kuthoda
Hansye
Heroulee
Jetowar
Donaree
Barra Hingonah
Kuwari R.
Sehoree
Hingonah
to Agra
British Camp 28th Dec. 1843.

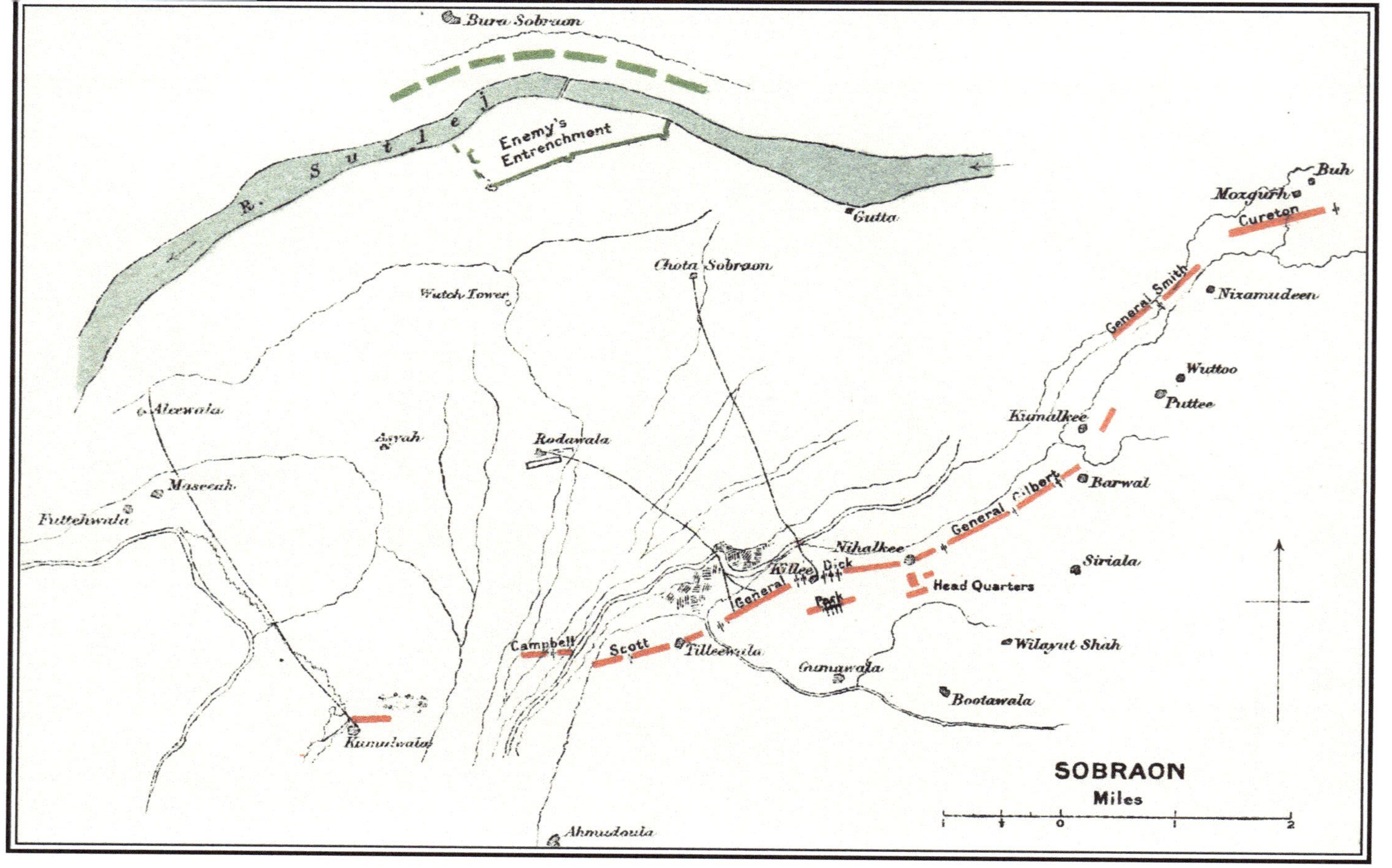
Bura Sobraon
R. Sutlej
Enemy's Entrenchment
Gutta
Buh
Mozgurh
Cureton
General Smith
Nizamudeen
Chota Sobraon
Watch Tower
Wuttoo
Puttee
Kumalkee
Aleewala
Assah
Rodawala
Barwal
Maseeah
Futtehwala
General Gilbert
Nihalkee
Killee
Dick
General
Peck
Head Quarters
Siriala
Campbell
Scott
Tilleewala
Wilayut Shah
Gumawala
Bootawala
Kunnuwala
Ahmudoula
SOBRAON
Miles
0
1
2

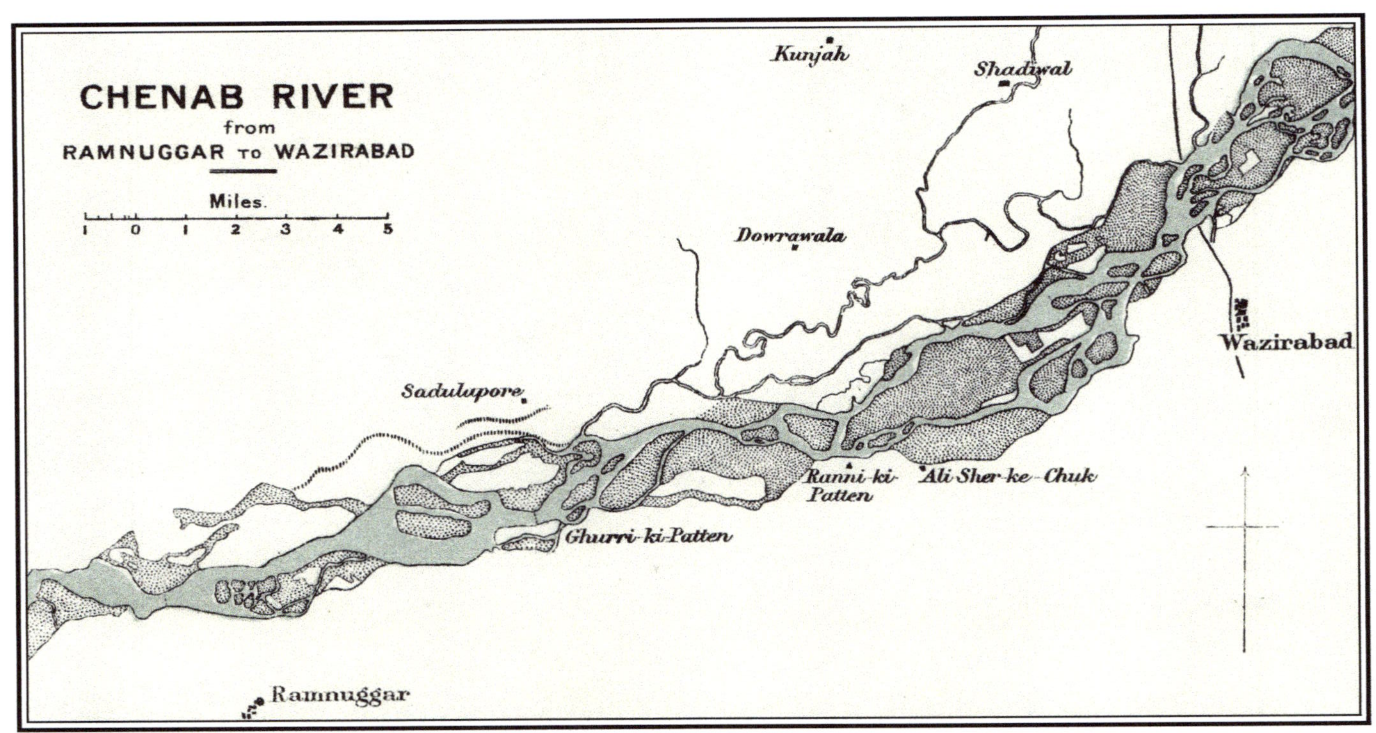
CHENAB RIVER
from
RAMNUGGAR TO WAZIRABAD
Miles.
1 0 1 2 3 4 5
Kunjah
Shadiwal
Dowrawala
Wazirabad
Sadulupore
Ranni-ki-Patten
Ali-Sher-ke-Chuk
Ghurri-ki-Patten
Ramnuggar

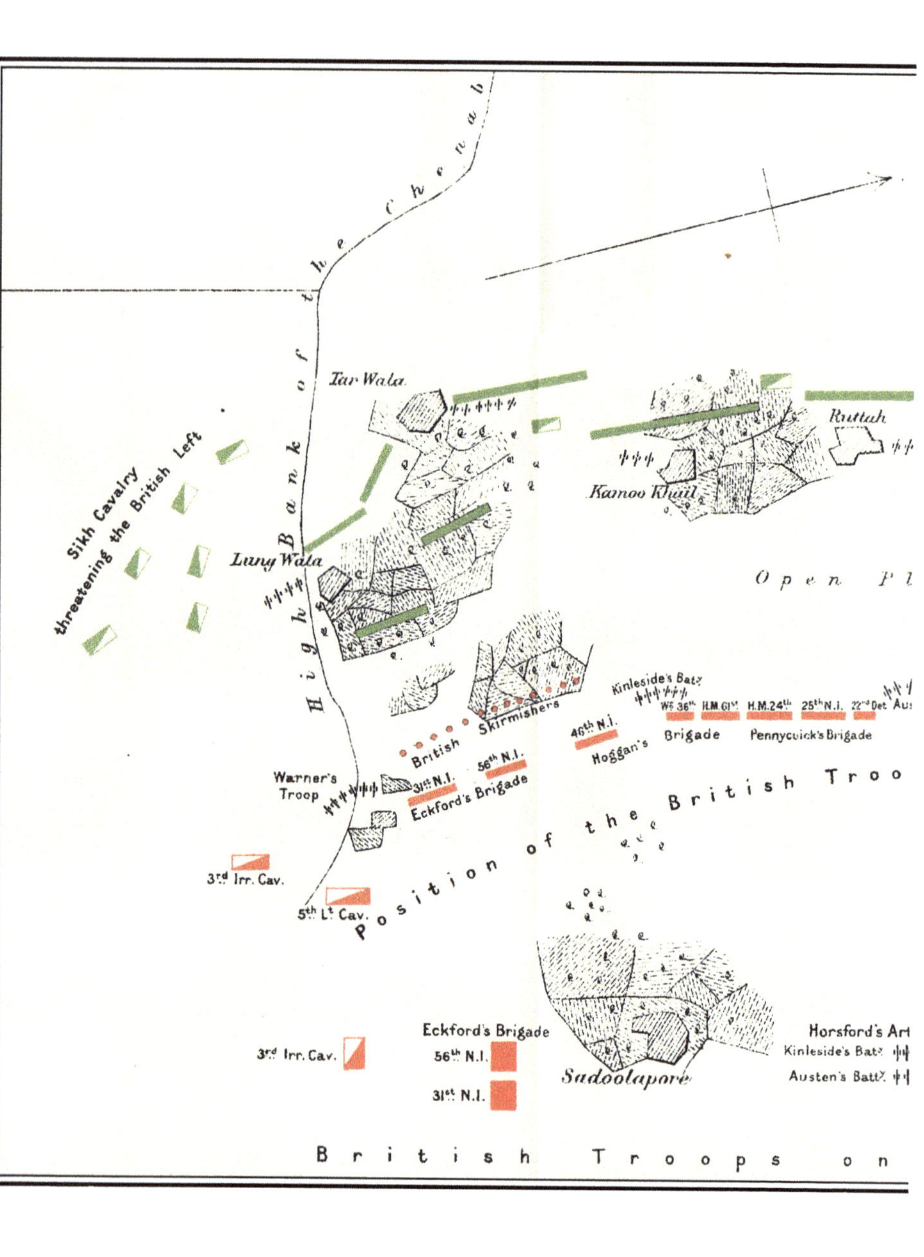

the Chenab
High Bank of the Chenab
Tar Wala
Ruttah
Kamoo Khuil
Sikh Cavalry threatening the British Left
Lung Wala
Open Pl
Kinleside's Baty
W's 36th
H.M.61st
H.M.24th
25th N.I.
22nd Det
Aus
Brigade
Pennycuick's Brigade
British Skirmishers
46th N.I.
Hoggan's
56th N.I.
31st N.I.
Eckford's Brigade
Warner's Troop
Position of the British Troo
3rd Irr. Cav.
5th Lt Cav.
Eckford's Brigade
56th N.I.
31st N.I.
3rd Irr. Cav.
Sadoolapore
Horsford's Art
Kinleside's Batty
Austen's Batty
British Troops on

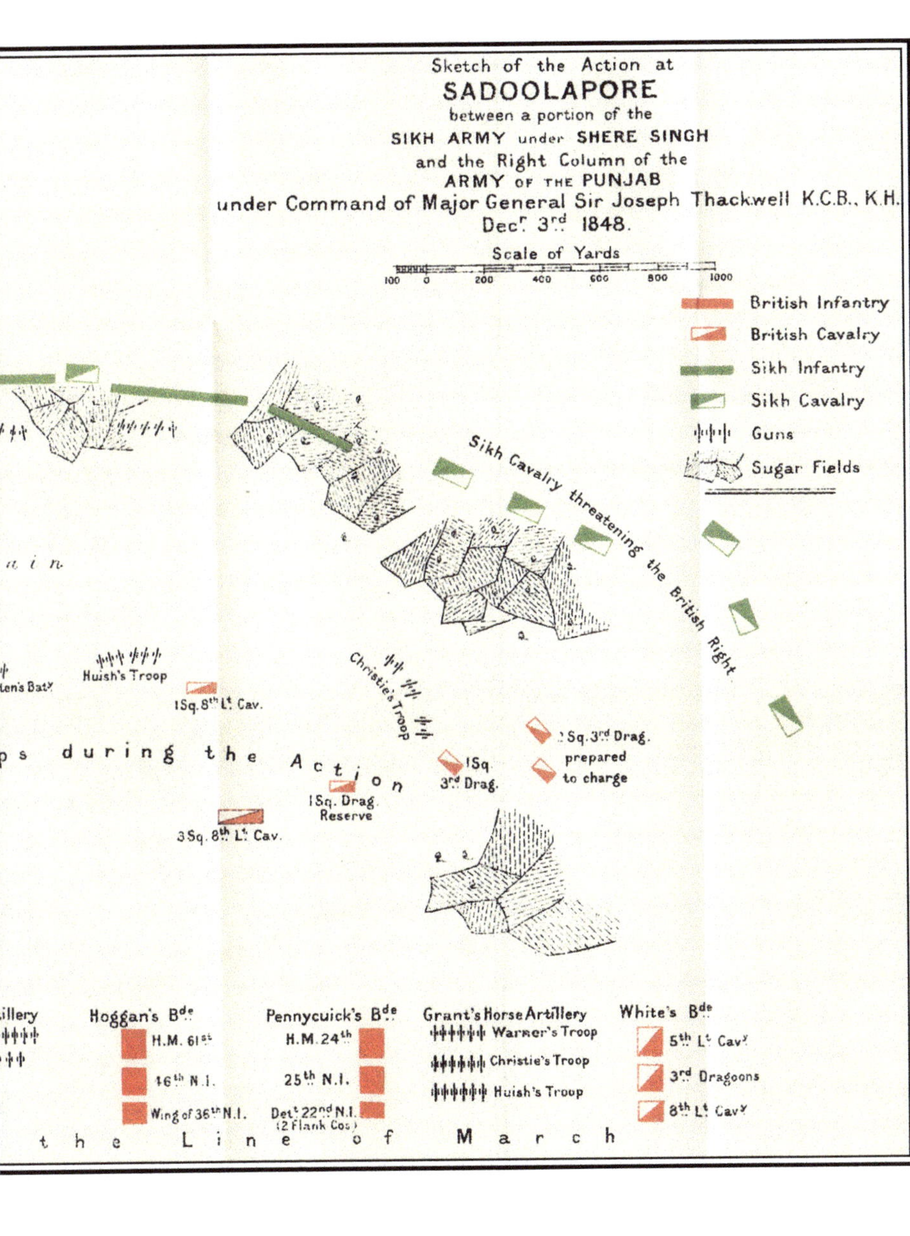

Sketch of the Action at
SADOOLAPORE
between a portion of the
SIKH ARMY under SHERE SINGH
and the Right Column of the
ARMY OF THE PUNJAB
under Command of Major General Sir Joseph Thackwell K.C.B., K.H.
Decr. 3rd 1848.
Scale of Yards
100 0 200 400 600 800 1000
British Infantry
British Cavalry
Sikh Infantry
Sikh Cavalry
Guns
Sugar Fields
Sikh Cavalry threatening the British Right
Huish's Troop
1Sq. 8th Lt. Cav.
Christie's Troop
2 Sq. 3rd Drag. prepared to charge
1Sq 3rd Drag.
1Sq. Drag. Reserve
3Sq. 8th Lt. Cav.
Hoggan's Bde
H.M. 61st
16th N.I.
Wing of 36th N.I.
Pennycuick's Bde
H.M. 24th
25th N.I.
Det. 22nd N.I. (2 Flank Cos.)
Grant's Horse Artillery
Warner's Troop
Christie's Troop
Huish's Troop
White's Bde
5th Lt. Cavy.
3rd Dragoons
8th Lt. Cavy.

ROUGH SKETCH

OF THE

BATTLE OF GOOJERAT

FOUGHT ON THE

21st FEBRUARY, 1849.

A B 1st Position of the Army when deployed into line from Brigade Columns.

C D 2nd Position. The Artillery being in action.

E 3rd Position of the left wing of Infantry, left brought forward, the Enemy's right wing of Cavalry having been driven into retreat towards the Jelum by the charge of the Scinde Horse and two Squadrons of Lancers.

G Position of the Sikh Army with their batteries well to the front, and strongly occupying the three Kalra villages with advanced troops at other points.

H White's Brigade of Cavalry at the Baradurree cannonading the Enemy's Infantry Columns.

J The Sikh Encampment with much baggage and tents left standing.

K The Sikh Army retreating in mass of columns, forced from the Jelum road and eventually from that to Bimber, by our Cavalry of the left acting on their flank whilst that of our right pressed them in the rear, and in the pursuit they were completely dispersed.

M British Camp after the Battle.

British Infantry. British Cavalry. British Guns.
Sikh ,, Sikh ,, Sikh ,,

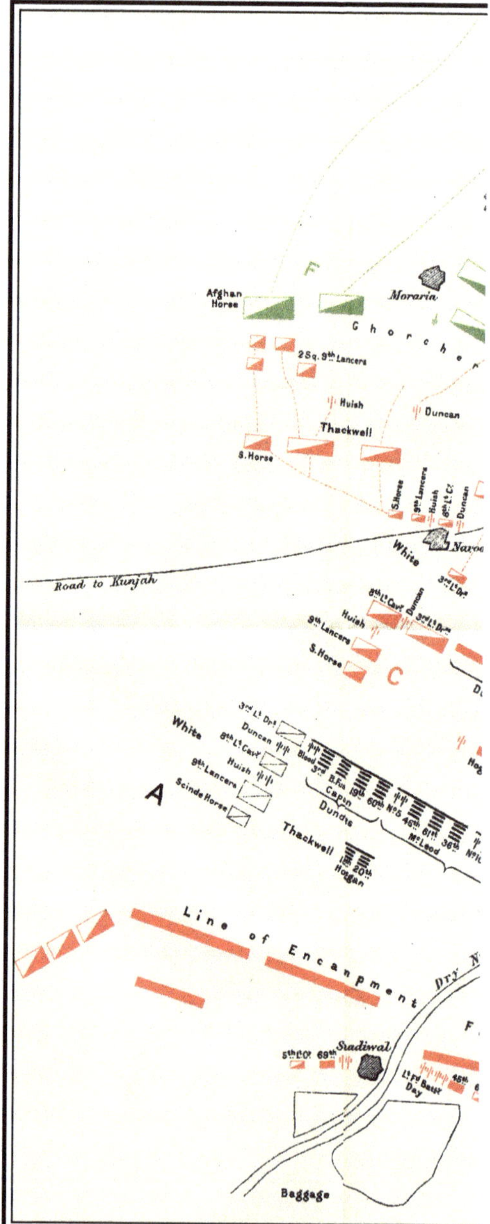

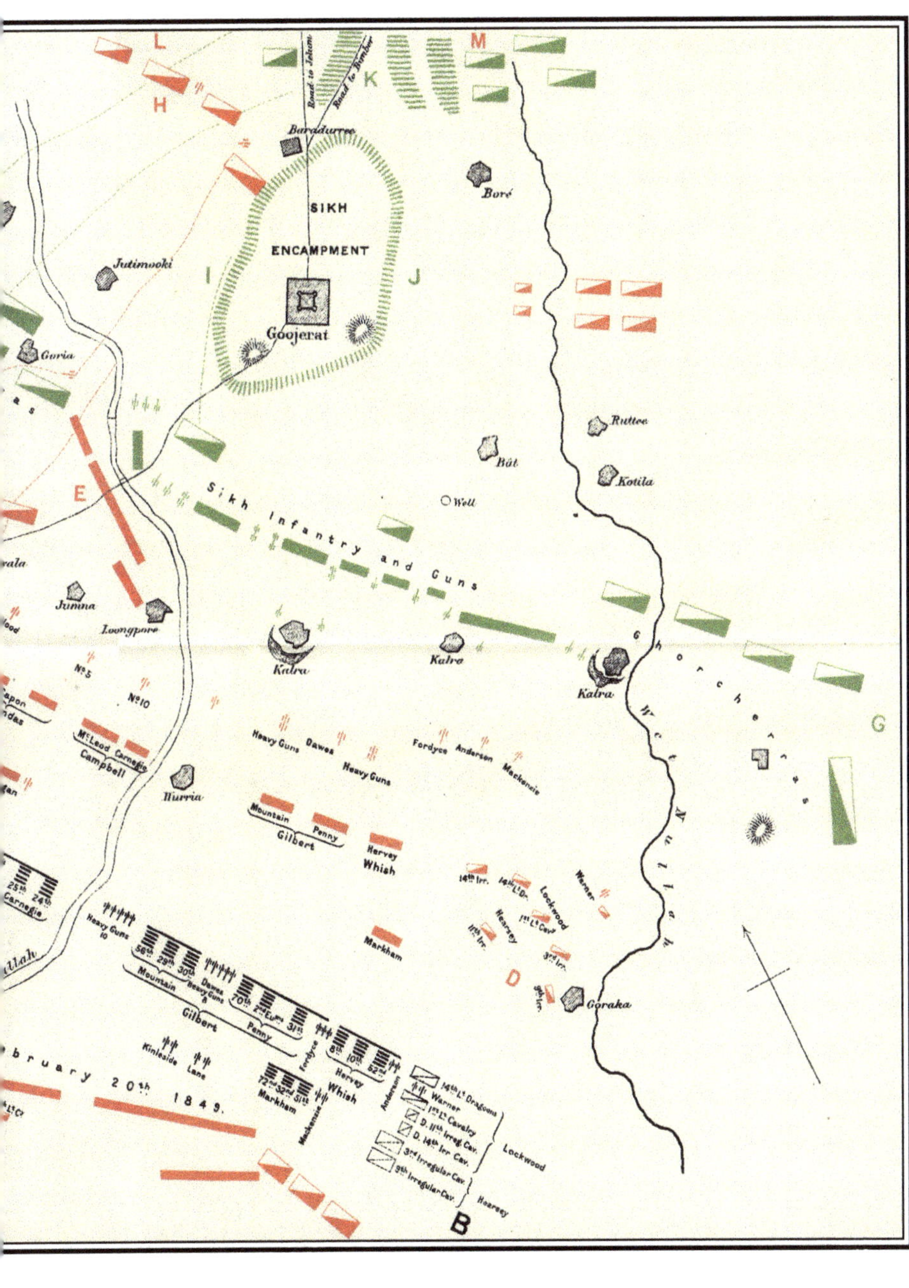

L
H
M
K
Road to Jelum
Road to Bimber
Baradurree
SIKH
ENCAMPMENT
Boré
Jutimooki
I
J
Goojerat
Goria
Ruttee
Bût
Kotila
E
Well
Sikh Infantry and Guns
Jumna
Loongpore
Kalra
Kalra
Kalra
Ghorcheras
G
No 5
No 10
McLeod
Carnegie
Campbell
Heavy Guns
Dawes
Heavy Guns
Fordyce
Anderson
Mackenzie
Hurria
Wet Nullah
Mountain
Penny
Gilbert
Hervey
Whish
14th Irr.
14th Lt Dn.
Lockwood
Warner
1st Lt Cavy
Hearsey
11th Irr.
3rd Irr.
9th Irr.
Markham
D
Goraka
25th
24th
Carnegie
Heavy Guns
10
56th
29th
30th
Mountain
Dawes
Heavy Guns
70th
2nd Eurns
31st
Gilbert
Penny
Kinleside
Lane
Fordyce
8th
10th
52nd
Hervey
Whish
72nd
32nd
51st
Markham
Mackenzie
Anderson
bruary 20th 1849.
14th Lt Dragoons
Warner
1st Lt Cavalry
D. 11th Irreg. Cav.
D. 14th Irr Cav.
3rd Irregular Cav.
9th Irregular Cav.
Lockwood
Hearsey
B

A
Sketch Map
of the
STATES
Bordering on the Indus River,
and of
AFGHANISTAN
shewing the Routes marched by the
Army of the Indus
Composed of
TROOPS FROM THE PRESIDENCIES
of
Bengal and Bombay
Under the personal Command of
HIS EXCELLENCY
Lieutenant General Sir John Keane G.C.B. & G.C.H.
Commander in Chief
Scale of British Miles to an Inch
By A. M. Becher Lieut 61st Regt N.I. D. A. Qr. Mr. Genl.
BELOOCHISTAN
SINDH
Shikarpoor
Bukkur
Khyrpoor
Larkhana
Bickaneer
Jessulmere
Jodhpoor
Nusserabad
Ajmere
Kishenghur
Beawr
Sukerkund
Hyderabad
Omerkote
Jurruk
Lyaree
Sonmeanee
Meerpoor
Sirohee
Bhooj
Mandavee
Moowa
Rahaunpoor
Nouanuggur
Baroda
Cambay
Gogo
Baroach
Surat
Diu
Damaun
Nassuk
Bassun
Bhewndy
Panwell
BOMBAY
ROUTE
BOMBAY COLUMN

HERAT
BALKH
KOONDOOZ
Khooloom
MOORAD BEG'S TERRITORIES
HINDOO KOOSH
CABUL
Ghuznee
Kandahar
Kilat i Ghiljee
Furrah
Girishk
Seistan
Gurumseer
DESERT
Shorawuk
Noosky
Kilat
Dadur
Pisheen
INDEPENDENT TRIBES
Peshawur
Attok
Cashmere
Jelalabad
Jhelum
Lahore
Umritsir
Ferozepoor
Leia
Dera Ishmael Khan
Dera Ghazi Khan
Mooltan
Bahawulpoor
Ahmedpoor
Khanpoor
Ooch
Indus River
Chenab
Helmund River
PUNJAB
ROUTE OF BENGAL COLUMN

Military Memoirs of Lieutenant-General Sir Joseph Thackwell

CHAPTER I

JOSEPH THACKWELL, the subject of these memoirs, was born on the 1st February 1781 at Rye Court, almost under the shadow of the Malvern Hills, and in that picturesque and historically interesting portion of the county of Worcester which marches with Gloucestershire, and which is known as Malvern Chase. He was the fourth son of John Thackwell, of Rye Court and Moreton Court, Lord of the Manor of Birtsmoreton and Berrow, and was descended from an old Worcestershire family. His mother was Judith, daughter of J. Daffy, of Maysington, a descendant of the Egyoke family.

There is little or no record of young Thackwell's boyhood; he appears, however, to have been educated chiefly in Worcester, and when barely seventeen and a half he was appointed to the Worcestershire Regiment of Provisional Cavalry, in which his commission as Cornet signed by Lord Coventry, Lord-Lieutenant of the county, bears date 18th August 1798. The Provisional Cavalry was a short-lived force, which had been called into existence just ten years previously when the alarm of invasion was at its height, and appears to have been in some degree a Home Defence Regular Cavalry for service in the British Isles only—being quite distinct from either the Yeomanry or the Volunteer Cavalry. Fortescue states in his "History of the British Army" that owners of horses kept for riding or carriage were "required to provide one trooper and horse for every ten of such horses, while those that possessed fewer were lumped together to provide their horsemen jointly. This Provisional Cavalry was entitled to pay if embodied, and was reckoned to comprise fifteen thousand men." The Worcestershire Regiment had been embodied in 1797, and when Cornet Thackwell joined them in Worcester the corps was commanded by Lieutenant-Colonel Hon. John Somers Cocks. Thackwell was posted on appointment to Captain Parker's troop—the regiment comprised but four troops—and his brother subaltern was a Lieutenant Thomas Garden.

On the 4th November the regiment marched from Worcester *viâ* Shrewsbury and Bridgnorth to Chester, and about this time Captain Parker appears to have resigned his commission, the command of his troop devolving upon Captain Thomas Webb.

During April and May of the following year the Worcestershire Regiment of Provisional Cavalry had its headquarters in Liverpool, and was there commanded by Major Bromley, the commanding officer being absent at Westminster attending to his parliamentary duties; at this time Thackwell was on detachment at Wigan, and later—on the regiment moving to Shrewsbury—he seems to have been detached in Liverpool and afterwards in Preston. On the 12th September 1799 Thackwell was promoted to the rank of lieutenant, and transferred to Major Henry Bromley's troop; and on the 7th of the following month the regiment moved by march route to Liverpool, under orders from Major-General O. Nicholls, and thence embarked for service in Ireland, where, "in greater measure than in other countries, the French Revolution had stirred inert minds to activity, and active minds to precipitation." The embarking strength of the four troops would appear to have been 187 men, since the pay-list of that date contains a credit for £294 10*s.* 6*d.* "embarkation money," at the rate of £1 11*s.* 6*d.* per man. The regiment disembarked at the Pigeon House Dock, remained for a short time in Dublin, and moved on the 14th November to Carlow.

At this period the British army contained an enormous variety of regiments—"regular regiments for general service, regular regiments for European service, regular regiments for home service, invalid companies and other corps for garrisons at home and abroad, militia, provisional cavalry, yeomanry, volunteers, associations of cavalry, and associations of infantry,"[1] coloured levies and foreign troops; and yet the military authorities of those days seem to have found the time for the in-

[1] Fortescue.

dulgence of a taste—one moreover which they have bequeathed in full measure to their successors in more modern times—for constantly and bewilderingly changing the titles of regiments. Thus the corps in which young Thackwell was now serving had commenced its existence in 1796 as the Worcestershire Regiment of Provisional Cavalry; barely three years later it became the Worcestershire Regiment of Light Dragoons, and finally in January 1800 its title was once again altered to that of the Worcestershire Regiment of Fencible Cavalry.

The regiment did not remain long in Ireland, and I have been unable to discover that it was ever actively engaged against the rebels; early in 1800 it again left Dublin, marched from Liverpool to Bewdley and thence to Worcester, where on the 12th April the regiment was finally disbanded, Joseph Thackwell being gazetted on the 23rd of that month Cornet by purchase in the 15th Light Dragoons, whose headquarters were then stationed at Canterbury, and joining a troop quartered at Ramsgate on or before the 15th May.

At this period the 15th King's Light Dragoons bore a reputation second to none in the British—or indeed in any cavalry. As Eliott's Light Horse the regiment had taken part, just forty years before, in its maiden battle, and the name "Emsdorff" commemorates their gallantry and their heavy losses during that engagement in the Seven Years' War, when going into action 450 strong they lost in killed and wounded 4 officers, 125 men and 126 horses, and "five Battalions of French were defeated and taken by this regiment with their Colours and nine pieces of cannon on the plains of Emsdorff July the sixteenth 1760."[1] But when young Thackwell was appointed there were officers still serving in the regiment who had taken part in an even greater feat of arms; Aylett was there who had led the three squadrons of the Fifteenth at Villers en Cauchies; there too were Calcraft, who had ridden with him, and Robert Wilson—this last in after-years the British Military Commissioner with the Russian Army during Napoleon's disastrous campaign of 1812. The regiment had even now but just returned from active service in Holland under H.R.H. the Duke of York, where their behaviour had earned for them the right to bear the word "Egmont-op-Zee" upon their guidons and appointments.

Truly a good school for a young cavalryman anxious so to train his mind and body that he might rise to high rank in this particular branch of the service!

When Thackwell joined the 15th Light Dragoons, then commanded by Lieutenant-Colonel George Anson, he was at first posted to Captain Sandford Lambe's troop, but in the following June he was transferred to that commanded by Major and Lieutenant-Colonel Aylett, with whom he seems to have remained until promoted lieutenant on the 13th June 1801, when he joined Captain James Mansfield's troop at Hertford.

During the autumn of 1800 and the winter of

[1] Inscription on a helmet issued to Eliott's Light Horse after the Seven Years' War, and now in the Officers' Mess, 15th Hussars.

1800–1 the Fifteenth were at Dorchester, but in the early part of April the regiment was ordered to the neighbourhood of Taunton in consequence of serious riots in Somersetshire, two troops being sent to Launceston and Bodmin. Just now the Fifteenth must have presented a very fine appearance; it was already 800 strong, but the ten troops were now each augmented to 90 men and horses, and in June the strength of the regiment was again further increased until its establishment was 950 rank and file, with 1,012 horses. It seems clear that already thus early Thackwell had shown the zeal and ability which were ever afterwards to distinguish him; for in September 1801, it being determined to attach two galloper guns to each regiment of light cavalry, he was sent in charge of a party to Brighton to go through a special course with Horse Artillery. In the spring of the following year, however, it seemed as though the career of our young officer was likely to be prematurely blighted; the preliminaries of peace had been already signed in London in the previous October, and the treaty of Amiens—which was really nothing more than a suspension of arms—was signed on the 25th March 1802. The immediate result of a peace was the reduction of the army, and the 15th Dragoons, which at the moment were only twelve men short of their establishment, was one of the first corps to suffer. In May 325 men were discharged and 244 horses were transferred or cast, while in June two troops were reduced, and the remaining eight were placed on an establishment of 64 men and 54 horses, when the two junior captains, lieutenants, and cornets were placed on the half-pay list. Among these officers thus cast adrift was Joseph Thackwell, and on the 25th June he had severed—fortunately only temporarily—his connection with the regiment with which, in the future, he was to see so much and so distinguished service.

Already in March 1803 the hostile attitude of the First Consul was the cause of an augmentation in the strength of the various regiments of the British Army; the Fifteenth was brought up to a strength of 75 men and 65 horses per troop, and was stationed at different points on the Kentish coast, and by July the establishment was further raised to a total of 85 men for each of the eight troops.

Thackwell appears to have been re-posted to his old regiment on the 20th April 1804, but seems for a time to have been supernumerary to the establishment, since it is not until August that his name is to be found on the pay list as attached to a troop. In this month, however, two additional troops, making a total of ten, were placed on the establishment, and Thackwell was then posted to that nominally commanded by Captain Leitch, who had been adjutant of the Fifteenth when Thackwell first joined. As Leitch, however, was absent on the staff at this time as aide-de-camp to the Duke of Cumberland, the troop—which was at Blandford, the headquarters of the regiment being at Winchester—was, for some time at least, commanded by Lieutenant Thackwell.

In the beginning of March 1805 the establishment of each troop was raised to 106 men and the same number of horses; and in July the regiment proceeded to Weymouth, where a very large force had been assembled, and where the Fifteenth appears to have been brigaded with the 3rd Hussars of the King's German Legion. The regiment remained at Weymouth—in Radipole Barracks—during the succeeding year, and here at last Thackwell had to hand over the troop he had so long commanded as a subaltern to a Captain Augustus Heiliger, who was brought in from the 2nd Light Dragoons of the Legion. It was while quartered at Weymouth that the Fifteenth were converted into hussars. On the 9th April 1807 Captain Loftus was promoted from the Fifteenth to a majority in the 38th Foot; but Thackwell's promotion to the vacant troop, which should have followed, was not notified until the 11th July, although the date of his promotion to captain was afterwards antedated to the earlier date. He was appointed to command I. Troop. Meanwhile the regiment had begun to hope that they were again to see service in the field, for on the 21st May an order was received from the Horse Guards to hold themselves in readiness to embark for foreign service.

For some reason, however, the order was cancelled, and it was not until the autumn of the year following that the wishes of all ranks were to be realised. In July 1807 the 15th Hussars had moved to Woodbridge Barracks and Wickham Market, and here the Hussar Brigade—7th, 10th, and 15th Hussars—was got together under command of Major-General Lord Paget. The spring and summer of 1808 were spent in Essex, and then at last, early in October, an order was received for eight troops, each of 85 men and horses, to be held in readiness for foreign service.

The reason for this movement was the events which had been transpiring in Portugal. On the 21st August the battle of Vimiera had been fought, and on the 30th a treaty, known as the Convention of Cintra, had been concluded at Lisbon; Sir Hew Dalrymple had been recalled to England; Sir Arthur Wellesley and, later, Sir Harry Burrard had quitted the army in the field; and the forces in the Peninsula had been left under the command of Sir John Moore, who had landed on the very day of the battle which closed that phase of the campaign. It was the 6th October before a despatch containing the first determinate plan of campaign arrived at Lisbon, where Sir John Moore then was.

> "Thirty thousand infantry and five thousand cavalry were to be employed in the north of Spain. Of these, ten thousand were to be embarked at the English ports, the remainder to be composed of regiments drafted from the army then in Portugal. Sir John Moore was to command, and was authorised to unite the whole by a voyage round the coast or by a march through the interior."[1]

Sir David Baird was directed to take under his orders the force destined to be added to that already under Sir John Moore in the Peninsula,

[1] Napier.

and of which the greater part was collecting at Cork and Falmouth—the cavalry at Portsmouth.

On the 20th the regiment began its march for the port of embarkation, remained a few days at Guildford *en route*, and finally embarked at Portsmouth on the 28th and 30th October. The embarking strength was 27 officers, 36 sergeants, 682 rank and file, and 682 horses. Captain Thackwell was the junior but one of his rank. The 7th and 10th Hussars, completing the brigade, had already set sail some days previously, and with them was the brigadier, General John Slade, who had served in the 10th until he had attained the rank of second lieutenant-colonel, and under whom the 15th Hussars had already served when at Weymouth.

The voyage across the Bay was stormy, and twenty-two horses died; but between the 12th and 15th November the men and baggage of the regiment were safely landed at the pier at Corunna—the horses being towed ashore at the sterns of the boats—and on the 21st the 15th Hussars commenced their march up-country.

The strength of the Fifteenth marching out of Corunna is given in the "states" as 527 rank and file—total of Hussar Brigade, 1,538.

CHAPTER II

Sir David Baird, with the bulk of his division, had arrived at Corunna on the 13th October, but his disembarkation had been very greatly delayed by the action of the Spanish authorities. He had expected to find every arrangement made for the reception and accommodation of his troops, but he speedily learnt that not only was the arrival of the British force entirely unexpected, but that the Junta of Galicia, then sitting at Corunna, did not consider itself authorised to receive the English troops, or to permit their disembarkation without the sanction of the Supreme Government, and the projected landing was consequently delayed to obtain this permission.

Sir John Moore had written to Sir David from Lisbon on the 12th October, the day before the ships conveying the reinforcements from England had anchored at Corunna, stating that he had decided to march the troops with him through Spain and Portugal, in preference to embarking them and joining Sir David at Corunna by sea; that he meant to move upon Almeida and Ciudad Rodrigo by three different roads; that his march thence would be on Burgos; and that "at some intermediate place, which shall afterwards be settled, our junction must be made." Sir David Baird was enjoined to place his troops in the most convenient cantonments in and about Corunna, and equip them for the field. On the 22nd Sir John Moore wrote again announcing that most of his regiments had started, but in the meantime the courier sent to Madrid had returned to Corunna conveying to Sir David permission to land in the event only of its being found impracticable to proceed by sea to Santander; if disembarkation did take place it must be in detachments of not more than 300 men at one time, which were then to be successively pushed on into Castile from the port of disembarkation. It was only by objecting in the strongest terms to this division of his force, that Sir David was eventually enabled to canton his troops in the towns and villages on the two principal roads leading to Leon and Castile, pending their proper equipment for field service. As a matter of fact, however, the leading troops of the force under Sir David Baird—the light brigade with Brigadier-General Craufurd—was actually on its march towards Astorga on the 28th October, six days only after the permission to land the troops had arrived at Corunna.

The cavalry—whose embarkation at Portsmouth had been delayed owing to the dearth of horse-transports—were fortunate in that they were able to disembark at Corunna very shortly after arrival. Marching by the great Madrid road, General Slade's brigade moved through Betanzos, Monte Guitirz, Lugo, Nogales, Villafranca and Bembibre, and closed up with the force under command of Sir David Baird, whose advanced posts were in front of Astorga. General Slade, writing in his diary of this march, states that the Hussar Brigade halted at Betanzos on the 20th, Guitirz on the 21st, Lugo on the 22nd, and Nogales on the 24th, but these dates do not, however, appear to agree with those mentioned in the records of the 10th and 15th Hussars. The marches were made by night, and the cold was very great, while between Travedelos and Rogalos the road was nothing but a track through the mountains with a precipice on one side.

Much uncertainty had existed with respect to the movements of the force under Sir David Baird. Late on the night of the 29th November Sir David had received a letter from Sir John Moore announcing the defeat and dispersion of the Spanish army under Castanos, and stating that it was in consequence his intention to retreat upon Portugal, while Baird was directed to fall back upon Corunna, whence he was to sail for the Tagus. At this time Moore estimated that the French had 110,000 men in Spain; it was certain that the Spaniards had no longer anything like an army in the field; while the British force, even if united, was not of sufficient strength to contend single-handed with the armies of Napoleon. Baird then, in pursuance of his instructions, withdrew his infantry to Villafranca, but decided to retain for a few days at Astorga the cavalry brigade which had but lately arrived at that place. On the 5th December Sir John

Moore wrote to Sir David Baird ordering him to return with his whole force to Astorga, and stating that it was his intention to do "all that the wishes of his country and duty demanded" to support the resistance which the people of Madrid seemed ready to offer to the advance of the French. In this letter Sir John Moore asked that two regiments of cavalry should be sent to him; Sir David, however, decided to send forward the whole of the Hussar Brigade which reached Benevente on the 6th, and on the 8th December arrived at Zamora on the Duero, up the left bank of which river it advanced through Toro to Tordesillas. On the 15th it threw out outposts towards Rio Seco to cover Sir John Moore's movement from Salamanca, joining hands with Sir John Moore's cavalry, consisting of the 18th Hussars and the 3rd Hussars of the King's German Legion under Brigadier-General Charles Stewart. Moore's projected advance upon Valladolid was now given up, and on the 16th the army, marching by Toro, continued its movement upon Mayorga for the purpose of effecting a junction with Sir David Baird's corps and attacking Soult.

At Mayorga on the 20th December the long anticipated junction of the forces of Moore and Baird was at last effected.

Mayorga was within three leagues of Sahagun, where it was stated that between seven and eight hundred of the enemy's cavalry were posted under General Debelle. The Fifteenth had only marched into Mayorga late that evening but were ordered to co-operate with the 10th Hussars and four guns in the endeavour to cut off the French detachment. For this purpose the 10th Hussars with the guns, under General Slade, were to move along the right bank of the Cea direct upon Sahagun, and about daybreak push into the town, whilst the Fifteenth with Lord Paget advanced by the opposite bank of the river. The 10th and 15th, with the guns and a small party of the 7th Hussars, accordingly left Mayorga at midnight on the 20th December. General Slade writes in his diary:

"A more dreadful night the troops could not be exposed to, as it was particularly dark, a severe frost, with sleet falling, and the snow drifted in many places to the depth of four feet. Many horses fell, and one man had his leg broken."

Between five and six in the morning of the 21st the advanced guard of the Fifteenth fell in with a French patrol and took five prisoners, but the rest escaping, owing to the extreme darkness, the alarm was given. The regiment now advanced with all possible expedition to cut off the retreat of the enemy, and a little before daybreak approached Sahagun, where the French were discovered formed up. Between the two forces the ground was broken by a hollow way, and, as the Fifteenth drew near, the enemy retired towards a bridge on their left. Lord Paget moved the regiment in columns of divisions and trotted along parallel to the enemy's line of march, but a good deal behind them. The French frequently endeavoured to cross the head of his column, when he changed the direction to prevent his flank being turned. At length the French halted and formed line, when the Fifteenth, having passed the enemy's left flank, also halted, wheeled into line and charged with resistless impetuosity. The hostile cavalry was absolutely borne down and unhorsed by the superior activity and weight of our Hussars; many were killed and wounded, and the whole body was overwhelmed and scattered; two lieutenant-colonels, eleven other officers and 154 men were taken prisoners, twenty were killed and 125 horses captured, while the casualties in the Fifteenth, considering the numbers opposed to the regiment—double its own force—were few—only two rank and file and four horses being killed, whilst the Colonel and Adjutant, 18 rank and file, and 10 horses were wounded.

Lord Paget expressed to all ranks of the Fifteenth his grateful thanks for the gallant way in which they had behaved; Colonel Colquhoun Grant and Lord Paget were each given a medal; the conduct of the Hussars received warm praise from Sir John Moore in his despatch, dated Benevente, 28th December, to Lord Castlereagh; and the regiment was subsequently honoured with the royal authority to bear the word "Sahagun" on its appointments in commemoration of this action.

On the 21st, Baird's division occupied the town from which the French had been driven, and here, too, Sir John Moore established his headquarters; the cavalry was pushed out towards the Carrion and Saldanha.

The army was to have advanced at eight o'clock

on the night of the 23rd December to attack Soult in his position on the Carrion River; but, previous to the hour named, Sir John Moore received information from the Marquis de la Romana that Bonaparte, having become acquainted with the menace of the forward movement of the British, had not only turned against them the troops which were marching on Portugal, but was himself advancing from Madrid with an overwhelming force to cut off the English armies from Galicia. In consequence of this information, Sir John Moore determined to abandon the proposed movement against Soult and to lose no time in regaining the neighbourhood of Astorga, whence the retreat of the British force might in some degree be assured.

"From the points occupied by the British army, two principal roads lead on Astorga. That on the north, crossing the Esla by a bridge at Mansilla de los Mulos, and passing through Leon, is the more direct line of communication" (but this road was occupied by the Spanish troops under la Romana). "The southern line crosses the Esla by the bridge at Castro Gonzalo, about a league in front of Benevente, through which town it passes. It was along this road that Sir John Moore and the principal part of the army began to retreat; whilst Sir David Baird was directed with his division to take an intermediate direction by cross roads, leading to Valencia de Don Juan, a town situated on the Esla, about equidistant from the bridges before mentioned, at which place the river is passable by a large ferry-boat, and, in a dry season, at a ford in the neighbourhood."[1]

[1] Hook.

At this moment the English army was almost hemmed in by the converging forces of the French. In front, Soult had been strongly reinforced until he alone was now in superior strength to the British, and there was danger lest he might slip round Moore's left and cut him off from Corunna; Junot, with the army liberated by the terms of the Convention of Cintra, was moving up from Burgos, and had reached Palencia; Lefebvre was marching from Salamanca; while Bonaparte himself, moving up like a whirlwind from Madrid, across the deep snows and the wild hills of the Guadarama, was already close upon Tordesillas.

Lord Paget sent forward strong patrols, and the cavalry preserved a bold front while the infantry withdrew. General Slade chronicles in his diary:

"On the 25th, at three in the morning, we marched back to Sahagun; but how different the treatment to that we received on advancing! No more ringing of bells, no longer did the air resound with 'Long live the English!' All the shops were shut, and not anything to be got. I may truly say it was the most unpleasant Christmas Day that could be spent."

On the 26th the whole army was in full retreat towards the coast. The cavalry reached Mayorga on this date, and here the 10th Hussars had what Napier describes as "a hardy action," adding:

"The English cavalry had been engaged more or less for twelve successive days, with such fortune and bravery that above five hundred prisoners had already fallen into their hands, and their leader being excellent, their confidence was unbounded."

On the 27th the Fifteenth reached Valderas, and on the following day Benevente, and of these two marches General Slade remarks that—

"we were thirteen hours on horseback—and a most severe day it was. I should conceive that not less than forty horses sunk from fatigue, and had to be shot. . . . Marched to Benevente. It rained the whole of this day, which after the frost made the roads almost impassable. No ploughed field, after the breaking up of a frost, could be in a worse state. The exertions of the artillery, and the difficulties they encountered, exceed everything I could have conceived."

On the 29th the Reserve and Craufurd's Brigade quitted Benevente; but the cavalry remained in that town, having guards at the fords of the Esla: on this day General Lefebvre Desnouettes, with five squadrons of Hussars of the Imperial Guard, forded the river above the town and attacked the picquets commanded by Lieutenant-Colonel Otway, of the 18th Hussars. The result was most creditable to the 10th, 18th, and 3rd German Hussars, led by Brigadier-General Stewart; but a few orderlies, one of whom was killed, were the only men of the Fifteenth who were engaged, the rest were with the column, but moved up to the support of the cavalry. The French did not, however, cross the Esla again, and the Fifteenth continued its retirement on La Baneza. The cavalry arrived at Astorga on the 31st; and here, as recorded by General Slade—

"it was reported that the French were advancing upon us. The Hussar Brigade assembled at our alarm posts at 8 p.m., where we remained till midnight." (Thackwell says until two in the morning.) "We then marched to Bembibre, arriving at 2 p.m. the following morning. The many delays were owing, in the first instance, to the depth of the snow, and also from the frequent interruptions we met with from carriages being overturned, artillery wagons burning, etc."

The mountains were covered with snow, and the cold was intense, and some idea of the sufferings of the troops may be gathered from the statement in the records of the 10th Hussars, that "in the Tenth alone, Captain Darley and seventeen private soldiers died of fatigue, while sixty horses were destroyed to prevent their falling into the hands of the enemy."

Part of the Fifteenth covered the retreat from Bembibre on the 2nd January, and with all its exertions could not keep in its front the many stragglers, who had, therefore, to be left to their fate. The French cavalry had entered the town before our picquets had well left it, and they pressed much upon the rear during this march, the Fifteenth having three horses killed, and two or three men and horses wounded. That night the whole regiment bivouacked on the high road near a large dwarf wood three or four miles in front of Calcabello—two troops on some open rising ground, and the remainder nearly a mile further to the rear. Thackwell states that here "several infantry stragglers, who had been miserably gashed by the French cavalry, crawled to our fires in the night, and were sent to the rear under protection."

Next morning the enemy showed a disposition

to advance; but losing a few men killed, wounded, and taken, they desisted. At this time the strength of the Fifteenth was under 300 men and horses, and many of the latter were utterly exhausted, lame from want of shoes, and could scarcely be urged to a faster pace than a trot. In the evening the British retired fighting through the town; the French dragoons closed twice with the skirmishers of the Fifteenth, and had several men killed, but did not succeed in taking a single prisoner from the regiment. From Villafranca the route lay over the Monte del Cebrero, traversing a country so broken and intersected as to prevent cavalry acting, or indeed of moving at all except by the road. The regiment, therefore, continued its retirement during the night, and on the morning of the 4th left a squadron of 80 rank and file, mounted on the freshest horses, at Nogales to act with the infantry rear guard.

On the 6th Moore halted at Lugo and offered battle, but the offer not being accepted the march was resumed at night—the Fifteenth remaining on picquet until 5 a.m., keeping the fires burning to delude the enemy into the belief that the British Army was still in position, and then retiring in torrents of rain to Betanzos. On the 11th the army reached Corunna, and on the 15th the enemy in considerable force drove in the outposts and took possession of the wooded heights overlooking the British position; of the picquet of Hussars, 100 strong, several horses were wounded on this occasion. A patrol under Thackwell, who was in command of the cavalry picquets, was sent by Sir John Moore to ascertain if the French were extending to their left. It proceeded about five miles to our right front without meeting the enemy, and part of the picquet remained out till the morrow, but was not engaged in the battle.[1]

Of the horses of the 15th Hussars, thirty only were permitted to embark, some were given up to the Commissariat Department, and the remainder were slaughtered on the beach, the regiment embarking on the afternoon and night of the 16th January. Towards the end of January, the Fifteenth were disembarked at Portsmouth, Plymouth, and Falmouth, landing with only 29 horses out of the 682 which they had taken with them when proceeding on active service little more than three months previously.

"The distance from Sahagun, the point at which Moore's retreat began, to Corunna, where he expected the British transports to be waiting for him, was, in a direct line, about 160 miles; the actual march of the troops was probably about 220 miles. . . . During this retreat of eighteen days it will be seen that Moore's forces actually halted four days, and it seems difficult to understand how a British Army, unshaken by defeat, and splendidly led, should practically have fallen into ruin in a period of time so short. But the march from Sahagun to Corunna was, for suffering and horror, like a tiny section of the Moscow retreat. The track lay through the savage Asturian hills. It was winter time—tempests raged almost incessantly. Every stream was swollen, every ravine was choked with snow."[2]

[1] The only cavalry taking any part in the action appear to have been forty men of the 15th Hussars, under Lieutenant Knight, forming the escort of Sir John Moore.

[2] Fitchett.

Of the march from Astorga to Villafranca, Lord Londonderry says:

"The condition of the army was at this time a most melancholy one; the rain came down in torrents; men and horses were floundering at every step; . . . the shoes of the cavalry horses dropped off, and the horses themselves soon became useless. It was a sad spectacle to behold these fine creatures urged and goaded on till their strength utterly failed them, and then shot to death by their riders, in order to prevent them falling into the hands of the enemy."

Writing of the march of the 1st January, he says:

"At this time the enemy's cavalry, though they seldom sought an opportunity of coming to blows with us, pressed closely and incessantly upon our rear; we rode frequently many miles in sight of each other; and from time to time our rearmost dragoons would exchange pistol shots with their leading files."

But throughout the retreat, as in the advance, the work of the mounted portion of the force was singularly brilliant, and few will be found to disagree with the historian who says that—

"the part taken by the cavalry under Paget in this retreat was very gallant. They faced with cheerful courage and tireless hardihood the vastly superior French cavalry, which pressed on the British rear, and never failed to overthrow them in the actual shock of the charge."[1]

[1] Fitchett.

CHAPTER III

During the next four years the 15th Hussars remained in England, and Captain Thackwell was thus debarred from taking any part in the events of that momentous period of the war in Spain and Portugal. Instead, there was the usual round of home soldiering—change of quarters, reviews on Hounslow Heath and Wimbledon Common; the Fifteenth being on at least one occasion brigaded under their old commander, Lord Paget, with the regiments—the 10th and the 18th Hussars—in company with which they were again, ere long, to see service in Spain.

Twice the monotony of home service was broken by employment of an unusual character. In April 1810 the regiment was ordered to London from Hounslow and Hampton Court to aid in suppressing the riotous assemblages of the populace, which took place when the House of Commons ordered one of its members, Sir Francis Burdett, to be taken into custody and lodged in the Tower for a breach of privilege. Burdett had made himself unpopular with the Government by vigorously supporting freedom of speech and Catholic emancipation, as well as by protesting against the suspension of the Habeas Corpus Act, against the existing prison discipline, and against the enormous taxation. He was apprehended at his rooms, and conducted to the Tower by the Fifteenth and a detachment of the Life Guards.

Again, on the 15th November 1811 the regiment was hurried up from Colchester, where it was then quartered, to Nottingham in consequence of the disturbed state of the manufacturing districts. Ned Lud was a Leicestershire imbecile who had acquired some local notoriety by destroying machinery, and his name was given to the riots in which the popular discontent expressed itself in the Midlands about this time. General distress being caused by the progress of the industrial revolution, the anger of the Luddites was directed against the new machinery, much of which was destroyed, and the whole of the regiment was employed for several months at the end of 1811 and the beginning of 1812 in keeping order in Nottinghamshire, Yorkshire, and Lancashire.

It was while quartered in November 1812 at Manchester and the neighbouring cities and towns that the Fifteenth received orders to hold six troops, each of 90 men and horses, in readiness for foreign service. On the 15th of the following month the regiment marched for Portsmouth *viâ* Chichester and Arundel, was inspected by General Hammond on the 13th January and embarked, under command of Colonel Colquhoun Grant, on the 15th and 16th, landing at Lisbon between the 3rd and 7th February.

The four years which had elapsed since Thackwell had quitted the Peninsula had been—

"years of struggle—of advance and retreat, of triumph and of disaster; shining threads of victory interwoven with black threads of calamity and hardship. If those years had seen the glories of Talavera and of Salamanca, of Busaco and of Badajos, they had also seen the black days of the retreat to Torres Vedras, and the later retreat after the failure at Burgos."[1]

But the Fifteenth reached Lisbon in time to take part in a campaign as glorious and as successful as any which had preceded it, and as epoch-making as any which were to follow under the same leadership. Wellington had utilised the months which followed the retreat from Burgos in reorganising—

"the allied army with greater strength than before. Large reinforcements, especially of cavalry, had come out from England, the efficiency of the Portuguese was restored in a surprising manner, and discipline had been vindicated in both services with a rough but salutary hand. . . . Nor had the English general failed to amend the condition of those Spanish troops which the Cortes had placed at his disposal. By a strict and jealous watch over the application of the subsidy, he kept them clothed and fed during the winter, and now had several powerful bodies fit to act in conjunction with his own forces."[2]

At the opening, then, of the campaign of 1813 Wellington was able to command the services of something like 90,000 fighting men, 40,000 of

[1] Fitchett. [2] Napier.

whom were British and all of whom were in a high state of efficiency.

"The relative strength for battle was no longer in favour of the French; their force had been reduced by losses in the secondary warfare, and by drafts since Wellington's retreat, from 260,000 to 230,000. Of the last number 30,000 were in hospital, and only 190,000 men, including the reserve at Bayonne, were present with the Eagles; 68,000, including sick, were in Aragon, Catalonia, and Valencia; the remainder, with the exception of the 10,000 left at Madrid, were distributed on the northern line of communication from the Tormes to Bayonne."[1]

These troops were strung out across Spain from Clausel's army in the Asturias on the north-west to the force under Suchet on the east in Valencia, and the success of the French in the struggle which was impending depended upon King Joseph being able to concentrate his scattered armies rapidly upon any one point. Wellington had decided—

"to operate with his left, ascending the right of the Duero to the Esla, crossing that river to unite with the Galicians, while the rest of the army advancing from the Agueda should force the passage of the Tormes. By this combination, which he hoped to effect so suddenly that the King should not have time to concentrate in opposition, the front of the allies would be changed to their right, the Duero and Carrion turned and the enemy thrown in confusion over the Pisuerga. Then moving forward in mass the English general could fight or turn any position taken up by the King; gaining at each step more force by the junction of the Spanish irregulars until he reached

[1] Napier.

the insurgents at Biscay; gaining also new communications with the fleet and consequently new depôts at every port opened."[1]

To prevent the concentration of the French armies by which alone King Joseph might have offered an effectual resistance to the design of the British, Wellington held Suchet to the eastern coast by despatching an expedition to Tarragona, while Clausel in the Asturias was rendered immovable by the partisan warfare which was set going in his neighbourhood.

The Fifteenth was intended to form a Hussar Brigade with the 10th and 18th Hussars, and was the first regiment of the brigade to land in Portugal, the Eighteenth arriving at the beginning, and the Tenth not until the middle of February, and it was only on 4th April that the brigade[2] started to join Lord Wellington, the Fifteenth leading. The following extracts from Captain Thackwell's diary describe the main incidents of the march and of the ensuing compaign:

"*4th April.*—The three right troops of the regiment marched at 11 o'clock from Belem to Saccavem, distant three leagues."

"*5th.*—Marched at 8 o'clock, and passed a small river in front of Saccavem on a bridge of boats . . . passed through the right of the Lines of Torres Vedras, within half a league of Alhandra; the road is closed in with a work containing five or six guns, and the line continues over a very steep and broken ridge; several batteries and a deep ditch connect it with the Tagus; and two or three batteries from the islands therein flank the road from Alhandra."

[1] Napier.

[2] Commanded by Colonel Colquhoun Grant.

Azambuja was reached on 6th April, and here and in the vicinity the brigade was closed up and remained for several days, not leaving till the 20th, when the Fifteenth moved on and reached Santarem. Marching by Thomar, Espinal—where "the country," remarks Captain Thackwell, "is getting very rugged and mountainous and the roads very bad,"—and Celorico, Freixados was reached on the 11th May, and here a halt was made for the whole brigade once more to close up.

"18*th May.*—The three regiments of Hussars were inspected at 10 o'clock this day, six miles in front of Freixados on the Almeida road, by the Commander of the Forces, who was pleased to express his entire approbation of the fine appearance of the brigade, but particularly of that of the 15th Hussars, which was considerably stronger than either of the others, each squadron having 80 files in it; my squadron had 81 including half-squadron officers. The brigade merely marched past by half-squadrons, ranked off, trotted past by divisions and advanced in parade order."

(On the 26th April, the "state" of the Anglo-Portuguese Army shows the strength of the Hussar brigade as follows: 10th Hussars, 505; 15th Hussars, 521; 18th Hussars, 504.)

To Lieutenant-General Sir Thomas Graham, who had only the month previous rejoined the army in the field from sick-leave in England, had been entrusted the direction of the force which was to turn the French right by the Tras-os-Montes—a region which King Joseph and his military advisers had judged impassable for the movements of an army.

"It was shaggy with forests, horrent with snowy peaks, scarred deep with leaping mountain torrents. Three great rivers had to be crossed; hill-crests, white with winter snows or buffeted with angry winds, had to be surmounted, and many a mountain pass, that never before had echoed to the tramp of disciplined battalions, had to be threaded."[1]

General Graham had under his command some 40,000 men, comprising Ponsonby's, Grant's, Bock's, Anson's and D'Urban's cavalry brigades, the First, Third, Fourth, Fifth, Sixth, and Seventh Divisions, Pack's and Bradford's brigades of Portuguese infantry, and a Spanish Division under Colonel Longa.

"19*th May.*—Marched this day to Corriscada."

"21*st.*—Marched to St. Amaro, and thence to the river Duero—two small leagues, the country very hilly and bold, the road tolerable. The two left squadrons passed the river and encamped on the right bank, the right squadron encamped on the left bank in a beautiful olive grove and had abundance of green forage. The difficulty of getting the pontoon train over the river and up a tremendous hill on the other side prevented our crossing to-day."

"22*nd.*—The right squadron passed the Duero at Barca de Focinho at daybreak this morning and marched to Torre de Moncorvo—a league and a quarter. This is a very pretty town, pleasantly situated among the mountains. The 18th and 10th Hussars also crossed to-day and marched to the same place. The river Duero is beautiful beyond description, and about 130 yards across at the ferry, the current in places very rapid. On the

[1] Fitchett.

hills on the right bank are three small field-works thrown up by General Silviera, to prevent the French crossing the river in 1811."

(" By the 24th May, after trying marches over nearly impassable mountain tracts, Graham's wing was occupying a line on a front of forty miles extending from Braganza on the left to Miranda de Duero on the right. The force was formed in three columns. On the left, the cavalry brigades of Anson and Ponsonby, the First Division, and Pack's brigade held Braganza. In the centre, Bock, D'Urban, the Fifth Division, and Bradford were at Onteiro, the Third Division at Vimioso on the right, the Fifth and Seventh Divisions, with the 18-pounder battery, occupied Molhadas and Miranda de Duero, which the Hussar Brigade and Fourth Division, who formed a link between the two wings, were rapidly approaching.")[1]

" 26*th May*.—Marched to Sindim, a very pretty town situated on a commanding eminence. Saw from a hill in the neighbourhood a long way into Spain, also Forcatello, a large town near the junction of the Tormes with the Duero, also Miranda de Duero three leagues up the right bank of the Duero."

" 27*th*.—Marched and encamped with the right of the brigade on Castes, and the left on Brandilanes in an oak grove in rear of a rivulet; passed Villa Garcia and Constantin. The Fourth Division of infantry were encamped on the left of the road, midway between these villages and several brigades of artillery. On this day we found the outposts of the army; plenty of green forage, no enemy near. The infantry passed our encampment in two columns at half-past 10 o'clock in the direction of the River Esla, also some of the artillery."

[1] Butler.

"*29th.*—Marched at 4 a.m. and crossed the River Aliste at the ford of Mugo; the 10th Hussars formed the right column and passed the wooden bridge on the Aliste—the left column passed through the village of Mugo, turned to the right through Carvajales and encamped within two miles and a half of the ford of Almenda on the Esla, the whole brigade on a woody ridge with a marshy rivulet in front—not easy to surprise. The infantry and artillery encamped three miles in rear near Carvajales. . . . At 2 p.m. rode with Colonel Grant with a party of 10th Hussars to reconnoitre the ford of Almenda, and found a French picquet of about 60 cavalry composed of Lancers, Heavy Cavalry, and Chasseurs. Colonel Grant and his Brigade-Major endeavoured to discover the ford on the left, whilst I moved along the bank to the right and discovered the ford on the right, and proved the practicability of it by nearly passing over—the river broad; the current very rapid; the water at the deepest part near four feet, and the banks rising in hills. Obliged to return by a party of the enemy moving rapidly towards me. Joined by Colonel Grant, Brigade-Major Jones, and Colonel Ross (20th Foot), pointed the ford out to them and crossed the river with the two first. The enemy moved towards us, and we retired. This passage might be easily defended with heavy guns—but if the enemy have only light, two 18-pounders will cover the passage of the ford and secure the formation on the heights. I was the first man of the British Army that crossed the Esla at the ford of Almenda."

"*31st.*—The intention of Lord Wellington to pass the army to the left bank of the Esla this day was evident from the Hussar Brigade having received orders to be in readiness to march at 1 a.m. Accordingly it marched at that hour, left in front;

my squadron (although the right one) formed the advanced guard, a brigade of nine-pounders accompanying the brigade. At daylight the head of the column reached the ford of Almenda, and no enemy was visible. I received my instructions from Colonel Grant to trot up the hill, and to attack any description of force that might be opposed to me, and he would take care to support me with the brigade. The squadron entered the river attended by some companies of the 51st Regiment, and a corporal of the 18th Hussars acted as guide for the ford. The ford was much swollen and the river was very broad and rapid ; from the imperfect light a ledge of rocks in the ford could not be discovered, and many of the horses fell and threw their riders into the water, who were saved with difficulty, owing to the infantry holding the horses by the stirrups and preventing them recovering themselves. The whole brigade, however, got safely over, with the exception of a horse drowned, but between 20 and 30 of the Fifty-first and Chasseurs Britanniques were drowned. The remainder of the army crossed by a pontoon bridge thrown over a mile higher up. The right squadron, on crossing the river, advanced up the hill at a moderate trot, the right division forming the advance under Lieutenant Finch; the fourth division formed the reserve under Captain Wodehouse. On reaching the top of the hill I discovered the enemy's picquet, consisting of about 60 men, formed in the village. I immediately formed my two right divisions in line with intervals, whilst the two left formed the support. The enemy commenced firing; I instantly ascended the hill to the right to reconnoitre, and the enemy commenced retiring. I immediately advanced to the attack, but the orders given not being obeyed, the divisions in passing the village got into confusion, which from the enemy's rapid retreat and my rapid pursuit could not be remedied. I continued the pursuit nearly three miles, taking care to reconnoitre the country as I advanced, and at last came up with them and succeeded in making prisoners of about 50 men—nearly 20 of whom contrived to make their escape in the woods after they had been passed. Many were very badly wounded, and some were badly wounded who made their escape. I had never more than ten men in action, and sometimes not above four or five. The enemy during their retreat were reinforced by thirty men more, some of whom added to the number of prisoners. I stopped the pursuit within about a league and a half of Zamora, on discovering myself not supported and seeing in my front nearly 200 cavalry formed on a hill to cover the defeated body. The enemy commenced skirmishing, and I gradually retired to a position behind a rivulet covered by my skirmishers. This I maintained until the enemy displayed two squadrons on the hill in front. I then commenced my retreat through the wood, and took up a position in the rear of it, where the brigade soon made its appearance. . . . The prisoners were all of the 16th Dragoons."

"*2nd June.*—A most infernal night. Joined the regiment off picquet on the Toro road at half-past four on its march to Toro; near that town heard the French had just retired; the Tenth and Eighteenth advanced at a trot, and the enemy's rear was found formed midway between Toro and Morales, and consisted of part of the 21st Dragoons, supported at Morales by the remainder of the 21st and 16th Dragoons. The British artillery fired nine rounds upon their rear with little effect; they were charged by the 10th Hussars and a squadron of the Eighteenth at Morales and pursued half a league beyond under cover of six pieces of artillery, formed in a most commanding position, and the 5th and 12th Dragoons. The enemy's artillery did some execution among the Tenth, particularly when upon the causeway that leads across a morass. They however passed and drove the enemy over the first hill. During this time the Eighteenth and Fifteenth formed the reserve, and were extremely steady under a fire of round shot and shell. One of the latter dropped in the centre of my squadron without doing the slightest damage. The Tenth were ordered to retire, and the position being reconnoitred was considered as too strong to be attacked. The loss of the enemy in this affair was upwards of 200 men; that of the Tenth and Eighteenth was one officer killed and 27 men killed, wounded, and missing. Returned to Morales for the night."

In this action Colonel Grant was wounded, and Lieutenant Woodberry, 18th Hussars, relates that a Frenchwoman, whose husband, a French officer, was killed in the charge, was captured among the prisoners, disguised as a man in civilian clothes.

"*5th June.*—Marched to Penaflor; the 14th Dragoons and 1st Germans bivouacked near Penaflor, as did the Fifth and Seventh Divisions of infantry."

"*6th.*—Marched to within two leagues of Duenas, and bivouacked opposite St. Cecilia, where a flag of truce arrived concerning the prisoners taken by the 10th Hussars; the officer reported that the French headquarters were at Duenas. The whole of the Allied Army is so well concentrated that it could form in battle order in less than two hours."

"*7th.*—Marched to Villa Bona; the army

marched in five columns, our brigade with the centre one, the Fifteenth forming the advance guard. . . . Crossed the River Carrion at Palencia without opposition, the French having retired at 6 o'clock; we arrived at 9.30 a.m. . . . Joseph Bonaparte left Palencia yesterday."

"8*th*.—Marched to Tamora, left the Third Division at Mozon; the Fifth and the First were in front of Amusco, being the left column, the Sixth and Seventh were a league in rear of Tamora, and the Fourth and Light Divisions were with us. The right column marched on the Burgos road and consisted of the Second Division and Portuguese."

"9*th*.—The left column marched to Santillana, and five Divisions were at Pena and Tamora. . . . Crossed the canal of Castile, which we first saw at Palencia. . . . Lord Wellington's Headquarters were this day at Amusco, yesterday at Amusco also, on the 7th at Palencia, on the 6th at Castro Monte, on the 5th at La Mota, on the 4th at Toro."

"10*th*.—Crossed the River Pisuerga, of considerable breadth, fordable below the bridge. . . . The brigade formed the outposts with infantry columns in the rear. Headquarters this day at Molga de Juso."

"12*th*.—The brigade assembled between Ovillo and Castillio, and formed the advance of the centre columns; heard that the enemy were in force. Passed through Isar and ascended the opposite heights; the heavy brigade marched through Hormillos, and upwards of twenty squadrons of the enemy's cavalry were assembled on the hills. Dispositions were made for attack by the Hussar Brigade, Colonel Ponsonby's brigade and General D'Urban's Portuguese brigade, and the Fifteenth were ordered to attack the right. We advanced for that purpose, but the enemy retired and formed a junction with a large body of infantry on our right, and the whole filed over the River Urbel by the bridge of Tardajas under the fire of our guns, which did little execution. Had two divisions of infantry been up, the left wing of the enemy's army would have been destroyed. The 3rd Dragoons and 14th Light Dragoons attempted a charge, but without success. The French loss was about a hundred men killed, wounded, and prisoners. Bad generalship displayed by the French, which our want of infantry prevented us profiting by. The enemy showed upwards of 30,000 men on this occasion. The brigade retired to Isar for the night. Men on their horses from 3.30 a.m. to 4 p.m."

"*15th June.*—The troops that passed the Ebro to-day were the Hussars, 1st Germans and 14th Light Dragoons, 12th and 16th Light Dragoons, General D'Urban's Portuguese brigade, and the 3rd, 4th, and 5th Heavy Dragoons with twelve pieces of cannon. The Fourth and Light Divisions encamped on the other side of the river. The country becoming very mountainous and the road rugged. The descent to the river is very steep and the road bad—upwards of a mile in length through a tremendous ravine; the river is about twenty yards broad and fordable in several places, but the bed is rocky and the current very rapid; the bridge has five arches. The village of Arenas is in a valley half a mile across bounded by immense cliffs and mountains, the river's course being generally through a ravine bordered by stupendous rocks 600 or 700 feet high. After passing the river, the road runs up the left bank for three miles, and forms one of the strongest passes in nature. The scenery is romantic beyond description."

"18*th*.—Marched to Villa Alta and encamped for the night, three-quarters of a league in front of it. Longa's corps of guerillas had stopped at Villa Alta, because the French were in their front. A patrol of the 12th Light Dragoons had fallen in with the enemy the evening before, and on this day about 8,000 men took up a position at Osma —half a league in front of Berberana, on the Miranda road. They were skirmished with by the 12th and 16th Dragoons, and soon after attacked by the light battalions of the First and Fifth Divisions in front, whilst part of the Light Division turned their left. These movements obliged them to retire, with the loss of upwards of 300 men prisoners, besides many killed and wounded. Some of the latter reported that 1,100 men were entirely dispersed in the woods."

"19*th*.—Marched through Osma, and half a league from it turned to the left and followed a cross road direct to Vittoria, and encamped in advance of Subijana de Morillos, the road for nearly the last two leagues through a rocky valley. The French had about 10,000 men in this position, which was attacked and carried by the Fourth Division without material loss, probably owing to the Light Division driving an enemy's division before it at some distance to our right. The Fourth Division in front and General D'Urban's cavalry formed the advance posts."

"20*th*.—The prospect of a halt to collect the columns, as General Sir R. Hill's army marched considerably to the right yesterday morning. The French reported to be in position near Vittoria, to the amount of between 60,000 and 80,000 men."

The French left rested on the heights which came to an end at Puebla de Arlanzan, their line extending across wooded and broken country in front of the village of Arinez, while the right of their centre was posted on a steep height covered with artillery—which commanded the valley of the Zadorra. The right was stationed near Vittoria, intended to defend the passages of the river near that city, and to cover the right centre of the army, while there was a reserve in rear of their left, at the village of Gomecho. The attack on the French position was to take place as follows: the right column of the Allied Army, consisting of General Hill's corps with a brigade of the Spanish Division under Morillo, was to attack the enemy's left, and gain possession of the heights above Puebla, and afterwards of the village of Subijana, when the Fourth Division was to cross the Zadorra at Nanclares, the Light Division at Tres Puentes, and the Third and Seventh by a bridge higher up. These four divisions, forming the centre of the army, were to attack the heights on which the enemy's centre leaned, while the left, commanded by Sir Thomas Graham, and consisting of the First and Fifth Divisions, two Portuguese brigades and Longa's Spaniards—supported later on in the day by the army of Galicia under Giron—was to move across from the Bilbao road for the purpose of forcing the passages of the Zadorra, at the villages of Gamara Mayor and Abechuco, where the British had three divisions of infantry with a strong body of cavalry.

Wellington's orders for the movement of the Army on the 21st June direct that—

"The Light Division will move at daybreak, and

proceed by the road along the River Bayas, through the pass leading to Subijana, and thence by Montevite and the camp of the Fourth Division to the village of Nanclares.

"One squadron of the 15th Hussars will act with the A.G. of this division. The remainder of the 15th Hussars and Major Gardiner's troop of Horse Artillery will follow the rear of the Light Division. The Fourth Division will follow in rear of the 15th Hussars.

"The 18th and 10th Hussars will follow the Fourth Division."

"*Monday, 21st June* 1813.—Marched at half-past eight o'clock in the direction of Vittoria. A short distance from Nanclares halted to give time to the right and left columns to advance, on account of discovering the French army in position on the left bank of the Zadorra. . . . The enemy's army consisted, according to reports, of 6,000 cavalry, 70,000 infantry, with a numerous artillery advantageously posted on the salient angles of the position. At 9 a.m. Sir Rowland Hill's column commenced its attack on the enemy's left, and the sharpshooters began to gain ground. The centre column advanced by the right bank of the Zadorra to Tres Puentes—a village nearly opposite the enemy's right; this movement gave them some uneasiness, and they advanced a battery of Horse Artillery to oppose it, but did not prevent the division forming on the other side of the river, and a battery placed above the village soon silenced that of the enemy and occasioned him some considerable loss. At this moment Sir Thomas Picton's division made its appearance on the enemy's right, and commenced a cannonade and a brisk attack along the right bank of the Zadorra and crossed it above Tres Puentes. Till this time the enemy had defended himself most obstinately on his left and centre, but the Light Division, having in the most gallant style carried the hill that covered his centre, made him retreat from these points. He however took up a strong position on a ridge in rear of the village of Berostiguela, which he filled with infantry. This however was carried by the Light Division at the same time that his left was obliged to give way to the steady advance of Sir Rowland Hill's troops; his position on the ridge still remained, but this after some opposition was taken by the Fourth and Light Divisions, and many of the guns were captured. In this attack the Portuguese infantry behaved with the utmost gallantry and suffered severely.

"At this time the cavalry (with the exception of a squadron of the Eighteenth led by Captain Turing, who, without orders, charged a column of infantry which killed him and several men) had not been engaged, but suffered some loss from the enemy's artillery. At 6 p.m. the right and centre of the army were on the height above Vittoria, and the left of the enemy's army was retreating under the heights before mentioned—the centre in great confusion through Vittoria, and the right, which had made a most gallant resistance at the village of Gamarra against Sir Thomas Graham's column of the First and Fifth Divisions, partly by Arazua, and partly by Gomez. Lord Wellington now ordered the Fifteenth to pass the flats, leaving Vittoria to the right, and endeavour to cut off the enemy's retreat from Aranjuez. On descending, we found this operation very difficult, from the intersected nature of the ground, the dykes about which were filled with French horses. This, however, was surmounted, and the regiment moved to the attack of a regiment of heavy cavalry, which advanced to cover a retreating column of infantry, and many of both were sabred and the column broken. The cavalry retired under the protection of a regiment of Hussars and one of Dragoons, partly rallied and moved to the attack, which was met with the cool determination of the Fifteenth, and the enemy were beat back with loss. At this moment six squadrons of the enemy's cavalry, Hussars and Dragoons, made their appearance in our rear; these were charged by the squadron of reserve under Captain Cochrane, which did not prevent their endeavours to cut us off; but the Fifteenth, having in part changed its front, advanced to the charge and the enemy were driven back to the village of Gomez. During these operations the enemy's infantry succeeded in effecting its retreat, and not being supported it was judged prudent to discontinue the pursuit. Had either of the regiments of Hussars been in second line, 2,000 men would have been the fruit of our attack; as it turned out a general, a colonel, several officers, and about 130 men were made prisoners, besides sixty or seventy more, who were picked up by the people in our rear, and in addition to killed and desperately wounded left on the field. An infantry standard was also taken by us, and the enemy's loss altogether was 12,000 men, 151 pieces of cannon, most of his baggage and cattle, his military chest, etc.

"We encamped for the night in a wood half a league in front of Gomez."

At Vittoria Captain Thackwell received a severe contusion on the right shoulder from the hilt of a sword, from a thrust.

On the night of the battle the brigade bivouacked a short league in front of Vittoria, moving off next day with the main column towards Pampeluna.

The French left a considerable garrison in this town and retired towards the pass of Roncevalles, followed by the light troops. The left column under Sir Thomas Graham advanced by the pass of Adrian upon Tolosa, whither General Foy had retired with about 12,000 men; General Hill's corps was left to blockade Pampeluna; while the Third, Fourth, Seventh, and Light Divisions, with the Hussar Brigade, and Ponsonby's brigade of Heavy Cavalry, marched by Tafalla and passed the River Arragon by the bridge of Cappanoda in the hope of intercepting General Clausel's corps of upwards of 12,000 men. This general had advanced on the 22nd June to the neighbourhood of Vittoria; but finding the French to be totally defeated, he fell back upon Logrono, followed by the Fifth and Sixth Divisions, the Household Cavalry, and D'Urban's Cavalry Brigade, and thence by forced marches upon Tudela, which he reached on the 27th. Pressed by Mina's and Sanchez's cavalry, he recrossed the Ebro and marched on Zaragoza, eventually evading further pursuit by retiring through the pass of Jaca.

On the 30th June—all hope of intercepting Clausel being at an end—the Hussar Brigade recrossed the Arragon River and was stationed for three weeks at Olite. While here a change was effected in the composition of the cavalry brigades: the 10th and 15th Hussars forming one brigade, under Major-General Lord Edward Somerset; the 18th Hussars were brigaded with the 1st Hussars of the King's German Legion; and Colonel Grant was transferred to the command of the brigade consisting of the 13th and 14th Light Dragoons.

"In this campaign of six weeks, Wellington marched with 100,000 men 600 miles, passed six great rivers, gained one decisive battle, invested two fortresses, and drove 120,000 veteran troops from Spain."[1]

[1] Napier.

CHAPTER IV

"TEN days after the battle of Vittoria, Marshal Soult, under a decree issued from Dresden, succeeded the King as lieutenant to Napoleon. . . . Travelling with surprising expedition he was enabled on the 12th July to assume the command of the three beaten armies, now reorganised in one under the title of the *Army of Spain*. . . . At this period General Paris was still at Jaca, but Clausel had entered France, and Soult, reinforced from the interior, had nine divisions of infantry, a reserve, and two divisions of cavalry, besides light horsemen attached to the infantry. Including garrisons, and twelve Italian and Spanish battalions not included in the organisation, he had 114,000 men, and, as the armies of Aragon and Catalonia had about 66,000, 180,000 men and 26,000 horses were still menacing Spain." [1]

The army of Spain was posted as follows: Clausel was on the left at St. Jean Pied de Port; in the centre was d'Erlon about Ainhoa, while Reille was on the right at Puerto de Vera. The Reserve was behind the Bidassoa about Irun, and the cavalry divisions were respectively on the Nive and the Adour.

Wellington's dispositions were as under: the

[1] Napier

right was formed by Hill's corps, of which, on the outer flank, Byng's brigade with a Spanish division held the southern issues of the passes of Roncevalles and Ibaneta; the rest of the Second Division held the pass of Maya and the Fourth Division formed the support to the above; while Picton, eighteen miles south of Maya, was the general reserve to Hill's corps. Twelve miles west of the pass of Maya was the Seventh Division at Echallar; the Light Division held the village of Vera, the Sixth Division at Estevan forming a central reserve. The First Division, with the troops of Giron and Longa, held the line of the Bidassoa from Vera seawards; the Fifth Division, with the Portuguese, besieged San Sebastian on the left, while Pampeluna on the right was blockaded by the Spaniards. As the two fortresses were fifty miles apart no little work was thrown upon the cavalry in keeping open the communications.

Wellington's "theatre of operations was a trapezoid, with sides from forty to fifty miles in length, and having Bayonne, St. Jean Pied de Port, San Sebastian and Pampeluna, all fortresses in possession of the French, at the angles. The interior, broken and tormented by savage mountains, narrow craggy passes, deep water-courses, precipices and forests, appeared a wilderness, which no military combinations could embrace, and susceptible only of irregular and partisan operations."[1]

The allied forces formed practically three distinct armies—the one blockading Pampeluna, the other besieging San Sebastian, while the third

[1] Napier.

army, or centre of the allies, "was indeed an army of succour and connection; but of necessity very much scattered and with lateral communications so few, difficult, and indirect as to prevent any unity of movement."[1]

On the 24th July Soult collected the right and left wings of his army, with one division of the centre and two divisions of cavalry, at St. Jean Pied de Port, leaving General Villate with the reserve on the great road to Irun in front of Sir Thomas Graham's corps, and on the 25th attacked Byng's post at Roncevalles with between 30,000 and 40,000 men. The Fourth Division moved to the support of General Byng, and these troops maintained themselves during the day; but their position being turned in the afternoon, they fell back during the night to Zubiri. On the same day the enemy with two divisions had attacked the Second Division at Maya, which, in consequence of the retrograde movement from Roncevalles, retired to Irunta and on the 28th to Lizasso. Lord Wellington was not informed of these attacks till late at night on the 25th, and at first intended to concentrate his army on the 27th towards Zubiri; but Picton and Cole, not thinking they could hold their ground until that time, retreated to the position of Huerta and arrived there early on that date. In consequence of this event, the Hussar Brigade under Lord Edward Somerset marched at daybreak on the morning of the 27th and arrived on the right of the position at Huerta about five o'clock in the afternoon. Through the valleys of the Argu and

[1] Napier.

Lanz lead the direct roads from Roncevalles and the pass of Maya to Pampeluna, and between them and in front of Villa Alba is a broken mountain of considerable height and extent. Upon it was posted the Fourth Division, Byng's British and Campbell's Portuguese Brigades—their left at a chapel behind Sorauren in the valley of Lanz, and the right on a height which defended the high road from Roncevalles. Morillo's division of Spaniards and the troops which could be spared from the blockading corps before Pampeluna were in reserve. At an angle with the mountain above mentioned is a low ridge which joins the Argu on its left bank and there forms a strong rocky pass, but it is not difficult of access, as it extends to the hills beyond Olaz, and in front of this ridge was a small rivulet having steep banks. On this position were placed the Third Division and the Hussars, supported by Ponsonby's brigade of cavalry, this being the only ground on which the latter arm could act with advantage. The enemy formed on the mountain between the Lanz and Argu Rivers, and one division and a large body of cavalry on that in front of the Third Division. They attacked the hill commanding the Zubiri road, but were repulsed by a battalion of Portuguese and a Spanish regiment. The thunder was tremendous and rain fell in torrents towards night.

Early in the morning of the 28th the Sixth Division occupied the mountain on the right of the River Lanz, and extended in rear of the left of the Fourth Division. The Fifteenth was to-day posted in the first line between two brigades of the

Third Division; the Tenth and Eighteenth, with a brigade of Horse Artillery, were on the extreme right; the heavy cavalry supported the Hussars. Next morning early the enemy's right advanced in great force from Sorauren along the Lanz valley, but was assailed with a most destructive fire in front, flanks, and rear from the left of the Fourth and from part of the Sixth Division—which latter had most opportunely arrived—and was repulsed with enormous loss. To extricate it a serious attack was made on the left of the Fourth Division, and the battle became general along the line, but the enemy's repeated attacks were repulsed with heavy casualties. All the regiments of the Fourth Division charged with the bayonet—some of them even four times. The enemy having upwards of 2,000 cavalry and a numerous infantry in front of the Third Division, made demonstrations to attack its position, but these ended in partial skirmishing by their light troops and in detaching a body of cavalry over the rivulet to feel our right: these were driven back by the Hussars stationed there. On the 29th the regiment covered the right of the position; the Seventh Division was on the left of the Sixth, and two of the enemy's divisions had followed Hill's corps to Ostiz. The brigade of Household Cavalry arrived and took post on the right. Finding an impenetrable barrier to the relief of Pampeluna in the position of the British right and centre, the French hoped by driving back Sir Rowland Hill's corps on the left to turn it, and thus gain their object. For this purpose, on the night of the 28th, Soult

4

drew from his left more than half the infantry and on the following day a brigade of cavalry, reinforced his right with one division, and during the night of the 29th occupied in force the crest of the mountain on the right of the Lanz, opposite the Sixth and Seventh Divisions, withdrawing all his troops from the left. There was only some skirmishing on this day.

On the 30th Lord Wellington became the assailant whilst the enemy attacked Sir Rowland Hill. The Seventh Division carried the crest of the mountain in its front, and the Sixth then got possession of the village of Sorauren. The Third Division advanced at 8 o'clock in the morning, turned the left of the enemy's centre by the Roncevalles road, whilst the Fourth attacked the hill in front, the crest of which was gained by 12 o'clock, and the enemy fell rapidly back towards the frontier. Their guns were sent to St. Jean Pied de Port after the battle of the 28th, and their cavalry was employed in carrying their wounded men to the rear.

General Hill repulsed the enemy with great loss on the same day; on the morrow he defeated their rearguard of two divisions in the pass of Donna Maria, and on the 1st August the army was in the same position it had occupied on the 25th of the preceding month.

The infantry won immortal honour in these battles, but, fought among mountains and defiles where the cavalry could not act, this arm had little opportunity of sharing in the glory of their comrades or of reaping advantage from the enemy's retreat.

The Fifteenth remained for a short time in the villages of Elcano and Sagasetta, then marched to Artagona and then to Lurraga, General Sir Stapleton Cotton[1] and the Cavalry Headquarters being at Tafalla.

On the 18th October the right squadron under Captain Thackwell marched to Salinas de Pamplona to assist in the blockade, apprehensions being entertained that the garrison would attempt its escape in the direction of Jaca. A squadron of the 10th Hussars was placed under his command, and at a village in the rear of the blockading troops they remained on this duty till the town surrendered on the 31st, and the garrison, amounting, sick including, to about 4,000 men, marched out at 2 o'clock on the following day and laid down its arms. To support the attack on the enemy's fortified position on the Nivelle on the 10th November, the Fifteenth advanced through the mountains on the 4th to the banks of the Bidassoa; but the broken and hilly nature of the country, and the difficulty of procuring forage, rendered it expedient that it should remain at St. Estevan on the Bidassoa, whence it soon returned to the vicinity of Pampeluna. Marching again on the 15th December by Tolosa and the bridge of Irun, it entered France, and was quartered at Campo and the villages on the right bank of the Nive—and continued on outpost duty in front of Urcuray during the month, watching the valleys of Macaye and Mondionde and the road to St. Jean Pied de Port. The right of the army stationed in this quarter

[1] Afterwards Field-Marshal Lord Combermere.

consisted of two divisions of infantry, the Spanish division under Morillo, and two brigades of cavalry; these were several times threatened with attack by the enemy, but nothing of importance occurred.

In consequence of the bad roads and wet weather the regiment experienced much difficulty in procuring forage and had frequent skirmishes with the enemy's picquets and detachments in obtaining it. But as the foraging parties of the troops investing Bayonne were also permitted to enter the valleys of Macaye and Mondionde for the same purpose, the forage was soon exhausted, and shortly after the middle of January the horses were grazed, when practicable, and fed, until the opening of the fresh campaign, with chopped furze pounded with a mallet. Notwithstanding, however, the want of the usual rations of hay and the very small allowance of corn which was issued, the horses kept their condition remarkably well, a result attributable to the unremitting attention of the officers and the exertions of all ranks.

The "state" of the Allied Army for the 16th January 1814 shows the Hussar Brigade as having a strength of 1,438, the Fifteenth having 466 effectives, while the 7th Hussars, which had joined the brigade from England during the winter, were 513 strong.

"Early in February a sudden frost fell on the moist plains around Bayonne and turned the leagues of liquid mud into stone. And the frost, which made the earth rigid, set loose all the streams of war."[1]

The campaign opened on the 14th of that

[1] Fitchett.

month, the right of the army driving in the enemy's position on the Joyeuse River and the St. Jean Pied de Port road, and afterwards defeating General Harispe and Paris at Garris. The right squadron of the Fifteenth under Captain Thackwell moved out on the 14th, and had one man and two horses wounded in assisting to drive back the enemy's picquets; the movement of this squadron was intended to favour the observations of the general commanding the cavalry as well as to preserve the communications between the right and left columns. The squadron was then pushed out by St. Martin and Oregue in front of the Third Division, and established a picquet on the 17th, without interference by the enemy, beyond the heights of Came on the right bank of the Bidouze River—watching the roads to La Bastide, Leren, and Peyrehorade. The remainder of the regiment joining the advanced squadron, they watched the enemy's posts on the Gave d'Oleron. The right of the army having crossed the Gave, the Third Division occupied a position on its banks opposite the *tête-de-pont* at Sauveterre, where the enemy had a corps of 5,000 to 6,000 men, and on the 24th this division, with the Hussar Brigade, made a feint of passing the Gave d'Oleron at a ford near Sauveterre to favour the passage of the Second, Light, and Portuguese Divisions at Villenave and of the Sixth near Montford. The ford near the bridge was deep and impracticable, and although a few cavalry and infantry passed the river it was found expedient to withdraw them owing to the impossibility of affording support, and owing also to the superiority of the enemy. In withdrawing, the infantry sustained some loss; the Hussars did not suffer, although much exposed to the enemy's shells. During this demonstration, which was never intended to develop into a real attack, the enemy blew up the bridge covered by their *tête-de-pont*, and the columns on the right passed the Gave, and moved by the road leading from Sauveterre towards Orthes.

On the 25th the Hussars passed the Gave d'Oleron by a good ford below the bridge of Sauveterre, and occupied for the night the villages in front with outposts towards the bridge of Berenc . . . the Third Division were in and in rear of the town. Whilst these operations were taking place, the left of the army, consisting of the Fourth and Seventh Divisions, and Colonel Vivian's brigade of Hussars (18th Hussars and 1st Hussars King's German Legion) which had been in observation on the lower Bidouze, passed the Gave de Pau near Peyrehorade on the morning of the 26th, and advanced along the road leading to Orthes, the enemy having retired from that position in consequence of the movements on their left. As the column approached, the Hussar Brigade, with the Fifteenth in front, passed the Gave de Pau by a ford below Berenc, followed by the Third Division. The enemy's cavalry picquets were driven back, but the passage was effected a few moments too late to cut them off on the Peyrehorade road; the only result therefore was some skirmishing, in which one man and horse of the Fifteenth were killed and three rank and file and two horses wounded.

The French army had taken up a strong position; their left in the town of Orthes and on the heights above it; their centre on a continuance of the same heights, and their right on a salient occupying the village of St. Boes. The Fifth and Light Divisions crossed the Gave de Pau at Berenc at daylight on the 27th, and about ten o'clock the Fourth Division, supported by the Seventh, and by Colonel Vivian's brigade of Hussars, attacked the enemy's right, and soon after the Third and Sixth Divisions with the Hussar Brigade attacked his left centre, the Light Division, part of which was in reserve, maintaining communication between the columns on its flanks. These attacks at length dislodged the enemy from the heights and gained the victory. The Second Division forced the passage of the river above Orthes, and with a brigade of cavalry was directed on the great road between that place and St. Sever. The enemy had defended his several positions obstinately, and retired in good order, but threatened on his left, and, the British beginning to close on the last position held by his centre, he commenced a precipitate retreat on the river called the Luy de Bearn, and this soon became a flight, the fugitives spreading all over the face of the country. In these operations the Fifteenth was close in support of the infantry of the centre, and experienced some loss from the enemy's fire, but the ground was too broken to allow of its charging. The 7th Hussars were more fortunate, being at the head of the brigade during the enemy's retreat, and the leading squadron charged the rear of the French on the road to Sault de Novailles. On reaching the hill beyond this village the French right wing was discovered retreating in the greatest confusion over the meadows towards the river; the 7th were ordered along the road leading to Sault de Novailles, and turning to the left soon came up with the flying enemy. The Fifteenth continued to advance at a trot along the main road in the hope of closing on the confused mass in front, but within less than a mile of Sault de Novailles the brigade of cavalry attached to the Second Division debouched upon the road in its front. The enemy's left and centre passed the river in the greatest confusion, and it was thought that had the pursuit continued many prisoners might have been captured; it was, however, ordered to be given up, and some guns on the heights on the left of the town, covered by the river in front and part of the rallied French infantry, rendered any attack inexpedient, even had not the closing day prevented it.

The Fifteenth had one man and two horses killed, six men and five horses wounded.

On the 1st March the regiment formed the advanced guard of the centre column of the army from the bivouac on the river to near Caceres, crossing the Adour by a ford just below the broken bridge of St. Sever. Within a league of Grenade, the right squadron under Captain Thackwell, being in advance, began to skirmish with the enemy's rear guard, which defended for a short time the passage of a broken bridge over a deep rivulet. On approaching the town the right half-squadron pressed the enemy rapidly through it upon their support by a continued attack, charged and drove 250 to 300 men of the 13th Chasseurs à Cheval along the road, for more than three-quarters of a mile, upon two companies of infantry posted in the enclosures of a farmhouse near the road. The fire from these at less than a hundred yards distance checked the pursuit, the more as the left half-squadron had been ordered to halt at the entrance of the market-place of Grenade nearly a mile in rear. The troop in advance did not consist of more than 45 men, owing to men being left with prisoners and patrols which had not rejoined, and it therefore became necessary gradually to withdraw from the heavy mass in front. This gave the French cavalry encouragement—they rallied, and, flanked by their infantry, commenced to charge; but the Fifteenth rear division, fronting, and galloping to the attack, drove them back for more than a hundred yards. The enemy repeated their charge and were again repulsed; but freed from the fire of assailants who could not be reached, the advanced guard made no further retrograde movement, and their opponents then retreated upon Caceres. Several prisoners who had been captured, contrived, however, to escape, and the enemy were able to recover a number of their wounded.

On the arrival of the head of the division, about half an hour afterwards, the squadron continued its movement to within a short distance of Caceres, when the enemy's rear guard of infantry with three guns posted on a thickly wooded eminence in front stopped its further advance.[1] These were soon, however, driven from their positions by a few rounds from the artillery of the division and the advance guard of the infantry, and outposts were then established for the night. In this affair the right troop had one horse killed and six men and six horses wounded, and Captain Thackwell had the good fortune to have his conduct approved by Lieutenant-General Sir Stapleton Cotton, and to be recommended by him for the brevet of Major, and the following cavalry divisional orders were issued in approbation of the conduct of the right squadron of the Fifteenth.

"Lieutenant-General Sir Stapleton Cotton requests that Major-General Lord Edward Somerset will express to the officers and men of the 15th Hussars his gratification at witnessing the gallant and soldierlike conduct of that part of the regiment which was engaged with the enemy yesterday.

"*(Signed)* J. ELLEY, COLONEL, A.A.G."

"Major-General Lord Edward Somerset has much pleasure in making known the cavalry orders to the corps composing his brigade, and joins the Lieutenant-General in expressing his perfect approbation of the conduct of the brigade on the 27th

[1] One man of the advanced squadron, Robert Walton, had a narrow escape from a shell which carried away the cloak from his back (the men having cloaked owing to the heavy rain) and burst without doing any further damage. The advance guard was here most fortunate, for many round shot and shell were fired at it, but only one horse—a French one taken a few hours before—was wounded.

ult. The Major-General has also to return his thanks to the 10th and 15th Hussars for their gallant attacks on the enemy's cavalry on the two following days, and feels convinced that, with troops thus disciplined, the most complete success may be expected to attend their future operations against the enemy.

"*(Signed)* C. JONES, M.B."

The French having been repulsed at Aire on the 2nd March by the Second Division with considerable loss, and by the Sixth Division at Caceres, the outposts were established near Plaisance, but soon fell back in front of St. Germier, in consequence of the enemy concentrating at Conchez and threatening the right at Aire. Lord Wellington's advance had been delayed owing to the heavy rain, the rapid current of the Adour preventing the laying down of pontoons, and owing to the necessity of repairing the bridges—all of which had been broken down by the enemy.

Some changes now took place in the regiment. On the 10th March Lieutenant-Colonel Dalrymple arrived from England and displaced Major Griffiths, who had commanded the Fifteenth since Colonel Grant had left it for a brigade. Five other officers also joined, and the establishment of the regiment being increased by two troops, ten sergeants, two trumpeters, and 148 rank and file joined headquarters on the 15th with 160 horses.

The French having retired upon Lembege on the 15th, it was determined to drive back their picquets in front of St. Germier, and on the afternoon of the following day the centre squadron of the Fifteenth was formed in column of divisions—the road not admitting of a greater front—and advanced supported by the right squadron under Captain Thackwell. On approaching the enemy the leading division charged with the utmost vigour, driving his advanced squadron back in confusion upon the remainder of the 13th Chasseurs à Cheval, composed of some 300 men. The attack was none the less continued by the centre squadron of the Fifteenth, and the enemy, giving ground, retreated as rapidly as his close formation would permit, and was pursued for about two miles. He at length gained the village of La Cassade, on the road to Plaisance, where, being secured by hedges and walls, it was deemed best to discontinue the pursuit. The right squadron was here ordered to the front to find the outposts for the night, and the enemy continued his retreat, a picquet watching the approach to the former place, and, recrossing the river on the outposts advancing at daybreak next morning.

Major Griffiths now took over command of the right squadron, Captain Thackwell being transferred to the charge of the left.

The regiment continued in the advance, and on the 20th the army moved in two columns from Vic-en-Bigorre and Rabastens upon Tarbes, where Marshal Soult was in position with his right upon the heights near the windmill of Oleac, and his centre and left retired, but with a strong corps occupying the town of Tarbes. The right column of the Allies advanced by the road leading from Vic-en-Bigorre, drove the enemy from Tarbes, and was then disposed for the attack of his left, whilst the Sixth Division, passing the village of Dours, attacked his right, and the Light Division, supported by the Hussars, drove the French troops from the heights above Orleix. The enemy opposed these attacks but feebly, and retired over the narrow river towards Tournay. The Hussars were pushed rapidly on in pursuit of his right, but the enemy avoided the only ground where cavalry could act, and unfortunately gave no opportunity for the attack.

The French Army continued its retreat upon Toulouse, followed by the British, and on the 25th March the Fifteenth was on outpost in front of St. Lys on the La Touche River. On the day following it had a squadron on duty at Tournefeuille, which had to resist an attack of the enemy's infantry in a situation where cavalry could not reach them; and on the 27th, Captain Thackwell's squadron, after gaining possession of St. Simon, found it to be untenable against infantry and therefore withdrew.

The heavy rain and the melting of the snow in the Pyrenees rendered it impracticable to lay a bridge over the Garonne before the 4th April, although a demonstration to pass the river was made between Toulouse and Muret on the 28th March; but the attempt was abandoned on account of the roads towards the Ariége being impracticable for cavalry and artillery. By the 1st April the enemy had withdrawn most of his troops into Toulouse, and his advanced posts were not more than a mile from the bridge, on the left bank of the river. The army under Wellington was stationed as follows: the Fourth Division and Vivian's Hussars were at St. Martin on the left; the Sixth Division and 10th Hussars at and in front of Tournefeuille; the Third Division was in reserve at Plaisance; the Light Division and 15th Hussars at and in front of St. Simon; and the Second Division was at Portel and Muret, with four brigades of cavalry. On the morning of the 4th the Hussars marched to La Chapelle, and passed the Garonne by a pontoon bridge of seventeen boats, without opposition, being followed in the course of the day by the Third, Fourth, and Sixth Infantry Divisions, two brigades of cavalry, and four of artillery. These troops took up a position in front with the right on the Garonne and the left on the Ers River; Captain Thackwell's squadron found the outposts at Gagnac.

The enemy's picquets in front of Feneuillet were driven in on the 8th, and on the same day the pontoon bridge was removed to the Château de Gagnac, Freyre's corps of Spaniards crossing to the right bank of the Garonne, followed on the next day by the Light Division.

Toulouse is encompassed on three sides by the Garonne and canal of Languedoc, and, in addition to its ancient walls, *têtes-de-pont* covered all the bridges over the river leading to the town, and these were in numerous places defended by artillery and by musketry from the old ramparts. On the heights between the town and the Ers River the enemy had constructed five redoubts connected in parts by lines of entrenchments extending to Montaudran. All the bridges over the Ers—the banks of which were steep and impracticable—were broken down except that at Croix d'Orade. Soon after 5 o'clock on the morning of the 10th, the British columns were put in motion to attack this formidable position occupied by the enemy in considerable force. The Second Division was to act on the left of the Garonne, and the pontoon bridge was laid over the river nearer to Toulouse, to afford a readier communication with the Third Division—intended to press on the *tête-de-pont* and fortified houses covering the canal bridge nearest the Garonne, supported by the Light Division and Bock's brigade of cavalry. The Spaniards were to attack the north-west part of the position supported by the Heavy Cavalry; the Fourth and Sixth Divisions, with the Hussars of Somerset's brigade, were to turn and attack the right; while Vivian's brigade of Hussars was to watch the movements of the enemy's cavalry on both banks of the Ers beyond the left.

At 8 o'clock in the morning, the Spaniards began to skirmish with the French, and, although annoyed by a galling fire from five or six field pieces on rising ground, began to form in two lines in front of the enemy's left, and about ten o'clock attacked with vigour; but, the right being turned, were soon obliged to retire with great loss, and were rallied with difficulty under cover of the cavalry and the Light Division, which moved up on their right in support. The Fourth and Sixth Divisions had crossed the Ers at the bridge of

Croix d'Orade, and about eleven o'clock, formed in three lines, the Fourth Division leading, carried the village of Montblanc, and moved up the Ers until they outflanked the enemy's right, then re-formed and advanced to theattack of the heights. The Hussars, left in front, quickly followed the movements of this column and did not suffer much, although the road led them within less than 300 yards of the enemy's guns, which from the redoubts were beating time to the din of small arms. A column of the enemy's cavalry, which had begun to descend the heights to attack the right of the Sixth Division and the brigade of nine-pounders, was repulsed by this timely movement, which brought the brigade in rear of the Sixth Division ; these had most gallantly taken the first redoubt with guns, and lodged themselves on the heights. The Hussars crossed, exposed to a heavy fire from the enemy's 2nd and 3rd redoubts, and six guns on an eminence near the canal, in the hope of falling on his infantry in retreat towards the canal and town ; but as it was seen that no advantage could be reaped from this movement, they fell back under cover of the heights.

The artillery of the left column had remained behind ; it was now brought up, the Spaniards were re-formed, and the Sixth Division moved to the attack of the 2nd and 3rd redoubts defending the enemy's centre, carried both and took a great number of prisoners. But before these could be disarmed and the redoubt nearest the town properly occupied, a heavy column made a desperate attack, retook it and endangered the brigade; the check was, however, only momentary, for all opposition gave way before the supporting body of troops, and the French were driven back to the town with great loss. The position of the Sixth Division commanded the two redoubts on the enemy's left, and these, on being cannonaded and threatened by this division along the heights and by the Spaniards in front, were evacuated by 5 o'clock in the afternoon. The whole range of the heights was now in our possession, but the enemy kept up a useless fire until late in the evening from guns on the canal covering his right, and from a six-gun battery on rising ground before the town.

During these operations the Second Division forced the enemy from the entrenched suburbs within the old town wall, while the Third Division drove them into the *tête-de-pont* on the canal bridge; but in attempting to carry this work, the division was repulsed with loss by a superior force, aided by the garrison of a fortified convent near to it.

On the night of the 11th the enemy evacuated Toulouse and retired towards Castelnaudary and Carcassone by the only road left them between the canal and the Ariége River, but Generals Harispe, Baurot, St. Hilaire, and 1,600 men were made prisoners, and guns and stores of all descriptions were taken. At 9 o'clock on the following morning the Hussars followed the enemy's movements to Baziége, near which place the Hussars of the King's German Legion had the last affair with the enemy, for on the morrow intelligence was received that Bonaparte had abdicated in favour of Louis XVIII., and hostilities ceased. The brigade, however, marched on the 17th from the vicinity of St. Sulpice to Puylaurens, in consequence of Soult not acknowledging the decree of the Provisional Government; but a few days afterwards—

"the Duke of Dalmatia, who had now received official information from the chief of the Emperor's staff, notified his adhesion to the new state of affairs in France—and with this honourable distinction, that he had faithfully sustained the cause of his great monarch until the very last moment."[1]

The Fifteenth went into cantonments in the villages in the neighbourhood of Toulouse, the headquarters of the Hussar Brigade being at Villandrique. On the 1st June the regiment commenced its march for Boulogne, where the British cavalry was to embark for England, and, moving by way of Cahors, Limoges, Orléans, Mantes, Gisors, Abbeville, and Montreuil, arrived in the vicinity of that place on the 11th July, having marched about 650 miles. The greater part of the regiment landed at Dover on the 17th, the remainder the next day, and the whole assembled at Hounslow, where it was quartered.

[1] Napier.

CHAPTER V

When the Fifteenth reached England from France, the strength of the twelve troops of which the regiment was composed amounted to 910 men, exclusive of officers, with 746 horses, but on the 9th August four troops were struck off the establishment, 350 men were discharged, and 288 horses were cast, the establishment being now fixed at eight troops, each of 69 men and 56 horses. On the 11th August the regiment marched to Liverpool and there embarked for Ireland on the 8th September. A few days were spent in Dublin, after which headquarters and three troops were quartered at Clonmel, the remainder of the Fifteenth being at out-stations.

But with the New Year the peace of garrison life in Ireland was dispelled. Napoleon had left Elba on the 26th February, and landed three days later in France, but it was not until the 6th March that the news even of his departure reached Vienna and the Congress which was there in session. The Congress, which was "perishing of mere strife among its own members, strife bred of unsatisfied greed and fast-kindling jealousies,"[1] became at once united at the news and acted with decision.

[1] Fitchett.

"The powers, therefore, acted as though Napoleon were once more the master of France. They signed a declaration on 13th March declaring him beyond the pale of civil and social relations. . . . On 25th March the League of Chaumont was formally renewed, the four great Powers binding themselves to contribute 150,000 men each, and not to lay down their arms without joint consent, and only when Napoleon should be unable to give further trouble."[1]

In England preparations were at once put forward for the speedy assembly in the Netherlands of a mixed force of 106,000 British, Hanoverians, Belgians, and Nassauers, under the command of the Duke of Wellington.

Under date of the 19th April Captain Thackwell's diary contains the following entry: "The day cloudy, with some rain. The order to prepare six troops for foreign service received this day." On the 4th May the regiment marched for Cork —where a number of small brigs had been taken up for the passage of the Fifteenth to Ostend —and sailed on the 13th. Ostend was reached on the morning of the 19th, and here the vessel in which Captain Thackwell sailed was run ashore on the sands near the town, and at 1 p.m. disembarkation commenced by lowering the horses into the water and swimming them ashore. No time was wasted on landing, for at 3.30 the same afternoon the regiment marched up the left bank of the canal fourteen miles to Bruges, and thence by Eckloo to Sleydinge within six miles of Ghent, where Louis XVIII. had established his court.

[1] Fitchett.

Moving on the 26th, the Fifteenth passed through Ghent and occupied several small villages about St. Gooritz, St. Marie Anderhove, and Mickelbecke Elste; they now found themselves brigaded with the 7th Hussars and the 2nd Hussars of the German Legion, under their old commander, Major-General Sir C. Grant, and took part on the 29th in a grand inspection, by Lord Wellington and Marshal Prince Blücher, of the whole of the British Cavalry—46 squadrons with 36 Horse Artillery guns, and a rocket brigade—commanded by Lord Uxbridge—near the village of Schendelbecke.

From Captain Thackwell's diary the weather at the end of May and beginning of June seems to have been very unsettled, rain falling nearly every day. Friday 16th June was, however, fine and warm, when an order was received to march with the greatest expedition. Starting at 7.30 a.m. the Fifteenth moved by Grammont, Braine le Comte, and Nivelles—a distance of 15½ leagues—and finally bivouacked at about midnight near Quatre Bras—too late to take part in the action fought there that day, and hearing there of the defeat of the Prussians at Ligny. The following entry occurs in Captain Thackwell's diary dated the 17th June:

"A very fine morning, everything quiet. Saw the ground on which the action was fought the preceding day. The right rested on a wood of considerable length, and this was occupied by the Brunswickers and Belgians. Nothing particular to mark the position. Many Cuirassiers lay dead. Sent by the Earl of Uxbridge to our picquet on the right, and to post another watching the forest of Hautain le Val. In consequence of the retreat of the Prussians, the infantry fell back about 10 a.m. to a position in front of Waterloo on the Brussels road. At nearly 3 p.m. the cavalry, which had remained in position, retired through Genappe in three columns. The column on the main road was followed by a large body of cavalry and some guns, and these were occasionally charged by the Life Guards and 7th Hussars; the last regiment suffered severely. The right column, consisting of the 13th Light Dragoons,[1] 15th Hussars, 2nd Light Dragoons of the German Legion, and the Duke of Cumberland's Hanoverian Hussars, was not pursued. I commanded the rear guard. On the Nivelles road near Lillois some French squadrons of light troops cut in and made some baggage wagons with wounded and cattle prisoners; these were checked by Captain Wodehouse's squadron, half a squadron of the 13th and some of the Germans, and a few prisoners were made. At three this afternoon the rain fell in torrents, which continued at intervals. The fields were perfect swamps. The Thirteenth and Fifteenth bivouacked in a field of rye on the right of the village of Mont St. Jean; fortunately there were some infantry huts standing, which afforded a little shelter from the torrents of rain which fell during the night. No rations or supplies of any description."

At 4 o'clock on the morning of the 18th the Fifteenth marched to their position in the front line at the angle in the rear of Hougoumont, with three troops detached on the right of the Nivelles

[1] The 13th Light Dragoons was attached to Grant's brigade in place of the 2nd Hussars of the German Legion, which had not yet joined.

road, in front of the right flank of the army, which at this point was partly refused.

During the earlier part of the action the Fifteenth suffered some loss from the cannonade in front, as well as from a heavy battery on the high ground overlooking the Nivelles road, and occupied the same position until 2 o'clock, when the brigade received orders to charge ten squadrons of Lancers posted on the heights on the enemy's left of the Nivelles road, covered by a deep ravine. For this purpose the Fifteenth, followed by the Thirteenth, moved off to their right, exposed to a sharp fire of round shot until they became sheltered by the inequalities of the ground. While making dispositions to cross the ravine, a tremendous shouting from the Lancers drew attention to a large body of Cuirassiers and other cavalry which seemed to carry all before them on the open ground between Hougoumont and La Haye Sainte. Grant's brigade at once made for them—this movement bringing the Thirteenth in the front line, the Fifteenth following in support. The French cavalry was driven back 200 or 300 yards, but its numbers being much superior to those of the English brigade, the latter's flanks were enveloped and it was obliged to fall back and rally behind the infantry.

From this period until the enemy was finally driven from the field, the regiment made various charges—sometimes attacking hostile infantry, again engaged with the Lancers, or driving back the Cuirassiers. Major Griffiths had been killed, Lieutenant-Colonel Dalrymple had had his left leg taken off by a round shot,[1] and the command of the Fifteenth devolved upon Captain Thackwell, who had already had two horses shot under him. It was about 7.30 p.m. that, in leading a charge upon a square of the Guard, Thackwell was shot through the bridle hand; he instantly placed the reins between his teeth, but a few seconds afterwards he was again shot in the left arm, shattering the bone between the elbow and the shoulder, and fell to the ground.

In the battle the Fifteenth had 3 officers and 25 men killed, or died of wounds, and 7 officers and 45 men wounded.

It may perhaps not be out of place to include here a copy of a letter written many years after the battle, in reply to a printed communication from Lieutenant Siborne, then engaged in constructing his well-known model of the field and battle of Waterloo, asking for information on certain specified points in connection with the position and formation of the 15th Hussars, at the period of the battle represented on the model.

"GLOUCESTER, *20th* Dec. 1834.

"SIR,

"I must plead a severe indisposition in excuse for having so long delayed replying to your communication of the 18th November relative to the position of the 15th Hussars in the attack of the right of the British by the Imperial Guard, about 7 o'clock in the evening at the Battle of Waterloo, and regret that from the lapse of time,

[1] The same round shot then passed through the body of Sir Colquhoun Grant's charger.

Photo Debenham, Gloucester.

p. 72

Captain Thackwell leading the Last Charge of the 15th Hussars at Waterloo.

and the circumstance of my having been severely wounded about that hour, I am unable to transmit that full account of movements and positions, which might have been rendered at an earlier date. However, as you have the advantage of Lieutenant-Colonel Wodehouse's statement of the occurrences of that day, and may procure one from Lieutenant-Colonel Hancox, who commanded the regiment at the close of the battle, and who resides near Nottingham, I trust the barrenness of my information may be supplied from those channels of authentic source and that the King's Hussars may be placed in the situation that belongs to them.

"Before I reply to your queries I had better here state that the squadrons of the regiment were not more than 52 to 55 files each, including officers, and that one squadron and one division of another were detached from the regiment on the morning of the 18th of June and did not rejoin it, except for a short period, during the day. This detached body was posted in observation in front of the valley leading to Braine le Leud, and as its operations were confined to skirmishing, its loss was trifling.

"1st. With respect to the query 'What was the particular formation of the 15th Hussars at the moment (about 7 p.m.) when the Imperial Guard, advancing to attack the right of the British forces, reached the crest of our position?' I beg to trace the formation of the 15th Hussars at A in the plan. On its flanks were British infantry in square, but I am not certain what regiments, as part of Lord Hill's corps from the second line, were then in the first line, and I am not sure whether the site of the traced position is not a little too much in advance. The 13th Light Dragoons were either to the right or the right rear of the Fifteenth; but as the troops were at this time closely concentrated in this part of the position, it was a difficult matter to distinguish particular corps.

"2nd Query. 'What was the formation of that part of the enemy's forces immediately in front of the 15th Hussars?' At B I beg to trace, according to the best of my recollection, a body of about 1,000 infantry in square, supported by a large body of Cuirassiers and other cavalry. This square was charged by the three troops and a half of the 15th Hussars, as it was halted in fine order, about the time of the advance of the Imperial Guard, or a little before; but as I was then severely wounded, I did not observe in what manner these troops were supported on their flanks, or how their retreat was conducted, but very large masses of cavalry were in their rear. With regard to the crops growing on the fields on the 18th June, I beg to state, as well as I am able to remember, that there was no fallow land in the vicinity of the Waterloo and Nivelles road, on the right, in the direction of Braine le Leud, nor on the left, as far as the crest of the position and Hougoumont. The crops were for the most part wheat, rye, and oats, or clover or grass hay, and particularly the latter in the hollows and ravines on the left of the track leading from the Nivelles road towards Braine le Leud; however, towards the enemy the crops were so much trodden down that the surface looked more like broken stubble than a golden harvest.

"In the early part of the day the position of the 15th Hussars was in the plan at C, that of the right squadron at C2, and that of a picquet at C3; but after the battle had begun until about half-past 2 p.m. it was at D, and for nearly an hour afterwards it was at E, whence it moved with the 13th Light

Dragoons to about F, for the purpose of attacking ten squadrons of Lancers posted in line in rear of a deep ravine at G. It then joined the right squadron of the regiment; but owing to the impetuous attack of the French cavalry on the right centre of the British position, the intended attack on the Lancers was given up, and the regiment, leaving the right squadron where it was originally posted, retraced its steps to the vicinity of the position A, and was immediately engaged in the attack, by charge or skirmishing, of Cuirassiers and other cavalry; and this lasted until the enemy's cavalry found it could make no lasting impression on this part of the position.

"The enemy's cavalry and infantry moved in column, both in advance and retreat, the former being at about quarter distance, and I understand when the British line advanced that three troops of the 15th Hussars charged a body of infantry as well as some Lancers.

"The position of the regiment being in rear of Hougoumont, the masses of infantry which would have closed on its post were intercepted by the troops defending that place and none of the enemy's infantry to the best of my recollection passed its enclosures, and the first I saw of that force in the immediate front of the Fifteenth was the column charged by my squadron; but I witnessed the advance of many heavy masses of infantry which attacked Hougoumont, although soon after the firing began the distant movements of the enemy's columns were from this point of the position but indistinctly seen owing to the smoke, which hung lazily on a surface saturated with rain. The left of the enemy's infantry extended to the Nivelles road nearly in line with G in the plan, whence a heavy fire of artillery was kept up, for the chief part of the engagement, upon the angle of the British position.

"Begging you will believe I shall be happy to give any further explanation, I have the honour to be, etc. etc. etc.

"Jos. Thackwell,

"*Late Lieutenant-Colonel Com. 15th Hussars.*"

On the 9th of the July following the date of this communication Lieutenant-Colonel Thackwell submitted to Lieutenant Siborne, for Major-General Sir Colquhoun Grant, this account of the proceedings of the 5th Cavalry Brigade at Waterloo:

"Having been requested by Lieutenant-General Sir Colquhoun Grant (owing to his inability in consequence of a severe family affliction) to reply to your printed letter of 28th October 1834, soliciting information with respect to the proceedings of the 5th Brigade of Cavalry in the Battle of Waterloo, I beg leave to state, in connection with a former communication relative to the movements of the 15th Hussars, that, with reference to the query, 'What was the formation of the 5th Cavalry Brigade, etc. etc.?':

"The brigade about this time was formed in line of squadrons at or about AA, A1, A2, in the plan, according to report and the best of my belief and recollection; but there might have been a square of our infantry between the Fifteenth and the Thirteenth, and their position might have been 50 or 80 yards more to the left. I must, however, here remark that the relative situation of the three regiments was not the same throughout the day, the Thirteenth being for a great part of the conflict on the left of the Fifteenth. The squadrons at this time did not probably amount to 30 files each. I should also perhaps explain here that the 13th Light Dragoons had joined the brigade that morning in consequence of the 2nd Hussars of the Legion not having returned from the frontier; that the 15th Hussars had a squadron in a ravine at A1, and a picquet at A2, from the beginning to the end of the battle, which detachments suffered some loss by cannon fire and skirmishing; and that the 7th Hussars were very weak, having suffered most severely on the debouch of the French cavalry from Genappe the preceding afternoon.

"With regard to the second query, 'What was the formation of that part of the enemy's forces immediately in front of the brigade?' The enemy's troops, cavalry and infantry, were in column, the former at perhaps half or quarter distance. It is a difficult matter at this distant period to trace the enemy's formation more particularly, but a large force of his cavalry was in this part of the field.

"In reference to the general proceedings of the brigade, I beg to state that it was under the crest of the position, in rear of the angle at Hougoumont, until about 3 p.m., when the 15th Hussars and 13th Light Dragoons were moved to the Ravine D between the Nivelles road and Braine-le-Leud, for the purpose of attacking 10 squadrons of Lancers in two lines, forming the left of the French Army at E. Whilst dispositions were making for the attack the Lancers began cheering, and on looking towards the position we had quitted, the cause of the cheering was discovered to be an impetuous attack by the French cavalry upon our infantry and guns, the limbers of which were going rapidly towards the Nivelles road. The French cavalry passing between the squares of the infantry were charged and driven back by the cavalry of the 3rd Brigade (Dornberg's) and the 3rd Hussars of the Legion.

"This attack was several times repeated; and Sir Colquhoun Grant, judging that the attack of the Lancers was only a secondary object, most judiciously took upon himself the responsibility of taking the two regiments back to the ground they had left, the Thirteenth leading, which regiment formed line to the front, and at or about the spot marked B charged a body of Cuirassiers, who were driven back for more than 300 yards to the low ground beyond C. The Fifteenth also formed to the front, to the left of the Thirteenth, and charged a mass of Cuirassiers, which were likewise driven back for a like distance upon heavy masses of cavalry, who, beginning offensive operations in front and on the flank, compelled the Thirteenth first and afterwards the Fifteenth to retreat to their own line, where the steadiness of the two regiments had the effect of checking any further serious attack upon this point for some time; but the skirmishers of the Fifteenth were employed against the Cuirassiers and other Cavalry in front, who were kept at some distance.

"More to the left, however, the enemy's cavalry made some demonstrations in advance and attacks. Between 6 and 7 o'clock a column of nearly 1,000 infantry advanced to within 150 yards, or less, of the first line, to near the spot F on the plan, supported by a large body of Cuirassiers and other cavalry about C. This square was charged by the two squadrons of the 15th Hussars and its further advance was checked. Shortly after some Light Cavalry and Lancers, who incommoded the front of the adjoining squares of infantry, were charged and driven back by the same corps. The 7th Hussars and the 13th Light

Dragoons were also most actively engaged, and on the general advance the regiments of the brigade made several attacks on the cavalry as well as infantry, in one of which Colonel Kerrison, of the first-named regiment, had a horse killed under him, and Major Griffiths, of the 15th Hussars, was killed in charging the latter; and these attacks contributed greatly to the loss and confusion of the enemy. Sir Colquhoun Grant had five horses killed and wounded under him. He was in Hussar uniform, and rode, at about 7 o'clock, a very fine, large, chestnut horse, which was wounded.

"The prompt, judicious, and fortunate movement of the brigade from the designed attack of the Lancers, to that of the Cuirassiers before mentioned, restored confidence to this part of the line, which seemed to be in danger, and may justly be considered an event of the utmost importance.

"I should imagine it was previous to the return of the brigade that the cavalry attack witnessed by Major Mercer occurred. However, I saw no cavalry in the ravine leading to Braine-le-Leud from the time we quitted the angle of the position to that of our return, except the Thirteenth and Fifteenth passing along it, although there might be cavalry at the spot stated by Major Mercer. It might have taken place afterwards, for I know, by report at the time, that a body of Cuirassiers passed between the squares of infantry to our left, and being unable or unwilling to return, retreated towards the Nivelles road, and passed the small post of the 15th Hussars; but those who escaped were said to be not more than thirty, some having been knocked down by the fire of the Fifty-first, the direction of which prevented their being charged by the above detachment. But a difficulty occurs here, as at this time a considerable part of Lord Hill's corps had joined the first line, and probably Major Mercer with it. I think I heard this body was charged, but never knew when, or by what corps. It is certain that no attack by cavalry had been made on the position near Hougoumont until after the two regiments had quitted that part of the field to move against the Lancers, and it is therefore probable some mistake may have occurred as to time, and that the circumstances mentioned may have originated in the same affair.

"I trust I need not apologise for this long statement, confiding in the hope that you have received Sir Colquhoun Grant's note intimating that he had desired me to transmit it after having perused and confirmed all the events of moment therein mentioned."

. . . . . .

Thackwell was left all night where he fell, but the following extracts from his diary show the events of the next few days:

"*Monday, June 19th.*—A very fine day; found by Assistant Surgeon Jeyes, and conveyed to the hospital. Arm very stiff and painful; at half-past 8 o'clock had it amputated close to the shoulder; the pain great, but I bore it with fortitude, and was greatly complimented on my heroism. The ball had shattered the bone from the shoulder to within an inch of where it was divided. Put to bed in billet No. 1,143, Section 1st, Place du Grand Sablon."

"*Tuesday, 20th.*—The day fine; a little better—wrote a letter to England."

"*Thursday, 22nd.*—A fine day; shaved myself."

"*Tuesday, 27th.*—A fine day; took a few turns in the square."

"*Wednesday, 28th.*—Walked in the Park for near two hours, rather fatigued myself. Called on the Colonel, who appeared in good spirits, chatted with him for more than half an hour."

. . . . . .

"*Saturday, July 29th.*—Rode to Waterloo, and ordered the monument on the officers of the Fifteenth who fell in the battle."

On Thursday, 10th August—little more than seven weeks after his severe wound—Thackwell was on his way to rejoin the regiment, and marching by Hal, Braine le Comte, Mons, Le Cateau, Cambrai, Bapaume, Amiens—where he witnessed the public entry of the Duke de Berri, and where he found General Dornberg and the 3rd Cavalry Brigade—Breteuil and Gisors to Lions la Forêt, where he rejoined the headquarters of the 15th Hussars.

(It is very noticeable all through Captain Thackwell's diaries how invariably when marching he studies the country passed over, from a military and especially from a cavalry point of view.)

In September Thackwell had a few days' leave in Paris, and shortly after his return, on the 3rd October, the regimental headquarters were moved to Gisors. From here he went to Dangu, where General Grant had his headquarters in a château belonging to General la Grange, and here he had a curious day's fox-hunting, of which sport he remarks: "The hounds bad—the woods are so large and the foxes so numerous that the practice is to shoot them whilst hunting. Killed one—chased a hare—bad sport."

On the 10th October the regiment marched by way of Rouen to Trouville, and remained here until the 13th December, when the Fifteenth marched to join a new brigade at Dieppe, moving by Ourville and St. Valery. Early in the New Year the regiment moved again, marching on the 11th January 1816 to Trevent, on the 21st to Avesnes le Comte, early in February to Bailleul, only half a league from the Belgian frontier, and on the 29th March to Bourbourg, the last station of the regiment during the Occupation.

A reduction in the strength of the British contingent in France having been decided on, the Fifteenth was amongst the first regiments ordered to England. On Sunday, the 5th May, the regiment marched to Hondschoute and handed over 151 horses to the 7th, 80 to the 18th Hussars, and 18 to the 11th Light Dragoons. The non-commissioned officers were allowed to retain their horses, of which only 76 were now remaining in the regiment. Here the Fifteenth appears to have been inspected by Sir Stapleton Cotton, now Lord Combermere, who, on Lord Uxbridge being wounded at Waterloo, had been summoned to France to command the British cavalry of the Army of Occupation.

On the 6th and 7th the regiment proceeded to Calais for embarkation, the dismounted men and a few horses sailing on the latter date. The remainder of the regiment was to have sailed on the 9th, but the wind was contrary, and the embarkation was not effected until the following day, and it was not until the 14th that the fleet of packets, with the 15th Hussars and 13th Light Dragoons, sailed for England, arriving at Dover at 8 o'clock the same night.

Next day the Fifteenth disembarked and marched to Canterbury, where the whole regiment was assembled, and where the medals granted for the Waterloo campaign were issued to officers and men. The following honours were also granted: the word "Waterloo" was added to the distinctions borne by the regiment; Colonel Dalrymple was honoured with the dignity of Companion of the Bath; Captain Thackwell was promoted Major *vice* Griffiths, killed in action; and Captain Hancox, who had brought the regiment out of action at Waterloo, was given the brevet rank of Major. All ranks had the further privilege of reckoning two years' service for the battle.

On the 28th May the regiment marched to Hounslow, where on the 1st June, having been newly mounted, it was inspected by the Duke of York, and on the 3rd marched for Nottingham, Birmingham, and Wolverhampton.

CHAPTER VI

Major Thackwell's squadron was stationed now at Birmingham, and here on the 28th October a riot took place towards evening. The troops were called out and cleared the streets, but many stones were thrown by the mob and one of these struck Major Thackwell on the head, rendering him insensible for several hours. Several of the rioters were lodged in the gaol, and by midnight all was quiet. The mob collected again in the course of the next morning, but the Riot Act was read, and by 2 p.m. the mob was entirely dispersed, and at night the streets were empty. Two more troops of the Fifteenth were brought into Birmingham, but their services were not required. At this time there was a very great deal of unrest all over the country, and both in this year and in the following the military were constantly called out in aid of the civil power.

In June 1817 Major Thackwell was promoted to the brevet rank of Lieutenant-Colonel in the *Gazette* of the 21st, wherein it was stated "the undermentioned officers to have a step of promotion whose former recommendations were overlooked for special services in the field." And now for some twenty long years Thackwell was to be condemned to a life, not of inaction, but of peace-soldiering. The army was reduced, the regiment was split up and occupied country quarters, and did not come together again until October 1821. But for Lieutenant-Colonel Thackwell the years were not wasted, and were devoted, in a manner somewhat unusual for those days, to the unceasing study of his profession. His correspondence shows how constantly he was in communication with leading cavalry soldiers upon questions of drill and manœuvre, and how greatly his views upon such matters were sought after and appreciated. Colonel Dalrymple[1] appears from this time forward until his death to have been but little with the regiment, all matters connected with its command falling in a large measure into the hands of Lieutenant-Colonel Thackwell. His diaries for these years are full of mentions of regimental field days lasting eight and nine hours, of exercises in reconnaissance and outpost duty, and in constant inspection of troops and squadrons at out-stations. While, however, he did not spare those under him, he spared himself still less; he was constantly in the saddle, and the long rides he took almost daily all over the country—forty to sixty miles a day—helped to keep him physically fit for the more active work of his profession which was still before him. On the 6th June 1820 Colonel Dalrymple died of abscess of the brain, and Lieutenant-

[1] Colonel Dalrymple was the second son of General Sir Hew Dalrymple of Cintra fame.

Colonel Thackwell succeeded him in the command of the King's Hussars.

Ernest Augustus, Duke of Cumberland, had been Colonel of the Fifteenth since 1801 and had always taken a very active interest in its affairs. All questions concerning it—not only in regard to appointments, exchanges, and promotions, but details of uniform and mounting—had always been referred to His Royal Highness, and in Colonel Thackwell's correspondence are many letters from the Duke of Cumberland or from his staff, showing how very close was the supervision the Duke always exercised over the regiment of which he was the Honorary Colonel. In January 1823 there is some question of reverting to a former pattern of bit, which has not the Duke's approval—"besides," he remarks, "the very look of them, as they resemble *coach horse bits* and therefore certainly very little becoming for a *Hussar*." On the question of remounts the Colonel cautions the Commanding Officer to "beware of purchasing cocktails; they are frightful, and what I have most studiously avoided buying." Writing in May 1823 His Royal Highness says, "I long very much to see once again the regiment which I hear from *all* sides is in very high order." He is very particular in regard to officers desirous of exchanging into the regiment, and when applications reach him he puts very searching inquiries to Colonel Thackwell on the subject. He is loth to consider any exchange with an infantry officer, remarking very properly on one occasion that "my object is to bring young men into the regiment who will rise progressively, and by being therefore thoroughly grounded in the system of the corps will, I hope, become good cavalry officers; you will therefore inform the officers in question that I cannot consent to the exchange."

On the 25th December 1827 the Duke of Cumberland, who had been Colonel of the Fifteenth for some six-and-twenty years, severed his connection with that regiment on being appointed to the Colonelcy of the Blues. There can be no doubt that His Royal Highness was very sincerely attached to the regiment he had so long commanded, and in several letters written at that time to Lieutenant-Colonel Thackwell he expressed the sorrow he felt at parting.

"The 25th," he wrote, "was for me a most painful day, as having had the honour and happiness of being Colonel of that distinguished regiment for twenty-seven years, I felt a sort of father's affection for the Corps." He writes to Colonel Thackwell of "an acquaintance of six-and-twenty years which has existed between yourself and me, and I believe you are the only officer or man I may say remaining in the regiment since I had first the honour of commanding it. I beg," repeats the Duke, "that you will assure the whole corps of officers of my sincere regard and faithful attachment. As to yourself, I need not say how much I regret that our military connection which has now subsisted twenty-six years, will cease naturally on the 25th December."

The Duke of Cumberland was succeeded in command of the Fifteenth by Major-General Sir Colquhoun Grant.

In August 1825 Colonel Thackwell and two officers of the regiment, Lieutenant Temple and Cornet Rose, went over to Berlin and attended some great manœuvres which were held in the vicinity. They left Hamburg on the 27th, travelling by post wagon, and the roads being very bad, only arrived at Berlin on the 29th in the evening, putting up at the Hôtel Stadt Rom. Two days after their arrival the manœuvres began, and Colonel Thackwell was out early and late. He describes very fully in his diary all he witnessed, and gives the Prussian cavalry every credit for steadiness in manœuvre, and for the thorough training of the horses, but finds their movements slow, everything being done at a steady trot, while even in the charge the gallop was only used for a very short distance. The infantry impressed him with their steadiness, but Colonel Thackwell adds a remark, which our military attachés have often uttered of late years in regard to German manœuvres—viz. that there was "unnecessary exposure to cannon and musketry fire in close order." The Dukes of Cumberland and Cambridge were both present at this time in Berlin, and Colonel Thackwell met and was presented to several of the men who had helped to raise Prussia after the disasters of Jena and Auerstadt—General Hake, von Knesebeck, Müffling and Gneisenau and others.

The officers of the Fifteenth left Berlin for Russia on the 25th September, and arrived at a town about thirty miles from Warsaw, on the Polish frontier, where the manœuvres of the

Russian army were to take place. Thackwell thought the men fine and the horses good, and the latter generally in excellent condition, while the Polish horses were much better bred than the Russian. Altogether some 1,300 cavalry and 7,000 infantry were seen here. On the 3rd October the three officers left for Berlin on their return to England; they returned thence by Leipsic, where they visited the battlefield, and by Frankfort to Calais, whence they crossed to Dover, arriving in England on the 9th November.

During the remainder of the present tour of service of the regiment in England, Colonel Thackwell was very frequently employed in the inspection of corps of yeomanry cavalry all over the country.

In July 1824 the Fifteenth embarked at Bristol for Ireland—no less than twelve vessels were then required to transport even a comparatively weak regiment of cavalry—and was quartered at Cork. Here General Sir John Lambert was in command, the Commander-in-Chief in Ireland being Lord Combermere. It was here that Colonel Thackwell seems to have met the young lady whom he afterwards married—Miss Maria Audriah Roche, the eldest daughter of the late Francis Roche of Rochemont, County Cork, of the Fermoy family, and niece of Colonel Roche of Trabolgan House, Co. Cork. In April of the year following the Fifteenth moved to Dublin, where Thackwell met Sir Hugh Gough, under whom he was to see so much service in the future, and who was then commanding the 22nd Regiment.

On the 9th July Colonel Thackwell was married

to Miss Roche at Trabolgan House by the Bishop of Cloyne, her uncle and guardian, Colonel Roche, giving her away—and on the 3rd September the newly married couple took up their abode in the Commanding Officers' quarters in Portobello Barracks, Dublin.

Almost throughout the whole of the remainder of Colonel Thackwell's command of the King's Hussars, the regiment was greatly split up, finding many detachments, and it was but seldom that the Fifteenth could be drilled or manœuvred together as a whole. None the less, the inspections by the commanding officer were frequent and close, and Colonel Thackwell was constantly travelling long distances by coach, chaise, or on horseback, to visit his troops and squadrons at out-stations. As a result of his unremitting supervision the regiment seems on every occasion to have been well reported on by the many distinguished officers by whom from time to time it was inspected—the Duke of Cumberland, Lord Hill, Lord Combermere, Sir Hussey Vivian, Sir Colquhoun Grant, and others—and it is very noticeable, from Colonel Thackwell's correspondence and diaries, how often commanding officers of other cavalry regiments made a point of being present whenever any special parade or inspection of the 15th Hussars was ordered. It is not only, however, in the maintenance of the high character of the regiment for drill and manœuvre that the hand of the commanding officer may be noticed. With the approval of Major-General Sir Colquhoun Grant, then Colonel of the regiment, Colonel Thackwell drew out and published a new set of Standing Orders, those formerly in use having been lost some years previously. He also appears to have conceived the idea of writing the history of the regiment, and some closely written bundles of manuscript in his handwriting, which are still extant, seem to leave no doubt, from the similarity of innumerable paragraphs, that for the greater part of the "Historical Record of the Fifteenth Hussars," written and published by Richard Cannon of the Adjutant-General's Office in 1841, that writer was very greatly indebted to the material Colonel Thackwell had collected, and much of which he had put into shape. It should be remembered that while Colonel Thackwell was in command of the King's Hussars, many of the actors in some of the most prominent achievements of the regiment were still alive, and were, some of them, in constant touch with their old corps. Sir William Aylett and Sir William Keir—or Keir Grant as he latterly called himself—are mentioned in Colonel Thackwell's diaries as dining with the regiment on "the thirty-fifth anniversary of the glorious action of Villers en Couché," where these two Paladins had won the cross of Maria Teresa; and there is, among the Colonel's correspondence, a letter from Major Pocklington, who brought the two squadrons of the Fifteenth out of action on the same glorious occasion, which is evidently written in compliance with Colonel Thackwell's request for some account of the service of the regiment in Flanders, and wherein Pockington tells, in modest terms, of those stormy but never-to-be-forgotten days. But besides these old officers of the 15th Light Dragoons, there were men still living who had served in the regiment at Emsdorff in 1760, and in the days when it was still known to fame as Eliott's Light Horse. Journeying through the Midlands in June 1827 Colonel Thackwell records in his diary that at Winslow, near Aylesbury—

"I saw poor old William Ovitts, aged eighty-six, a veteran soldier who was in the King's Hussars, and behaved nobly at the battle of Emsdorff. He seems decrepit, but has all his faculties, and would rejoice much in fighting his battles over again with the garrulity of old age. The Duke of Cumberland allows him £7 per annum, no pension having been granted him—but he seems poor, although well clothed. He lodges with one Mr. Jennings, a saddler."

There is a letter to Colonel Thackwell, dated Horse Guards, 19th February 1831, from Mr. Richard Cannon, wherein he writes:

"Being anxious to afford you every possible assistance in rendering the records of the 15th King's Hussars as creditable and as complete as they can be made, I have ventured to send you by this post the volume of Gazettes for the year 1762, to which you state you wish to refer for particulars of the Continental Wars at that period."

There is a copy also of a letter addressed by Colonel Thackwell, on the 11th August 1831, to Lieutenant-General Sir Robert Wilson—another of the Knights of the Maria Teresa Order—wherein he writes:

"That being in daily expectation of receiving an order from the Adjutant-General to transmit the revised history of the King's Hussars, he feels obliged to beg Sir Robert Wilson will have the goodness at his convenience to return him the narrative of the campaigns of 1799. At the same time Lieutenant-Colonel Thackwell ventures to hope Sir Robert Wilson will have the kindness to correct any error he may discover in the detail of circumstances, and add anything which may have been omitted, as he cannot but be extremely desirous that the account of the transactions of the regiment at these periods should have the authority of so distinguished a participator in them."

It seems greatly to be regretted that Colonel Thackwell was, for some reason or other, never able himself to complete the work he had undertaken, for there can be no doubt that he would have evolved an infinitely more interesting and a more vivid narrative than the dry-as-dust productions which, in the shape of "Records" of many regiments in the army, emanated some years later, wholesale, from the Horse Guards.

There seems no object in following the regiment from quarter to quarter—from Dublin to Ballinrobe, from Ballinrobe to Newbridge, back again to Dublin, thence to Kingston-on-Thames, and afterwards to Canterbury, and Hampton Court, and Brighton. At the last-named place Colonel Thackwell was greatly distressed at finding it necessary to destroy an old charger which had been given to him in 1807 as a four-year-old. He took the horse to Corunna in 1808 and rode him through Galicia, and at Corunna a round shot grazed his shoulder, but without doing him any injury.

"He was my favourite charger," writes his sorrowing master, "in embarking at Portsmouth in 1812; I rode him at the passage of the Esla, and in the battle of Vittoria, where he upset several horses which came in contact with him. I rode him also at the Pyrenees, at Orthes, Grenade, Toulouse, and in many skirmishes. At the latter battle a cannon ball passed under his belly. At Waterloo I rode him until he was wounded by a musket ball on the knee, from which he did not recover for a long time."

In April 1830 the Fifteenth moved to Nottingham and Sheffield, remaining in those districts for the greater part of two years; and it was while stationed in the vicinity of Nottingham that serious riots broke out, and Colonel Thackwell had again an opportunity of proving his ability to deal with a difficult situation.

On the 31st October 1831 the Duke of Newcastle wrote as follows to Colonel Thackwell:

"Since my return home I have had so many things to think of that I have omitted a very necessary part of my duty, namely, writing to inform you of my return, and to try to put myself in communication with you. . . . Pray let me hear from you on every occasion, either of what has occurred or respecting any measure of precaution that you may think necessary. You may rely upon my ready and cordial co-operation on all occasions.

"I should do more harm than good by appearing at Nottingham; I must therefore content myself with acting from home. I shall at all times be happy to see you here, or to meet you at any place where my presence will not raise a mob.

"I should be very glad to receive from you an impartial account of all that occurred on Sunday, Monday, and Tuesday at Nottingham and neighbourhood.

"Your communications, if requred, shall be strictly confidential."

"I remain, Sir,
"Your very obd. Ser.
"NEWCASTLE."

In reply, the following statement of all that took place during the riots in Nottingham and neighbourhood on the 9th, 10th, and 11th October 1831 was sent by Lieutenant-Colonel Thackwell to the Duke of Newcastle at Clumber:

"On Sunday the 9th about 12 o'clock a requisition was received from the County Magistrates and Mayor of Derby requesting that a troop might be attached to that place in aid of the Civil Power. The requisition was for a squadron, (which of course I could not think of), the ground being that the town was in the hands of a mob who were breaking windows and destroying the property of people who were obnoxious. Major Buckley, two subalterns, and 41 men and horses were instantly sent, communication having been previously held with the town magistrate, who did not apprehend any tumult, and thought the party might be spared."

"On the evening of the same day a great number of people assembled in the market-place at Nottingham, and proceeded to break the windows of obnoxious individuals. The Mayor, in endeavouring to quell this disturbance, was wounded on the head by a stone, thrown down and trampled on by the mob, his leg being much bruised; and, the number of the rioters increasing, the Military was called out, and succeeded in dispersing the crowds in various parts of the town and in the market-place. Small knots of the lowest rabble, however, still continued to glide through dark alleys and passages, and frequently succeeded in breaking windows before they could be interrupted. Mr. Wilkins, the Reformer, addressed the multitude late at night, entreating them to disperse, and about 2 o'clock two of the rioters were apprehended by the Civil Power. Soon after this, the streets being now nearly empty, the troops were withdrawn to the barracks, an officer's picquet of 20 men being left in the town. On Monday morning early the picquet was withdrawn; but after the meeting held at 11 o'clock to petition the King not to dismiss his ministers, a considerable excitement became apparent, although a communication from the Town Clerk led to the belief that everything would pass off peacefully. I had armed the recruiting parties stationed in the town with carbines, and had made disposable a few dismounted men who could be spared from the defence of the barracks; the whole of the troops were in readiness to turn out during the meeting, and a troop of Yeomanry was assembled near Wollaton. At half-past two everything was peaceable; the town clerk at this time sent a note to say most of the people had gone home, and the troops need no longer be in readiness; but between 3 and 4 a picquet was sent into the town, at the request of the Mayor, of the same strength as before, and soon after the whole disposable mounted force followed—the magistrates requesting that the mounted men might be brought into the town. The Riot Act was read in many places, and mobs which were parading the streets with flags and doing much injury to private property were frequently dispersed; it was, however, totally impossible to prevent their reassembling, as the whole force under my command consisted of only about 75 men and horses—many of the former recruits. Seeing that this force was inadequate to protect the town and county, I made application by express to the General commanding the District for a reinforcement, and having met with two county magistrates requested they would call out the Yeomanry.

"Soon after six o'clock intelligence was brought that a large mob was proceeding towards Colwick, but the magistrates of the town did not think themselves justified in allowing any part of the military to leave the town, nor, considering the weakness of my force, and the alarming appearance of the multitude which filled the street, did I consider myself at liberty to weaken my party by sending detachments away at this time. About half-past 7 o'clock, attacks were made almost simultaneously on the House of Correction and Nottingham Castle. I was out in the lower part of the town with a considerable part of my force at the time, and fortunately came upon the great concourse of people at the House of Correction, where part of them were forcing in the prison doors. These were driven into the surrounding streets and alleys, but as it was necessary to secure this place against further attacks, I sent for the whole of the disposable dismounted party from the barracks, and they were ultimately distributed for the protection of this place, the town and the county gaol, and the gasworks. In consequence of these occurrences at the House of Correction, the troops with me did not return to the market-place for some time, and when they did the castle was in flames. A party of mounted and dismounted men, with two county magistrates and myself, afterwards proceeded to the castle and dispersed the immense crowd in front; but it was judged quite useless to clear the castle yard, as there was but little chance of discovering the authors of the mischief among the great concourse of spectators. Colonel Wildman afterwards took a body of special constables and a party of Hussars to the castle, but I am not aware that any prisoners were made. The troops patrolled the side of the town next to the river until the mob had dispersed, but I believe it was owing to the extreme wetness of the night that more mischief was not attempted. It is worthy of remark, that whenever the military came in contact with the populace, the latter in being dispersed did not offer the slightest resistance by throwing stones or otherwise.

"This evening a great number of special constables were sworn in for the town, and the Mayor and magistrates exerted themselves to preserve the peace of the town, and afford assistance as far as their means would permit. Detachments were sent to Sharp Mill, the racecourse, and other places, but the aspect of things was too alarming to allow the chance of these small parties being overpowered, and they were recalled. An officer's picquet was left in the town, and the troops were recalled to the barracks at 2 o'clock in the morning.

"Early on Tuesday morning, a great crowd being assembled at the castle, Mr. Norton, a magistrate, requested an officer's party to assist the special constables in clearing the castle yard. An officer and 18 men were sent on this duty, but no magistrate being present when they first arrived, it was some little time before the mob were dispossessed of it. The magistrate shortly after left, and stones were for the first time soon after thrown at the military from the outside of the castle. About 10 o'clock I received a requisition from the town magistrates to take the troops into the town, and also to reinforce those at the castle. Proceeding with a party of twenty men, I met a very large mob on the Derby road opposite the park, and these men being in the act of pulling down some iron rails, I rode among them and dispersed them, but not so effectually as I should have done had there been any magistrate present. The troops marched into the town, but the chief part were afterwards sent to the barracks. Before 1 p.m. all were again ordered into the town, and various detachments were employed the whole afternoon in dispersing the mob in the market-place and other quarters. Early in the afternoon I received a requisition from a county magistrate for some men to assist a troop of yeomanry in attacking the mob which had gone to Beeston—that which had reunited after being dispersed in the morning; and as I could now calculate on the Yeomanry being assembled, I considered myself enabled to detach. I therefore obtained the consent of the town and county magistrates to proceed with one of the latter and 30 men for that purpose. Near Lenton, however, I met Major Rolleston, who stated that the mob, after having burnt Mr. Lowe's Mill at Beeston, had been rash enough to break into Wollaton Park, and had been there met and dispersed by a troop of Yeomanry, and 15 prisoners taken. We met the crowd near Lenton directing its course towards Nottingham, and the magistrate read the Riot Act; but as they did not remain together, the military were not then required to act. On returning towards the town I found one of the Holme Pierpoint troops of Yeomanry near the 'Admiral Warren,' who, having had some stones thrown at them by the mob from behind a wall and hedges, were firing their pistols to intimidate them. The mob also threw stones at the party of the Fifteenth which came up, but they were quickly chased away and a prisoner taken in the act of throwing was lodged in the county gaol. Soon after this the five troops of Yeomanry were placed under my command, and I was then enabled to dispose of them and the regulars, so as to protect both the town and the country. The prisoners taken in Wollaton Park had been lodged in the barracks, and towards evening I sent an officer's detachment of the Hussars to escort them to the county gaol. On the way the party was frequently pelted with stones, and in turning into Bridesmith Gate several stones were thrown which struck some of the escort. A pistol shot was then fired which wounded two people, and this at once stopped the hostile proceedings of the mob, and from that moment they melted away, nor did they ever again make head in any part of the town. I beg to state that on Tuesday the shot was fired by a soldier (not an officer), and the man wounded was not a special constable as reported, but a person who could be proved, were it necessary, to have thrown stones at the time. In the evening of this day I had recommended to the town and county magistrates to call out the out-pensioners to act as special constables; and this recommendation being attended to, an efficient civil force was now embodied.

"It was for a long time a matter of conjecture to what place the mob had retired. However, it was found at length that a strong party of it had retired to the meadows below the town to consult, and strong officers' patrols were at various times sent outside the town. The last appearance of an assemblage of people was at about 12 o'clock at night near the river, and those were dispersed by an officer's party with a town magistrate at its head. A picquet of an officer and 20 men was left in the town, and the out-picquets stationed near it and the inlying picquet at the barracks patrolled the country in various directions. The same precautions of strong picquets were taken by the troops, and various patrols were made from them into the country, during the nights of the 12th and 13th, after which the services of the troops of the Yeomanry were dispensed with. The picquet of an officer and 20 men of the Fifteenth was continued in the town until the 20th inst., and a strong inlying picquet was mounted at the barracks, from which the country had been patrolled by an officer's or smaller patrols in all directions, during the chief part of the night. The Wollaton and another troop of Yeomanry in part remained on duty for the nights of the 14th and 15th to patrol the villages of Bullwell, Busford, Wollaton and neighbourhood; but with the exception of an outrage or two in the direction of Plumtree on the 12th, everything has remained quiet in the town and neighbourhood up to the present period."

General Sir Henry Bouverie, commanding the district, left everything to Colonel Thackwell, as the General was unable himself to leave Manchester, and the riots, which at one time had presented a very ugly appearance, were completely crushed by the measures adopted. In writing to Colonel Thackwell on the 14th October, Major-General Bouverie says, "You appear to have managed everything in the most satisfactory manner," and again:

"I have not failed to call the attention of the Commander-in-Chief to the very judicious manner in which the duties which have devolved upon you and Major Buckley and the regiment under your command have been performed."

On the 19th October the General wrote to Colonel Thackwell as follows:

"Dear Sir,

"I have great pleasure in obeying the command of Lord Hill to signify to you the satisfaction which the conduct of the 15th Hussars under your command at Nottingham and under Major Buckley's at Derby have afforded him under such very trying and arduous services.

"I beg that you will make known his Lordship's approbation to your regiment in whatever way you may deem most desirable."

The letter from the Horse Guards speaks of Lord Hill's "sense of the steadiness, forbearance and superior discipline of the distinguished regiment." The civil authorities, too, were not behindhand in placing on record their appreciation of the services of the military: the magistrates of the County of Nottingham, assembled at Quarter Sessions, passed on the 17th October a resolution that—

"the most cordial and grateful thanks of this meeting are especially due and are hereby tendered to Colonel Thackwell for the admirable and judicious arrangements and disposition of the different military corps after the whole were placed under his sole command, by which the late most formidable tumults were effectually suppressed and the public peace preserved without the loss of a single life and almost without bloodshed."

The inhabitants of Nottingham raised a subscription, and at a meeting held on the 21st unanimously resolved that—

"a silver soup-tureen be presented to Colonel Thackwell and the officers of the regiment (with an appropriate inscription) for their prompt and efficient exertions in restoring tranquillity amongst us, where destruction to life as well as property must have been the inevitable consequence of a less determined spirit.

"That part of the subscription be employed in purchasing books with a view to establishing a permanent library for the use of the men, under the sanction of Colonel Thackwell."

Matters, however, were by no means quiet, and throughout that winter there were constant rumours that the more desperate characters of the mob were arming; that men were being sent into Nottingham from Manchester and Birmingham; that the troops would be attacked on their way to church; that rioters who had been arrested and were under trial would be rescued and their judges assaulted. A troop of Horse Artillery and 150 men of the Royal Irish Regiment were brought into Nottingham and placed under the orders of Colonel Thackwell, who was in constant correspondence with Sir Henry Bouverie at Manchester and with the Duke of Newcastle at the Home Office. It was a troublous year: Lord John Russell had but recently introduced his first Reform Bill, and the opposition with which this measure was met had everywhere aroused the passions of the populace; upon tumult followed disease, and there was a very serious outbreak of cholera in Sunderland in the spring of 1831.

Already for some time past it would seem that Colonel Thackwell had had an idea of leaving the regiment and going upon half pay, and there is some mention of such an intention in his diary as early as March 1829. The idea took shape, and in November 1831 he arranged an exchange to half pay with Lord Brudenell, formerly of the 8th Hussars—an exchange which was finally announced in the *Gazette* of the 16th March 1832, though Colonel Thackwell did not finally sever his connection with the 15th Hussars until the 30th May of that year. The news of his exchange seems to have aroused among his brother officers a general chorus of regret: one of them writes:

"I am sure I need not express to you the very deep and sincere regret I feel at your intention of leaving the regiment, of which you seem to be so essential a part that one can hardly imagine its existing without you. . . . I firmly believe that no officer ever retired from the command of a regiment with such a reputation as you have, or so justly regretted."

Another says:

"Not only to myself is your retiring cause of the sincerest sorrow, but to every officer in the corps. There is not one in the Fifteenth (for however short time he may have served under you) that does not deplore the loss of so distinguished and impartial a commandant.

"The officers, my dear Colonel, cannot let you depart from among them without presenting some token, some testimonial of the admiration and esteem they entertain for you as a soldier and a man, and through me they beg to communicate to you their having ordered a piece of plate, your acceptance of which will be a heartfelt gratification to them. In their name and for myself, my dear Colonel, allow me to wish to yourself and family health and happiness wherever you may go, whatever realms to see.

"With thanks for your many acts of kindness to me, and assurance of a sincere regard and attachment to you, believe me, in grief at the step you have felt called upon to take,

"Yours most truly,

"FREDERICK BUCKLEY,

"*Major, 15th King's Hussars.*"

In reply Colonel Thackwell wrote:

"Believe me I can scarcely find words to express to you how much I am obliged for the kind sentiments conveyed in your letter, and it is a matter of the deepest sorrow that anticipations of the uncertainty of the future have induced me to remove from the command of the regiment where I have spent so many happy days. If the King's Hussars have been so fortunate as to obtain the approbation of the Commander of the Forces during the time I have had the honour to command the regiment, it has been owing to the zealous, cordial, and unremitted attention to their duty of the officers it has been my good fortune to be associated with."

On the exchange being finally carried out, Thackwell congratulates Lord Brudenell "most sincerely on succeeding to the command of one of the best regiments in His Majesty's service."

That Colonel Thackwell felt very deeply leaving the regiment in which he had served so long, and with which he had taken part in so many stirring scenes, there can be no question. It is significant that his diary, in which it was his custom to enter with minute exactness every event and action of his daily life, contains no mention whatever of the severance of his long and honourable connection with the Fifteenth.

Those who have stood, like him, at the "parting of the ways" will understand and respect the silence in which he passes out of a happy regimental life.

CHAPTER VII

COLONEL THACKWELL would seem to have settled down, at least for a time, to the ordinary life of a country gentleman. He had inherited a small property called Normansland, near Dymock, in Gloucestershire, and shortly after leaving the 15th Hussars he purchased an adjoining small property called Woodend Farm, and later on another at Ragland in Monmouthshire. He found plenty of occupation in the improvement of these properties, and in shooting and fishing, of which all his life he had been passionately fond; and his diaries at this period contain many references to the long journeys he made, and to the great distances he was accustomed to walk and ride when engaged in these pursuits.

His family, too, was increasing, and the following are the names of the children who survived infancy, the eldest being born in 1827 and the youngest in 1842: Edward Joseph, Elizabeth Cranbourne, Anne Maria Esther, William de Wilton Roche, Osbert d'Abitôt, Maria Roche, Francis John Roche.

If, however, Colonel Thackwell had ever any idea of finally severing his connection with the service, he seems to have speedily renounced it, for we find him a constant attendant at Lord Hill's levées, and there are not a few references in his diaries to a wish for re-employment; he was frequently engaged in the inspection of the Worcester, Gloucester, and Monmouthshire Yeomanry Cavalry; he rarely missed an attendance at dinner at Apsley House on the recurring anniversaries of the Battle of Waterloo; and just after leaving the Fifteenth he was gratified by being gazetted to the Third Class of the Hanoverian Guelphic Order, or K.H., dated the 17th January 1834. Already towards the end of 1836 there is frequent mention in his diary of the opening and prosecution of negotiations for a restoration to full pay, while the idea of service in India appears to have begun to exercise a novel and increasing fascination upon him; on the 10th February 1837 we find that he "offered £4,000 for an exchange, taking no responsibility, and provided the regiment goes out in June." Succeeding entries reveal the fact that he was in treaty for an exchange with Lieutenant-Colonel Charles Stisted, commanding the 3rd Light Dragoons, that regiment being then under orders to proceed shortly to India; and on Friday, 19th May, Colonel Thackwell records: "Saw my appointment to the 3rd Light Dragoons, and to a colonelcy in the army in the *Gazette*." The very next day he proceeded to Canterbury, where the Third was then stationed, and made the acquaintance of his future brother officers; but he does not appear to have actually taken over command of the regiment until the 24th June, just prior to embarkation.

Of the children, the four eldest were left in England, Osbert only, who was then the youngest, accompanying his parents. Miss Emma Webb, a niece of Colonel Thackwell's, also embarked with them for India. The 3rd Light Dragoons sailed in four East-Indiamen: the *Mountstuart Elphinstone*, the *London*, the *Thomas Grenville*, and the *Lord Moira*, the headquarters embarking on the first-named, and the little fleet dropped down the river on the 18th July. Travellers to India often complain in these days of the accommodation afforded them, but Colonel Thackwell remarks that he had to purchase the *whole* of the cabin furniture necessary for the voyage, the passage money giving him nothing but the bare walls of the cabins he had engaged.

Just before leaving England Colonel Thackwell attended the dinner at the Duke of Wellington's commemorating the twenty-second anniversary of the Battle of Waterloo, and notes that "in consequence of the dangerous illness of our beloved sovereign the party broke up early, and the usual toasts were not given"; and three days later he records the death of King William IV., and the accession to the throne of "Princess Alexandrina Victoria, the only child of his late Royal Highness Edward, Duke of Kent."

The *Mountstuart Elphinstone* does not appear from Colonel Thackwell's diary to have touched anywhere until she reached Calcutta, where she arrived—in company with the *London*—on the 13th November, when the regiment received the unpleasing intelligence that they were to proceed thence on foot the whole way to Cawnpore, instead of—as was the more usual procedure—being conveyed a great part of the distance in boats. The *Thomas Grenville* and the *Moira* arrived nearly five weeks after the first party, and the troops on board having been disembarked on the 21st and 23rd December, the whole regiment was assembled in camp, where, on the afternoon of New Year's Day, it was inspected by Major-General Sir Willoughby Cotton, K.C.B., commanding the Presidency Division.

On the 4th January the 3rd Light Dragoons commenced their march for Cawnpore; Benares was reached on the 17th February, Allahabad on the 27th, and Cawnpore on the 10th March, where the Third took over 608 horses formerly belonging to the 11th Light Dragoons—which corps had left Cawnpore for England in December—also 199 non-commissioned officers and men who had volunteered from that regiment to the 3rd Light Dragoons.

At this time Cawnpore was one of the largest garrisons in Upper India, where the main strength was kept on the lines of the Ganges and Jumna; the troops held Cawnpore, Meerut, and Kurnaul in considerable force, while smaller garrisons occupied Allahabad, Agra, and Delhi, the latter place being the most advanced magazine of the army.

On the 19th May a General Order announced Colonel Thackwell's appointment to the rank of Major-General in India, dated the 10th January 1837, and on the 20th July Lord Hill wrote to inform him that he had been gazetted a Companion of the Order of the Bath. In October he assumed the command of the station and garrison of Cawnpore on the departure of Major-General Oglander.

The following extract from a letter received at this time from Major Charles Slade, 3rd Light Dragoons, then on the staff of Sir Henry Fane, shows that General Thackwell had already become as popular in his new regiment as he had ever been in his former one:

"Though now a Major-General, I shall always consider you our Colonel. I do not like the new title; it appears that you are separated from us, which I should be sorry to think."

For some time past affairs in Afghanistan had been arousing great anxiety both in England and in India. There had been something like a break-up of the Empire of Afghanistan: Shah Shuja had gone into exile; Dost Muhamed had made himself master of Kabul and Kandahar; while Herat, which had continued under the Durani dominion, had in 1837 been besieged by the Persians, who were believed to be acting in the interests of Russia. The danger was deemed pressing, for the capture of Herat would introduce the Russians to a vulnerable portion of Afghanistan. The Persians and the Russians entered into negotiations with Dost Muhamed at Kandahar, who in his turn attempted to obtain support from Lord Auckland, the Governor-General. Lord Auckland, however, showed himself cold and unfriendly to the Dost, who then turned to the Russian Envoy. Lord Auckland now determined upon the deposition of Dost Muhamed and the restoration of Shah Shuja, and wrote that "we owe it to our safety to assist the lawful sovereign of Afghanistan to the recovery of his throne." It was apparently at first merely contemplated that Shah Shuja was to attack the Dost, assisted by the forces of the Punjab—Peshawar and Kashmir being ceded to Ranjit Singh in return for his help. The scheme, however, was enlarged so as to admit the presence of a few British corps to give stability to the movement. But here the military authorities were necessarily taken into council, and the Commander-in-Chief, Sir Henry Fane, without expressing political opinions, at once declared that if British troops were to be sent, they must be in sufficient numbers to take care of themselves; and the Governor-General, unwilling to recede from his plans, and still under the fear of Russian aggression, was led to make more extensive combinations. A column was ordered from Bombay to move upon Scinde, while a Bengal force, collected at Ferozepore, was to march down the left bank of the Sutlej, and thence through the territory of Bahawalpore to meet the southern force. Sir Willoughby Cotton, whom we have already seen at Calcutta, was appointed to the command of the Bengal portion of the force, and passed through Cawnpore in September, *en route* to join his troops; and on the 24th November Major-General Thackwell was gratified by the receipt of the following letter from Sir Henry Fane:

"LUTULLA, *16th November*, 1838.

"MY DEAR THACKWELL,

"I have at length persuaded the Government that the Cavalry of the Army of the Indus should have a chief; and I have therefore named you, and an order goes to you to-day to join as soon as you can.

"My opinion is that you would do wisely to come up with your traps, and not think about horses; for as a part of the army does not proceed farther than Ferozepore, or even lower down the river, you will not find difficulty in mounting yourself.

"Captain Willin has overtaken me already, so your task will not be very difficult; and I don't think that the troops can be off from Ferozepore before the end of the first week in December.

"I am, yours truly,

"H. FANE.

"I think that you may bring up an aide-de-camp, if desirable."

Major-General Thackwell lost no time over his preparations, and on the night of the 15th he started by palki-dak for Ferozepore, accompanied by Cornet Edmund Roche—a brother of Mrs. Thackwell's, who had been posted to the 3rd Light Dragoons shortly before they left England, and whom General Thackwell had appointed to be his aide-de-camp. Meerut was reached on the 4th December, and here the travellers put up for the night with Brigadier-General McCaskill. They were at Umballa on the 6th, Ludhiana on the following day, and finally arrived at Ferozepore—547 miles from Cawnpore—early on the afternoon of the 9th. "Found the whole army present," writes the Commander-elect of the Cavalry, "and took up my quarters at Colonel Beresford's tent. Called on Sir Henry Fane, Generals Torrens and Churchill, and also Sir Willoughby Cotton." On the next day he "called on Colonel Skinner, who is a fine old fellow"—the individual meant being evidently Colonel James Skinner, C.B., the Father of the Indian Cavalry, who had served under de Boigne, and Perron, and Scindhia, and Lord Lake, who had chased Holkar from Delhi to Fatehgurh, and Amir Khan, the Pathan adventurer, from Bhurtpore to the Himalayas. At the date when General Thackwell met Skinner, the latter was commanding a body of horse in the Army of the Indus, and must have been a man of sixty years of age.

General Thackwell had pushed on so fast to join the army that he had greatly outmarched his baggage, which had not come up, but none the less he left Ferozepore on the 12th December, in company with Roche, and hurried on 37 miles to overtake at Baggeeke the cavalry brigade, under Colonel Arnold of the 16th Lancers, forming part of the Cavalry Division which Major-General Thackwell had been selected to command.

"To perform this march," says he, "I had Beresford's horses for 7 miles, his gig with two horses as relays for about 12 miles, two of his ponies for 12 miles more, and two horses from Major Cureton" (the Assistant Adjutant-General of the Cavalry Division) "for the last march, a horse I purchased from Sir Henry Fane, and a pony of Dr. Wood's."

The Bengal force, as originally warned for field service, had consisted of a brigade of artillery, a brigade of cavalry, and five brigades of infantry, the latter told off into two divisions under Sir Willoughby Cotton and Major-General Duncan. The cavalry brigade consisted of the 16th Lancers and the 2nd and 3rd Light Cavalry, while there were also with the army two *risalas* of irregular horse. The force which had at the same time been collected at Bombay was composed of a cavalry and an infantry brigade, the former consisting of part of the 4th Light Dragoons and the 1st Light Cavalry with the Poona Local Horse. At the same time another body of troops was raised for service across the Indus—

> "the force that was to be led by Shah Shuja into Afghanistan; that was to be known distinctively as his force, but to be raised in the Company's territories, to be commanded by the Company's officers, and to be paid by the Company's coin." [1]

For this purpose there were raised two batteries of artillery, two regiments of cavalry, and five regiments of infantry, the total strength of these being some 6,000 men; the Bengal Army numbered 14,000, and the Bombay force 5,000. Durand points out that—

> "the British Army, consisting altogether of 25,000 men, was thus in two bodies, separated from each other by the whole length of the course of the Sutlej and Indus, that is, by a distance of 780 miles; the main body at Ferozepore movable, but that on the sea-board of Scinde, paralysed for a time by the want of camels or other beasts of burden, immovable."

[1] Kaye.

When the plan of campaign had been formed the principal object was the relief of Herat, and for this reason an advance south-west from Ferozepore—by Shikarpore on Kandahar—was the line selected. While, however, the Army of the Indus was yet completing its concentration at Ferozepore, news was received that the Persians had retired from before Herat, and it was then decided that the force to take the field should be reduced in strength, while at the same time the line of advance originally chosen was still held to, although the main reason for its selection had disappeared. Instead of two divisions, the Bengal Army was now to consist of one division of three brigades under Sir Willoughby Cotton, while General Duncan, with two brigades, was to stand fast at Ferozepore and Ludhiana. By this measure the strength of the Bengal force was reduced to 9,500 men; while in addition to the danger of the long line of communications to which the Army of the Indus was committed by the route selected for the advance into Afghanistan, there was the further danger of the distance by which it was to be separated from any reserve, and finally the hostility of the Ameers of Scinde, who strongly objected to the military occupation of their country, and who had, moreover, from their position the power to cut off or withhold all supplies.

"Scinde was at that time under a confederation, much like Afghanistan, only that all the Ameers were of one clan; the northern part of the province was under Rustam Khan, of Khyrpoor, the southern, or lower part, being ruled by some of his kinsfolk, of whom Nur Muhamed of Haidarabad was chief. By direction of the Government, Burnes, who now held the substantive post of agent with the Khan of Khelat (the leading chief of Baluchistan), negotiated with the Khyrpoor Ameer, while the dealings with the Ameers at Haidarabad were in the hands of Colonel Pottinger, the accredited Resident in Scinde."[1]

On the reduction of the force, Sir Henry Fane decided to relinquish the command of the expedition into the hands of Sir John Keane, who was then at Kurrachee at the head of the Bombay Division; but as Sir Henry was to proceed by boat down the Indus, *en route* for the port of embarkation for England, his advice and assistance were still, as Commander-in-Chief, at the service of Sir Willoughby Cotton, to whom now fell the command of the Bengal contingent of the Army of the Indus.

The position at this moment of Sir John Keane, the new Commander of the army, was by no means pleasant; he was unable to move from the sea-coast without transport and supplies, for both of which he was mainly dependent upon the goodwill of the Ameers of Scinde, whose attitude, if not actively hostile, was anything but friendly, and whose strength interposed between him and the major portion of the army he was to command, and which was nearly 800 miles distant from him.

When General Thackwell joined his command at Baggeeke the army had already commenced its march, the 9,500 fighting men of whom it was

[1] Keene.

composed being accompanied by the enormous number of 38,000 followers, while their supplies were carried by 30,000 camels. The baggage of the army was indeed enormous; no stringent orders on the subject appear to have been issued either by Sir Henry Fane or by his successor, and the former had contented himself with merely cautioning all concerned against the provision of large tents or establishments. This caution does not, however, appear to have been regarded: Sir Henry Havelock, who was serving with the force as A.D.C. to Sir Willoughby Cotton, admits to having himself had eighteen servants, while it is said that one officer had so many as forty.[1]

The Engineer officers, with the Sappers and Miners, preceded the leading column by two marches, improving the road as they moved on; then followed the cavalry brigade with a troop of Horse Artillery, accompanied by Sir Willoughby Cotton and the Headquarters, and one after the other on successive days marched the 1st, 2nd, and 4th Infantry Brigades, the siege train, and the park —a *risala* of irregular horse being attached to each of the three last columns. The distance from Ferozepore to Bahawalpore is 230 miles, and it was divided into 18 marches; the line of march never deviated more than twenty miles from, and was generally quite close to, the Sutlej, down whose stream the sick, the hospital stores, and commissariat supplies were conveyed in boats.

The following extracts from General Thackwell's

[1] The 16th Lancers took a pack of hounds with them on this campaign.

diary give some idea of the country traversed, and of the difficulties of supply by which the column in general, and the mounted portion of it in particular, were beginning already to be assailed.

"13*th December*.—Marched to Bahuk, the road a track, but passable for all descriptions of force: the same flat country, but little vegetation and quantities of low jungle, showing that we are entering the desert."

"14*th*.—But forty maunds of grain had been collected instead of nine hundred."

"15*th*.—All through the desert, the same kind of road and soil—no vegetation except the prickly thorn and tufts of long grass scantily diffused, with a little dwarf shrub and a few patches of cultivation surrounding a few of the small villages. The horses had to-day half-rations of grain and *jowa*. Roche overtook me last evening by dinner-time with two of my camels, and the remainder came up to-night."

"24*th*.—Our march all last week was through or on the borders of the Great Desert."

"26*th*.—Marched to Khyrpoor; road as before, but the desert more marked with deep sand in very high sand hills, with patches of cultivation around several villages. But a thin scattering of jungle and some small groups of trees."

"27*th*.—At sunrise and until 8 a.m. cloudy and very cold with a few drops of rain; the natives say it has not rained for three years on our line of march from Ferozepore. Partly through a broad plain between the Sutlej and the edge of the desert close on our left. The view to the left as far as the eye can reach is over a wide waste of desert with only a few shrubs and thorns growing on it."

"29*th*.—Marched to Bahawalpore and on to

our encampment, SW., and distant from it two miles."

"30*th.*—Sir Henry Fane held a durbar in his tent to receive the Rajah or Nawab of this place."

"1*st January.*—Marched to Khyrpoor. Despatches received that the fort of Bukker" (commanding the spot where the army was to cross to the right bank of the Indus) "has submitted. The despatches state that Sir John Keane with the Bombay division has landed in the bay of Kurrachee distant from Haidarabad about 120 miles, passing through Tatta, and that the Chief of Scinde intends to give him battle. Opposite Haidarabad, for about 40 miles, there is no road on the right bank of the Indus; then for about 40 miles thence none to within about 40 miles of Shikarpore."

"5*th, Chanikhan.*—Two days ago we fancied we saw, or really did see, the snow-clad mountains of Afghanistan opposite Multan. To-day the lower range of brown hills of that country show themselves, distant about 30 miles."

On the 14th January the force entered Scinde and left the vicinity of the desert, the country becoming richer and more cultivated; already, however, desertions were taking place among the followers, and on the 18th January General Thackwell records that "the 3rd Light Cavalry lost 35 camels with their drivers"; and it became apparent that the army might before long suffer from a scarcity of carriage and supplies. Shah Shuja had ere this gone on in advance of the British force, and having crossed the Indus, by ferry, seven miles above Roree, had moved on to Shikarpore, which had been fixed upon as the place of rendezvous. On the 25th January the Headquarters of the Bengal force arrived at Roree, and on the 29th the fort on the island of Bukker was taken over by a British garrison, and the crossing-place for the army was secured.

The news, however, from Lower Scinde was disquieting: Sir John Keane had advanced on the 23rd January from Tatta up the right bank of the Indus, and on the 25th reached Jurrukh, two marches from Haidarabad, where he learnt that the Ameers were moving forward a large army for the defence of the capital, and that another force lay between Haidarabad and the troops under Sir John Keane himself. A collision seemed unavoidable, and Sir John asked Sir Henry Fane, who had not yet left Roree, to permit a portion of the Bengal force to move down the Indus to co-operate in any movement on Haidarabad, and so facilitate the junction of the Bengal and Bombay armies. Fane had, however, already anticipated this request, and orders for the march of some 5,000 men southwards had been given very shortly after arrival at Roree—the Cavalry Brigade, the Horse Artillery, and the 1st and 2nd Infantry Brigades being detailed to march under Sir Willoughby Cotton. About the same time the Shah's contingent occupied Larkhana, almost due south of Shikarpore, and the force under Cotton, moving on the 30th January, had reached Khandeeree, seven miles from Roree, when a despatch was received from Sir John Keane, stating that the Ameers had agreed to all the terms proposed by

Government, and countermanding the movement southward.

On the 9th General Thackwell's diary records that—

> "news arrived that Sir John Keane was encamped opposite Haidarabad on the right bank of the Indus on the 5th, and that the *Wellesley*, with the 40th Regiment, had knocked the fort of Kurrachee about the ears of the Beloochees for having presumed to fire a shot at her."

This last was the "Scinde Reserve Force," under Brigadier Valiant, comprising the 40th Foot and the 22nd and 26th Bombay Native Infantry.

Sir Willoughby Cotton now retraced his steps to Roree, which was reached on the 15th, and, the bridge over the Indus having by now been completed, the cavalry crossed on the 17th, and by the following day the whole of the Bengal force was on the right bank of the river ready, under its new commander, for the trying work which lay before it.

CHAPTER VIII

THE Bengal column marched to, and was concentrated in, Shikarpore by the 20th February 1839, and here, too, was found the Shah's contingent under the command of Major-General Simpson. Shikarpore was the first town of any importance between Roree and Dadur—ten marches distant and situated at the entrance of the Bolan Pass; Quetta was eight marches on from Dadur, and Kandahar was another fourteen from Quetta.

Macnaghten, the political officer with the force, had been informed that the Bolan Pass was occupied by the enemy, and it was resolved to push on as soon as possible to secure its possession; this determination, the wisdom of which is incontestable, greatly increased the difficulties of supply, already sufficiently heavy. In the early days of the assembly of the force, Sir Henry Fane, the Commander-in-Chief, had warned the Commissariat Department that "failure on the part of the political officers would not be held a valid excuse in the event of a deficiency of supplies for the army being felt in the course of military operations." Supplies, however, of all kinds, and

[1] Keene.

especially of forage, were scarce, and came in but slowly; between the river and Shikarpore, the camels, it is said, had dropped down dead by scores; and the Bengal Commissariat had now been directed to supply the Bombay column also with food and transport. The Chief Commissariat Officer was most anxious to remain at Shikarpore for at least three weeks, so as to collect the greatest possible quantity of supplies for the troops; and as he calculated that he should require at least 40 to 50 days' supply to take the troops up to Kandahar, and as the force had moved from Ferozepore with but 30 days' supply and the equivalent transport, it was clear that not only was there far more food and forage to be carried than before, but a greatly increased number of animals would be necessary for its carriage.

At the time when Cotton reached Shikarpore, the Bombay column was some fifteen marches in rear. It was still accompanied by Keane, who for some reason did not push on and join the major portion of his command; and Cotton's decision to move forward and secure the passage of the Bolan Pass appears to have been made without reference to Keane, who followed slowly in rear by way of Larkhana and Gundava, and did not himself overtake the Bengal portion of his command until early in April.

Orders for the commencement of the march to Dadur were issued by General Cotton on the 21st, and on the 23rd the Headquarters left Shikarpore with the cavalry brigade, and a wing of native infantry. Almost at once it was discovered that there would be great difficulty about water and forage for the cavalry and the transport animals. On the 26th General Thackwell notes in his diary:

"There is no grass, and the food for the camels is bad. There are fifty *cutcha* wells of about 2½ ft. diameter at the bottom, only forty-seven of which had water, and none had more than 1 ft. 10 in. in depth, giving about 50 cubic feet of water on our arrival. I have had the wells deepened, but the supply, though improving, is still inadequate."

The heat in the daytime was increasing, and the marches were now generally made at night. The Beloochee robbers were busy, and on the 3rd March, near Bagh, camels were driven off, the post was plundered, while grass and camel fodder were again running short. Sir Willoughby Cotton writes to the Commander of the Cavalry Division:

"At any time when you cannot get grass or *kurbee* for the horses, I authorise you to purchase a field or fields of the standing corn and serve it out to the corps."

But there was little cultivation, and green crops issued as forage were not suitable for horses in hard work.

The road, too, was very bad, with steep ascents and descents, and with deep and narrow nullahs constantly crossing the line of march.

The cavalry reached Dadur on the 10th March, and a detachment of all arms was sent on to secure the pass; the leading column moved on on the 16th, and by the 23rd had emerged from the defiles of the Bolan into the camp of Sir-i-Ab. The difficulties of the ascent of the pass had been very great: the road a mere track strewn with flinty stones; the wind, howling through the pass, was so violent, that men could hardly keep their saddles; the tribesmen swooped down upon the stragglers; there was no lack of water, but the streams were tainted with the dead bodies of camels; supplies, too, were running short, for everything had to be brought up with, or in the rear of the army, *not one single day's supplies being obtainable between Shikarpore and Kandahar* from the country traversed!

"How the army is to subsist," writes General Thackwell in his diary, under the date of the 23rd March, "Heaven only knows. All the horse-corn carried by the Commissariat Department will be expended by to-morrow, and we have about twenty days' supply of *atta*. It has completely ruined this fine cavalry, and will reduce us to a state of starvation."

The followers had been on half-rations since leaving Dadur, the cavalry horses had latterly been getting only six pounds of corn daily and no grass, and the guns of the Horse Artillery were only dragged through the last few miles of the Bolan Pass by harnessing eight horses to each gun, and calling on the assistance of the infantry; even then they were five hours marching nine and a half miles.

"26*th*.—Marched to Quetta—distance from Shikarpore, 226¼ miles."

"*27th.*—Only four maunds of grain in store after this day's issue of one seer" (2lb.) "for each cavalry horse, and only ten days' consumption of *atta* and wheat. In consequence the troops are to be put upon half-rations of flour and the followers upon a quarter seer. We expect, however, to receive from Dadur 600 camel-loads of grain, and from Sukker about the 7th or 8th prox. 1,700 camel-loads; but if the same destruction of camels takes place amongst them as since the 1st inst.—namely, the death of one-third—we cannot expect more than 700 will arrive, which, however, it is hoped, will produce 8,500 maunds."

On arrival at Quetta General Cotton had sent back Major Craigie, his D.A. General, to represent to Sir John Keane the state of the advanced portion of his force as regards rations for the troops and forage for the horses. He was instructed to point out that Quetta and the neighbourhood could furnish but the scantiest supplies, and that many days must elapse before the arrival of a large convoy from the rear. The Bengal force had just sufficient stores to take it to Kandahar; and while asking instructions of Sir John Keane regarding the course to be pursued, Craigie was to point out that by delaying at Quetta, supplies, which it would be difficult to make good, would be consumed, while time would be given to the adherents of Dost Muhamed to devastate the country and occupy positions between Quetta and Kandahar.

Major Craigie reached Dadur on the 27th March and here found Sir John Keane, and on the 3rd April he was back at Quetta conveying to General

Cotton peremptory orders to halt until joined by the Commander of the Force.

Keane had reached Gundava on the 21st March, and had intended to march thence on Quetta by Khelat; but after delaying ten days in a reconnaissance of the road, he decided to follow in the tracks of the leading column, thereby giving up the advantage of a line of advance which, though perhaps more difficult, would have been but little if anything longer, and which would have had the advantage of being one on which the forage and local supplies had not been consumed. By retaining Cotton at Quetta until his own arrival—

"for eleven days he kept a body of fighting men on half-rations, and a mass of camp followers on quarter-rations, uselessly consuming their scanty supply, which, even with such severe husbandry, was barely enough to enable them to reach Kandahar; and when about to enter a hostile country, having the character of furnishing formidable horsemen, he gave the finishing blow to the greater part of his own cavalry by thus confining them to a spot which was soon bald of forage."[1]

To return, however, to General Thackwell's diary.

"*Saturday 6th April.*—His Majesty Shah Shuja and Sir John Keane arrived here to-day. The former has 3,000 infantry, 800 cavalry, and a troop of horse artillery; the latter has a wing of the 1st Bombay Cavalry only. Some Beloochees from a mud fort fired upon a party of the 16th Lancers, who dismounted a few men, attacked them, seized

[1] Durand.

four matchlocks through the loopholes and killed five men, wounding and bringing away the sixth a prisoner, who was hanged. Seven men made their escape when they saw the Lancers rushing to the attack."

The column moved out on the following day, marching by the north-westerly route towards Kandahar.

"9*th April.*—Marched into the valley of Pisheen. Our short rations of one seer, and sometimes not any corn, are fast destroying the cavalry. Fifty horses of the three regiments were cast and shot at Quetta, and seventeen horses have died to-day of sheer weakness."

"11*th.*—Seventy horses died yesterday from want of food or mere exhaustion—10 of the Lancers, 58 of the 2nd Native Cavalry, and 2 of the 3rd. The Sappers and Miners marched to the Kojeh Amran Pass this morning; about thirty of the Kandahar Horse were there, who fired two shots and then retired."

On the 14th the passage of the Khojak commenced.

"The Camel Battery began to drag their guns up at about half-past 7 a.m., and their last carriage was not down before 9 at night. From 40 to 50 men were put to each, and they had to drag the guns up or ease down the ascent and descent for at least a mile. Severe work for all. The chief part of the baggage remained on the mountain or in the entrance of the pass all night, and some things were plundered by the robbers, who were so audacious as to exchange shots. I was most fortunate with my baggage, as we got it over the mountain and in camp by 4 p.m."

(General Thackwell mentions here a second aide-de-camp, Lieutenant Crispin of the 2nd Light Cavalry, but there is no record of when this officer took up his appointment.)

"15*th.*—No corn could be got for the cavalry, and but few could get water."

"19*th.*—It is reported that 2,500 of the enemy's cavalry are encamped 14 miles in our front; we have therefore increased our picquets, as, if the account be true, we shall probably be attacked to-night."

"20*th.*—All the troop horses have been without corn since the 15th."

"21*st.*—No water to be got the chief part of yesterday, and very little to-day."

"22*nd.*—Marched to Mehel Mandah, but owing, as was imagined, to there not being sufficient water for the whole body of troops, the cavalry continued its march through the pass of the Ghautee Hills, and diverged from the Upper Kandahar road to the left, and encamped on the Lower Kandahar road on the right bank of the Doree River—where there was plenty of good water and wheat and barley. Total, 23 miles. This long march in a meridian sun has caused the death of several followers, and of near 60 horses. The folly of it was very great. The brigade had been ordered to halt at Killa Futtoola, but Brigadier Arnold went to Sir John Keane and prevailed on him to suffer the cavalry to march, and when he got to Mehel Mandah he thought that there would not be water enough, and again prevailed on Sir John Keane to let him move on, and I could not find the latter soon enough to prevent it. In consequence we didn't get into camp until half-past 11, and not under canvas until near the evening, during which time the sun was at times absolutely roasting us."

"*24th.*—The Shah left our ground this morning and marched towards Kandahar, on hearing the people had deserted the chiefs and were ready to receive him; and in the afternoon some of his troops were admitted into the town."

"*27th.*—Marched at 3 p.m. to a camp within a mile and a half of Kandahar. The Cavalry Divisional Staff Camp is pitched on a verdant turf within little better than half a mile of the eastern angle of the town wall. This piece of turf delights the eye much more than anything I have seen since I entered the Ganges River."

The Bengal column on reaching Kandahar had marched 1,005 miles from Ferozepore. The cavalry and horse artillery were put on half-rations on the 24th March, so that they had been on a reduced scale, with scanty forage, for 32 days. But from the 30th March they had had no grain whatever, so that for the 26 days prior to their arrival in Kandahar the horses had been fed entirely on green forage, supplemented by very poor grass. It is evident that from the casualties among the horses, and the weakness of the survivors, the efficiency of General Thackwell's fine command had by this time been seriously impaired.

"*4th May.*—I inspected the 4th Local Horse and part of the *risala* of the 1st or Skinner's Horse. Many are in wretched condition owing to the severe work they have had; many, however, are fit for good service. The Bombay column marched into camp this morning; it consisted of two troops of European artillery, one company of foot ditto, about 280 of the 4th Light Dragoons, 190 of the 1st Light Cavalry, the 2nd and 17th Queen's Regiments, a wing of the 19th Native Infantry, and about 300 Local Horse."

"*8th.*—Shah Shuja-ul-Mulk was formally enthroned to-day, on which occasion most of the regular army was paraded at daylight to the north of the town on the fine sloping plain in review order. Four salutes of 21 guns each were fired from different stations, until His Majesty took his seat upon the throne, when 101 guns were fired, and the troops marched past in slow time and filed to their encampments."

"*9th.*—Brigadier Sale called at my tent. I regret I have not the command he is nominated to, viz. 100 European Light Infantry, the 16th Native Regiment made up to 1,000 men by the Shah's infantry, 120 of the 2nd and 3rd Bengal Cavalry, 300 of the Shah's horse, 2 nine-pounders, one 24-pounder howitzer, two 5-inch mortars with a proper number of European artillery with Sappers and Miners—upwards of 1,600 men. They march on the 11th, towards Girishk on the Helmand River, to drive the Sirdars into Persia, unless they come in on the terms offered. There is a report brought by a *hurkara* that Timur Shah, the son of the Shah Shuja, has defeated and dangerously wounded the son of Dost Muhamed at this end of the Khyber Pass."

Already the weather was getting very hot, and on the 16th May General Thackwell records:

"I and Edmund Roche dined with Sir John Keane, but he was too unwell to be at the table, suffering greatly from the heat, which in the hospital tent of the 19th Bombay Infantry has been as high as 114°."

"18*th*.—I rode to look at the Bengal Cavalry in their lines, and they seem to be picking up. . . . Various rumours afloat—one, that the Sirdars had left Girishk and with 100 men had gone to Meshed; another, that they still occupy a fort 25 miles beyond the Helmand River. A rumour that Dost Muhamed had advanced 60 miles on this side of Kabul with 100,000 men to give us battle; another, that his bodyguard intend to come over to the Shah as soon as he came near; another, that he was prepared for a start into the Hindoo Koosh as soon as we drew near; another, that Ranjit Singh had not fulfilled his engagement to the Governor-General of sending 5,000 men with the Political Agent Wade, in support of Timur Shah's advance to the Khyber Pass; and another, that Ranjit Singh intended to join Dost Muhamed in an attack upon us. It is said for certain that the Governor-General has refused to ratify the treaty entered into with the Ameers of Scinde and that Brigadier Douglas, Commandant of Bukker, had written for reinforcements."

Towards the end of the month General Thackwell was engaged in the inspection of the regiments of the Cavalry Division, and found the horses much improved in condition, and the remounts very promising.

"14*th June*.—The advance of the army upon Kabul, which was to have begun to-morrow, is postponed to await the arrival of a large convoy of camels (4,000) with grain."

"22*nd*.—The convoy of 4,000 camels came in to-day without any loss."

"*27th.*—The army began its march for Kabul at 4 p.m. . . . Cavalry Division and the 4th Local Horse, 18 guns, Sappers and Miners, and 1st Brigade of Infantry of Bengal."

But as the men who had come in with the hired camels of the convoy refused to proceed further, 20,000 maunds of supplies had to be left at Kandahar in charge of the garrison which remained there. The troops and followers were consequently again obliged to march on half-rations, but it was hoped that matters would improve on nearing Kabul, where, the season being later than at Kandahar, the crops would yet be uncut.

Kelat-i-Ghilzai was reached unopposed on the 4th July, but there had of late been constant reports of intended attacks by the enemy. "We were to have met 10,000 Ghilzais in position here," writes General Thackwell on this date, "but not a hundred have shown themselves, and they immediately retired.

"*18th July.*—Marched at 4.30 a.m. to Mooshakee. . . . The Ghilzai Chief with his 500 horse is not far from us, and the low range of hills on our left may give him a field to amuse our baggage to-morrow. Dost Muhamed's son said to be in position at Ghuznee. . . . For the last three marches we have been among the Hazarahs (a tribe), who are civil, and hail the crowning of the Shah apparently with joy. Their features are much broader than the Ghilzais, and give one a strong idea of the Chinese countenance."

"*19th.*—The enemy had a picquet of about 40 men on the hills to-day, but they retired before our advanced guard. At Ghuznee or before we are led to expect a fight, as it is supposed Dost Muhamed with the chief part of his army has arrived there. For some time the advanced guard has consisted of a regiment of native cavalry, four companies of native infantry, the Sappers and Miners, and two field guns. The Shah's column, with the 4th Brigade of Infantry, has closed up to-day, and Willshire's column will be within a march; this looks like business."

"*20th.*—From a hill two miles on the left of the cavalry camp I had a most delightful view of the plain between it and the high range. I had also a good view of Ghuznee, and a picquet of about 60 of the enemy's cavalry midway between it and our camp. A nephew of Dost Muhamed, with some followers, joined the Shah to-day. . . . A report that our camp is to be attacked to-night by Dost Muhamed's sons, and the troops will lie on their arms."

"*21st.*—The army lay down in position, but there was no attack or any alarm, although a stupid bugler of the Bombay Infantry sounded the 'assembly.' A report came in last night that the chiefs had evacuated Ghuznee, and a man came in this morning to say he saw them go away at six yesterday evening. Marched at 4.30 a.m. to a position three-quarters of a mile to the SSW. of Ghuznee, and at 4 p.m. to a position SE. and NE. of the town opposite to it and on the Kabul road and about one mile from the former.

"The army marched in four columns; the infantry, left in front, under and along the hills, the artillery on the main road, the cavalry in two columns of troops right in front at deploying distance, and the Shah's force in the rear, as well as Willshire's Bombay Infantry. The advance was beautiful, until within two short miles of Ghuznee, when a message from Sir John Keane, to the purport that 2,500 cavalry had passed through the mountains on our left to attack the baggage, caused the Bombay brigade of cavalry to be sent to the rear to meet any attack which might be made. This caused an opening in the line which the difficult nature of the ground, over which the right column passed, took some time to fill up. Strong patrols had been sent out to examine the numerous villages to the right and right front. Roche brought word from the several villages he reconnoitred that the chiefs were still at Ghuznee; but before he could be sent to communicate with Sir John Keane, hostilities had begun by a firing of matchlocks from the numerous villages and gardens, and shortly after the first cannon shot was fired from the town. The enemy were soon driven in within or around the town, and the two troops of Horse Artillery took up a position at about 450 yards from the walls, and opened a heavy fire which was soon followed by the camel battery. These knocked down some of the parapets and the upper part of the town at the southern angle and silenced most of the guns, which were but few; but, as it was found that the walls and profile of the town were too high to admit of a practicable breach, the artillery were withdrawn from their exposed situation after about half or three-quarters of an hour's fire, and the infantry followed, except a party with a squadron of the 2nd Light Cavalry, that went to the right of the town with the Chief Engineer to reconnoitre its weak points.

"Some casualties have occurred. Two officers wounded, and a man or two, and some horses have been killed, and several of both have been wounded. The Commander of the Forces was nearly killed by a cannon ball, and I had several narrow escapes

—one round shot at about 500 yards passed close over my head and dropped about twenty yards behind me. Roche and Crispin were also a good deal exposed. I had halted the cavalry within about 1,100 yards of the town, reinforcing the advanced guard with another troop, in order to repel any sortie by the enemy's cavalry, which was at one time expected. These withdrew when the artillery did, and on my return I found the cavalry in a sad jumble, intermixed with hundreds of camels loaded with baggage, Colonel Persse having withdrawn the Lancers and the 3rd Light Cavalry to the rear, leaving only the 2nd Light Cavalry in position, and the Bombay Brigade, which had by this time returned, was in a still more encumbered spot. A shot from the citadel, while I was having breakfast, carrying off the leg of a horse of the 2nd Cavalry, soon made the camel people glad to go to the rear. . . . At 4 p.m. we marched to the position before mentioned out of cannon shot. This afternoon's work has invested the town, but it would have been much better if the wild work of the morning had been avoided. A fatiguing day's work for men and horses—some poor fellows, I daresay, had nothing all day to eat."

"*Monday, 22nd.*—A quiet night. Having arrived so late in camp, the Engineers could not begin operations during the night, and therefore Sir John Keane will not realise what he told me yesterday—that he would be in possession of Ghuznee to-day *coûte que coûte*. . . . About 11 o'clock a multitude of men, infantry and cavalry, were seen to descend from the mountains four miles in rear of the Shah's and our camp, and it was supposed that a part of the Dost's army were the people. I rode to Sir John Keane, who ordered out the whole cavalry and six guns. On my arriving a little way to the rear I however found the enemy not nearly so strong as reported, and I therefore left the 2nd Light Cavalry to check any advance from the town, and sent word for the guns and Bombay Cavalry to remain on their ground. I took in support of the Shah's troops the 16th Lancers and 3rd Light Cavalry, and finding the Shah's infantry and part of his cavalry were attacking the enemy on their left, who were retiring up the hill, I sent a squadron of the 3rd and Christie's Cavalry to the right of the hill and two squadrons of the 16th Lancers to the left to cut off the enemy's retreat, and these flanking parties were afterwards reinforced by a second squadron of the 3rd and a third of the Lancers, when it was seen that the enemy's cavalry had all gone. The irregulars drove the enemy from the top of the mountain after a good deal of labour, but they retired to another pinnacle, and the Shah's troops were too exhausted to follow them up. Had they done so the Sixteenth would have cut them to pieces, and had their right been attacked instead of the left not a man would have escaped; as it was a good many had been killed and taken, and such a lesson had been taught them that they did not presume to follow the retreat."

"23*rd*.—The dispositions for the assault of Ghuznee were made last night, and at 3 a.m. the troops on the south side of the town began a fire of musketry on the parapet, which was soon followed by a fire on the other fronts, and of the artillery in position on the north side of the town, under cover of which the gate, the only one not walled up, was burst open by two successive petards and the troops rushed in to the assault. There were four light companies, under Colonel Dennie, of the four European Regiments, the 2nd Queen's, the European Regiment, and the remainder of the 13th Light Infantry were to follow supported by the 17th Regiment (the whole under Brigadier Sale). The attack succeeded at all points, and by sunrise the town, the Citadel, and Dost Muhamed's third son, Hyder Khan, were in our possession, although hundreds of the enemy still occupied houses and isolated points until the afternoon, when all surrendered to save their lives. The attack was well planned, parties of infantry being disposed on the other faces of the town to distract by their fire the attention of the garrison from the main attack. But very few of the defenders escaped, owing to the cavalry being judiciously stationed round the town very early in the morning. They had about 600 killed and 1,700 taken prisoners, most of whose lives were spared. . . . I entered the fort with Sir John Keane and the Shah over the dead bodies of the slain in the gateway, where all were obliged to dismount, and found that the citadel or palace had not been well defended. The zenana was there, and the poor women in sad distress, but they were protected. Many bodies had been sadly mutilated by shot and shell, hundreds of horses were galloping loose in the fort, fighting and desperately wounding each other.

"It was reported the Khan had escaped to a fort to the NE. and was surrounded by the 3rd Light Cavalry. Two guns and the European Regiment were therefore ordered to capture him, and I offered to go in command when breakfasting with Sir John Keane, and he was well pleased. However it turned out a false report, and I met the young Dost being brought a prisoner from the town. Went to Macnaghten's tent with Sir John Keane and the Khan, and there I saw a large body of cavalry to the south and got permission to march against them ; but as I was doing this with a few cavalry, I met one of their chiefs, and they proved to be the followers of Hadjee Khan, who had unaccountably remained in the rear—no doubt to have plundered us had we met with a reverse. The bullets yesterday kept a respectful distance, but to-day some matchlock balls were very near."

In Sir John Keane's orders issued on the capture of Ghuznee he writes in regard to the action of the cavalry:

"In sieges and storming it does not fall to the lot of cavalry to bear the same conspicuous part as the two other arms of the profession. On this occasion Sir J. Keane is happy to have an opportunity of thanking Major-General Thackwell and the officers and men of the Cavalry Division under his orders for having successfully executed the directions given, to sweep the plain and to intercept fugitives of the enemy attempting to escape from the fort in any direction around it; and had an enemy appeared for the relief of the place during the storming, H.E. is fully satisfied that the different regiments of this fine arm would have distinguished themselves, and that the opportunity alone was wanting."

CHAPTER IX

Major-General Thackwell gives the following description of the fort and town of Ghuznee:

"The town of Ghuznee stands on a hill gradually rising towards the north-west until it meets the precipitous hill on which the Citadel stands. The height of this natural hill above the country is about 20 feet on the eastern front, which front is about 430 yards long, but irregular in shape. The gate is within an angle, well covered by recent works of masonry, but the doors are very old and would not for an instant withstand the petard, as it was not, as supposed, walled up, nor was there any wall within it. The ramparts are about 25 to 30 feet high, but their profile in places is not 12 feet, and the road round them is continually obstructed by houses. There is a ditch of 15 yards wide, with gradual scarped counterscarp some 20 feet deep with probably 5 or 6 feet water, but a good roadway is left at all the gates. There is moreover a new *fausse braie* well loop-holed, constructed on a broad berm so that the débris of the rampart could not fall into the ditch. The south front is about 420 yards, the rise of the natural hill is about 30 feet, and the height of the rampart is about 30 feet above; it has no angle, but *demi-lune* projections at short distances, the same as on the east front.

It has the same kind of *fausse braie* and wet ditch as before described, and a large round tower, lower than the adjoining parapets, is at the SE. angle. It has strong brickwork with loop-holed parapet to protect it. There are two doors to the entrance about 12 feet from each other, but no wall across the inside. They would have easily been blown open, but neither of the doors could be seen from the outside. The river flows close to the gateway, a nice troutlike-looking stream, and on the right bank in front of the gate is a semicircular work covered by a wet ditch with a curtain about 40 yards long, and inside a raised work of some 15 feet high, with a diameter of some 15 feet. These works have a parapet, and are well loop-holed.

"The western face is not quite a straight line; it is in length, to the hill and rock on which the Citadel stands, about 330 yards, and thence to the NW. angle is 235 yards more. The natural hill rises from the commencement of this part to nearly 100 feet to the base of the hill on which stands the Citadel, and the height of the latter is at least 120 feet more. The rampart of the town is here much higher than before described, and it is here covered and flanked in like manner with wet ditch, *fausse braie*, and small semicircular flanks. There are two separate walls rising high in air covering the Citadel on this side with *fausse braie* and wet ditch for a short distance; a natural perpendicular rock rises at the angle, and the ground rises from the counterscarp into a broken ridgy hill running NNE. The north front is from the NW. to the Kabul Gate 180, and thence to the north-east angle 300 yards long—total 480 yards. The Kabul Gate has a very efficient covering of masonry. Its natural ground is about 40 feet high, with a rampart. It has a *fausse braie*, and a deep ditch, in part wet and in part dry. A road leads to the doors at the gate, which was visible from the country. This is by far the weakest front, although the Citadel has it under protection, as the ground is more hilly and broken to its left front than it is on either of the other sides, although it is broken on the east as well as rising and broken on the south.

"The ramparts of the enceinte are in some places about 10 feet broad, with a parapet six or eight feet high ; in others the houses are built close to them, and the flat roof is the rampart and parapet. There are very large houses of this sort on the right of the Kabul Gate, which quite overtop the walls, and to the left there are also houses connected with the rampart. In other parts the natural rise of the hill forms a solid rampart, and again in other places is so thin as to be easily breached. On the north-west portion of wall from the Gate a new work has been raised about 50 yards within the old wall, and overtopping it—the better to command the hill. On this platform was a 12-pounder, and a 6-pounder was nearer the gates ; two 4-pounders were at the SE. angle, and one other on the south front ; below the Citadel was the large gun and a 6-pounder, and there was also an iron 1-pounder, and these were all the guns I have seen. There were also a good many gingalls.

"The town cuts a most wretched appearance ; miserable streets of houses, and all filthy in the extreme. The streets are nothing but narrow, filthy lanes, and the stench and dirt are overpowering. The Citadel is 135 yards from the NW. angle of the walls on the west side, and on this raised space is a small, handsome building with a good deal of carved work about it. The continuation of the outer wall at the back of the Citadel is 60 yards; it then descends the steep obliquely for about 50 yards, and joins the town wall. Above this the Citadel is elevated at least 60 feet, and is irregular in form, the eastern front being about 170 yards, the SE. nearly 40, the western still more irregular, being about 200 yards, and the NW. about 40. The gate is on the east side, and ramps from the N. and S. lead to it. The wall around is very high and loop-holed, the entrance by two very thick folding doors—at least 6 inches—covered with very thick sheet-iron, highly embossed, and covered with tin or some fused white metal. There are many wells of water in the town, and the Dost calculated it would hold out until the winter began to set in, when we should be obliged to retire upon Kandahar. There were great stores of grain, bhoosa, and lucerne laid in for the cattle, and a great quantity of ammunition.

"I rode to Old Ghuznee and saw the tomb of Mahmood, surrounded with houses and gardens. It is a small mean building approached by passages resembling cloisters; the room containing the tomb is small and without ornament, a canopy of silk floats over the tomb, which is of fine white marble, sculptured all round in Arabic or Persian, with inscriptions that none of the interpreters can make out. The tomb is plain and nothing elaborate about it, except fine sculpture resembling a coat of arms, which does great credit to the remote age wherein it was done. The mace of this great warrior-king was not at the tomb; it was, it is said, sent into the camp a few days ago. The carving of the doors of sandal-wood is elaborate and rich, showing there were very good artists in India in that remote age. They were brought by the victorious Mahmood from the temple of Somnaut in Gujerat, which he plundered of immense riches."

On the 30th July the force marched from Ghuznee for Kabul in two columns, leaving behind a detachment of artillery, the 16th Native Infantry, some of the 4th Local Horse, 200 of the Shah's mounted men, and such sick and wounded as could not be removed without risk. With the first column were the Headquarters, the whole of the cavalry, and the Bengal portion of the army, while with the second—which moved on the 31st—were the Bombay troops under Major-General Willshire.

On the 3rd August, at Shekhabad, General Thackwell writes:

"I rode on the hills to the left, as we halted to-day to enable the Shah's column to join us. The view was a very pretty one over the Logur valley. When I got to the left front of the hills about two and a half miles from camp, I saw thick columns of dust rising on the Kabul road, and when I got back to the main picquet to give advice of it, I found that the report from the front was that Dost Muhamed's army had broken up, that he had left his artillery behind, and had gone to Bameean with 300 followers only, that 100 of his troops had just gone into camp to the Shah, whose column was marching in, and it therefore seems that the dust noticed proceeds from others of these miscreants, who have not the courage to defend their king and country. . . . Thus I fear the game is over, and the poor Dost is running for Balkh, unless he gets checkmated by the party of the 2nd Cavalry and 4th Poona Local Horse of 100 men with a corps of Afghans, which the Shah is sending to try and intercept his retreat to Bameean. This I don't think they will do, unless the Dost's usual good sense, as well as good fortune, has left him. Major Cureton also marched at 12 o'clock with 200 cavalry to Maydan to take possession of the Dost's artillery, and to move on Kabul if safe. The Dost's army was afraid to come to blows with us. He committed a great mistake in not having come to Kandahar with 6,000 or 8,000 men to have joined the Sirdars; had he done so, we should have had 15,000 men opposed to us there, which would have occasioned us loss, and retarded our operations. When he did not do this, he ought to have fought a battle in front of Ghuznee before his troops had been dispirited. If he had been defeated he would have retired under the protection of that fortress upon Kabul, and the attack on the place must have been made with diminished means. He would then have probably had more than 20,000 men in the field."

"*Sunday, 4th.*—Our camp—that is, the Indus Army and the Shah's—is along the Kabul road for two miles or more; the mountains on either hand are thrown about in wild confusion, and some with snow on their crests are 3,000 or 4,000 feet above our camp, which is not far short of 8,000 feet above sea-level. The appearance of these mountains is sublime—rocky, serrated peak upon peak, with an outline broken and tossed about in every variety of shape. The river is a fine rapid stream of transparent water, covering the valley in heavy rains and at the melting of the snows; a witness of this is in the remains of a dry arch with abutments on each side of the river, indicating that the centre had been swept away in ancient days. The road to and beyond the river rocky, and in places bad, several small streams having to be passed over rugged bottoms. The Bengal cavalry camp is in the gorge of the range of hills on the right—the main picquet to its right. The artillery and Bombay brigade in the preceding gorge, and the infantry in rear. A great number of the Dost's disbanded troops were on the road waiting to make their salaam to the Shah, and a multitude of the inhabitants were waiting with presents of fruit to hail the return of their legitimate sovereign, after an absence of 32 years."

"*5th.*—At Urghundee we passed the forsaken artillery of the Dost in position—though apparently an odd one. It consisted of between 20 and 30 pieces ; some good guns were there—a very long 12-pounder, a 24-lb. howitzer, and nearly a 50-lb. gun were the heaviest, all the others seemed sixes and fours ; but the carriages of all were wretched, clumsy affairs that would not have stood much marching. The guns were all in one position at about 350 yards behind a small deep nullah with perpendicular banks, where at two places only could horses pass in single file, but to the left this nullah could have been turned. An express arrived to the king this morning saying the Dost had halted a few marches off, to attend on his sick son Akbar Khan, and in consequence a subaltern's party of cavalry of the Indus was joined to a large party of Afghans and sent in pursuit at 12 o'clock to-day."

"*Tuesday, 6th.*—Marched at 3 a.m. by the northern road to a camp within three miles of Kabul and on its west. Views most beautiful and changing at every step."

"*7th.*—The Shah was to have gone into the town this morning at 6 a.m. We all paraded, but by some mistake the king was not ready, and the display was deferred until 3 o'clock, when he shifted his quarters to the town, attended by the Commander-in-Chief, the generals of division and their staffs, six guns, and two squadrons of British cavalry. He entered Kabul amid the acclamations of the people, and took up his quarters in the fortress of Bala Hissar, where there are two palaces, amid royal salutes and every demonstration of joy. News from Captain Outram that the Dost was ahead of him, that he occupied some forts with 200 riflemen, and that he could do no good against him with the troops he had, as his corps of 1,500 Afghans had dwindled to 700, and those could not be depended on. (Query.—Why did not a battalion of infantry march with this corps and another with two guns follow on the practicable road?)"

"30*th.*—It was fortunate that Ghuznee fell on the morning of the 23rd July, or we should have had to sustain a formidable night attack on our camp that very night, for it has been discovered by intercepted letters that the Dost's eldest son was to have attacked us with his 2,500 or 3,000 cavalry in conjunction with the three or four thousand fanatics whom we encountered on the mountains on the 22nd, and assisted by at least 4,000 horse and foot, then encamped as friends with the Shah under the leading of Hadjee Dost Muhamed. These would probably have been assisted by a sortie from the town, and our friend Hadjee Khan would not have been far away to lend a helping hand. Our picquets and patrols would, however, have given timely intimation of the movement of these people, and we should have been soon on the alarm post, and doubtless have repelled all these hostile attacks, yet they must have cost us a good deal of trouble and loss."

"*3rd September.*—The Shahzada Timur, with Major Wade, some Sikh Lancers and Light Dragoons, and some infantry arrived—the latter tall straight fellows, armed with English muskets and bayonets, and clothed in white linen dresses and trousers, with a red shawl wrapped round their heads. The cavalry have a similar headdress, and brown clothing, but their horses are miserable. There was a small party of the 20th and 21st Bengal regiments, some infantry of the Shah's, some irregulars got together by Wade, two of the Bengal 24-pounder howitzers, and two 4-pounder brass guns of the Shah's—in all, from twelve to fifteen hundred men."

"*5th.*—Sir John Keane said yesterday when he saw the 16th Lancers that the first charge he ever saw made was by Sir Colquhoun Grant at Vittoria, and he pointed out that it was at the left angle of the town. He gave great eulogy to the 15th Hussars, and I rejoiced because it was my squadron, led by myself—no Sir Colquhoun Grant or Major Griffiths being near—that charged and defeated the heavy dragoon regiment that came to attack us, drove them through a column of 800 infantry which we passed through by sabring and caused to lay down their arms—but I have told this story in another place."

About this time there were many robberies by the Afghans, and also attacks upon officers and men who were at any distance from camp, isolated or unarmed. Colonel Herring, commanding the 57th Native Infantry, was cut down and killed, several soldiers were attacked and wounded, and not a few followers were murdered, proving that if the Afghans were prepared to acquiesce in the re-establishment of their former sovereign, they did not approve of the continued presence in their country of the foreigners by whose assistance he had been reinstated.

On the 17th September there was a Durbar at the Palace in the Bala Hissar at which several officers, military and civil, were invested with the new Order of the Durani Empire just instituted by the Shah; the Governor-General, Sir John Keane, Sir Willoughby Cotton, Mr. Macnaghten, Sir Alexander Burnes, and Major Wade received the 1st Class of the Order; while several of the senior officers were invested with the Star of the 2nd Class: Major-General Thackwell received the decoration on the 10th October.

Preparations were now being made for the breaking up of the army of the Indus: on the 18th the Bombay portion of the force left Kabul *en route* for Quetta, and on the 2nd October orders were issued that the whole of the Bengal infantry division, the 2nd Light Cavalry, a light field battery, and a detachment of Sappers would remain in Afghanistan under the command of Sir Willoughby Cotton, the rest of the troops to return to India *viâ* the Khyber Pass.

That portion of the Bengal force returning to India marched on the 14th and 16th October. The Commander-in-Chief, Sir John Keane, moved with the 1st column, the 16th Lancers, a company of the 21st Native Infantry, and a *risala* of the 4th Local Horse; while Major-General Thackwell marched in command of the 2nd column, composed of the 3rd Light Cavalry, a troop of Horse Artillery, and four detachments of native infantry, numbering about 500 men: with him marched as a prisoner Muhamed Hyder Khan, late Governor of Kabul, and second son of Dost Muhamed.

The difficulties of the march India-wards began at once.

"*17th October*.—A severe frost in the morning; marched to Koord Kabul, ten miles. The road an ascent for the first three and a half miles, with small rocks and loose stones, to the Koord Kabul River, which runs down a pass and narrow defile, for about four and a half miles between very high mountains. This stream is crossed twenty-six times and the road is rocky and indifferent. If well defended it would be next to impossible to force this defile, as the sides are very precipitous, and in one place approach within forty yards. This is called the Koord Kabul Pass, and the mountain to the left is at least 5,000 feet high—nothing but a wall without a blade of vegetation, and this may be said of all the mountains around us.

"The cold in the pass was very severe, icicles hanging from the shaggy mane of a Turkestan pony, and from my stirrup irons, and, in fact, everything that got wet was frozen; one unfortunate follower died of cold in the pass."

"*21st*.—A fine day, but the water frozen in the rivulet—marched to Jugduluk, the road stony and uneven to the Jugduluk River, and then a gradual ascent along the stony bed through a close defile of three miles. This pass offers the finest scenery I have seen in this country. It is bounded by perpendicular rocks on either side from 100 to 600 feet high, and in one place so narrow that it barely took the breadth of the guns with about two feet to spare. In its windings it offers many fine views, and it is the closest pass for this long distance I have ever seen. A fine view of the Sufed Koh mountains with snow on their summits about 20 miles off to the SE."

"*23rd*.—A Jemadar, a Havildar, and one Sepoy were killed on the 22nd between Jugduluk and Surkhad, and one Sepoy was wounded; the robbers also fired upon a party of the 13th Light Infantry."

Jellalabad was reached on 27th October, Dhaka on the 2nd November, and Landikhana on the following day. The second column under General Thackwell arrived at Ali Musjid on the 5th: here an attack from the tribesmen was expected, but the General's diary records that—

"the night was passed in quietness, though the Khyberees are reported to be in force to the SW. of our camp. They had some days ago made an attack on Captain Ferris' detachment and were repulsed. They had also attacked some Punjabee troops on a hill *chokee* in our front near a round tower, for the recovery of ten or twelve thousand rupees they had left there when Colonel Wade ascended the pass. The Sikhs repulsed all the attacks, but at night they became panic-stricken, and retired from the fort, leaving 170 of their sick and wounded, all of whom were butchered."

"*6th.*—Marched to Kuddam; a close defile for the chief part of the way, and where it in the least opened, it was still commanded by high and broken and rocky mountains. The column, however, and baggage marched in close order, and all got in by 2 o'clock without any loss whatever, although a great number of the Khyberees lined the hills to the south; they, however, meditated an attack about 5 miles on the road from our last camp; but their ambuscade being discovered, a few of the European regiment drove them from the hill like chaff before the wind, killing two, and suffering no loss themselves. A large body of the Sikh troops lined the pass for the last four miles."

Peshawar was reached on the 8th November, and here the second column went into camp to the E. of the city, near the force which Sir John Keane had accompanied.

"The Governor of Peshawar, General Avitabili, gave a most sumptuous dinner to the officers of the first column yesterday, and to those of mine to-day. About fifty officers sat down to an excellent dinner, camp fashion, dressed and served in the best style. He must have a very good French cook, for the dinner was in every respect most excellent, and abundance of beer, sherry, port, and champagne cheered our hungry appetites and palates, which had not met with such things for so long. His house, built under his own superintendence, is more like a palace, and is tastefully arranged and ornamented. The gateway leading to it is also a handsome building, and there are several small temples tastefully ornamented within the enclosure: all were brilliantly illuminated on the occasion of the dinner, with glass oil-lamps as in England. Very good fireworks were let off before dinner, and afterwards there was an exhibition of nautch girls for some hour or more."

"10*th*.—I called on General Avitabili and saw some of his sword and target players exhibit. They were very expert and would be ugly customers if their hearts be in the right place. They are also expert in the use of the two-headed mace, which is a stave of four or five feet in length with a knob or ball at each end, which they hold in the middle and whirl in all directions, first with one hand and then the other in a most extraordinary manner. . . . They say Avitabili has amassed lakhs of rupees since he entered Ranjit's service about 1821, and talks of shortly going to Italy to enjoy the goods of his adventurous life in his own beloved clime."

"12*th.*—The Governor most kindly took me in his carriage, drawn by four mules, to see the fort and gardens. His style of travelling is princely: a guard of ten or twelve horse soldiers, six or seven outriders, and ten or twelve chuprassies on foot, and his beautiful mules constantly at full gallop. His two pages, interesting youths of eight or ten, are also in constant attendance on his person. We passed a gibbet having three bodies, and a string about 80 yards in length tied to the feet, which is attached to the post of the sentry on the town wall. By the motion of the string the sentinel would discover whether anybody was attempting to steal the bodies, when the shot from his fusee would probably send the intruder away."

"13*th.*—News has arrived from Mr. Macnaghten stating that Russian troops had arrived at Khiva, and that their agents have demanded Colonel Stoddert of the Chief of Bukhara, in consequence of which he had ordered the halt of the Bombay Army until the decision of Lord Auckland is given. An order is just arrived for the march back to Ali Musjid of the European and native drafts, a company of the Sappers and Miners, and two companies of the 20th and of the 21st Native Infantry to hold the fort until the arrival of the 37th and 48th Regiments of Native Infantry."

"14*th.*—The party marched this morning, but the ammunition was not ready to send after them. Pretty work this! The two guns and *risala* of Local Horse march to Jamrood at 12 to-night with some ammunition."

"15*th.*—The remainder of the ammunition to complete the infantry to 200 rounds per man was sent off to Jamrood at 2 o'clock to-day."

"16*th.*—All is reported to be quiet at Ali Musjid, and the party who marched from this under Captain Prole is stated to be still at Jamrood.

It is unlucky that the unhappy affair of the 12th should have occurred, as it has caused this irksome delay. On the 10th Captain Farmer, with two companies of the Sappers and Miners, two of the Twenty-first and one of the Twentieth, with a squadron of the 3rd Cavalry and two guns, and camels in sufficient numbers to carry grain for a month's consumption, marched for Ali Musjid. The artillery and cavalry halted at Jamrood on the 11th, and the infantry and convoy went on by the short road to Ali Musjid; they had some skirmishing and safely lodged their grain in the fort on the forenoon of the 12th. Farmer wanted to remain all night, but the political agent, Mackeson, overruled his wish by saying the Sikhs would not remain. They consequently commenced the march back between two and three in the afternoon by the same route, and were attacked by the Khyberees; the Sikhs ran away, some of the natives behaved well, others did not, and the whole came out of the defile *sauve qui peut* in the greatest confusion, losing killed, wounded and missing, as well as 436 camels. A disgraceful affair, wherever the fault lies."

"19*th*.—The loss of camels, *tattoos*, and horses during this campaign has been very great, and the column under my command lost in the march from Kabul to Jellalabad 1,600 camels by death and 200 more on the march to this place, and Captain Reddie states that between four and five hundred were taken by the Khyberees; but I think all these statements of loss are exaggerated by the surwans, who take every opportunity to cheat in every way they can. No doubt the camels were half starved at Kabul; and as a proof of their wretched plight, even in the middle of September nearly forty died in one night, which had come up from graze for the use of the Bombay troops—and these were selected as the best. Another cause of the mortality was the extreme coldness of the weather for four or five months, and the poisonous plants or shrubs they picked up at some of the encampments. The other column did not lose quite so many. If to these be added the loss of at least a hundred of the second column's private camels, we shall have a serious total.[1] . . . Everything quiet at Ali Musjid, but some firing is said to have been heard in the direction of Lieutenant-Colonel Wheeler's march."[2]

"22*nd*.—Yesterday an action took place in the pass with the detachments returning from Ali Musjid, accompanied by Lieutenant-Colonel Wheeler and the 37th and 48th Regiments. Two Europeans and a few Sepoys were killed, and from fifteen to twenty of the latter were wounded. The Khyberees are said to have had more than fifty killed and wounded: eleven dead bodies were seen by Colonel Wheeler together. The two European dead and one Sepoy were left behind, several camels were taken, and some officers lost all their baggage."

"23*rd*.—Called on General Avitabili and Sir Willoughby Cotton to take leave. They say Avitabili is a tiger in this government; he has been known to flay criminals alive and to break the bones of poor wretches on the wheel previous to hanging them in chains, and at our conference to-day very gravely wondered we did not put poison in sugar to send in traffic among the Khyberees! They say he has two wives, who reside in the top part of the house—beautiful Cashmerians.

"The Khyberees' rifle is a heavy affair, from 15 to 20 pounds in weight. It is longer than our

[1] The actual loss in animals in the whole army was 32,483.

[2] This officer was marching down from Jellalabad with the 37th and 48th Native Infantry.

musket, with a much smaller bore. It has a raised breech, with a hole through to take aim, and at an elevation is said to carry true to 450 yards. Two prongs work on a hinge near the muzzle, upon which it rests when the man kneels to take aim."

On the 23rd November the whole force marched out of Peshawar *en route* for Ferozepore, where the army of the Indus was to be finally broken up. The road traversed would seem to have been little more than a track, cut up by ditches, nullahs, and ravines. The Indus was crossed at Attock by a bridge of boats, but on arriving at Jhelum it was found that there had been a terrible mishap to the 16th Lancers, marching with the first column, in fording the Jhelum River. The ford would appear to have been almost V-shape—the apex pointing down stream; two squadrons of the Sixteenth crossed in safety :—

"The last squadron got too far down and became crowded, and some men got on a soft bottom, and then into deep water. The consequence was poor Captain Hilton, a corporal, and nine fine young men were drowned ; several followers were also drowned. Lieutenant-Colonel Cureton had to swim, being dismounted, and several officers and men had narrow escapes. Six horses also were drowned."

Ferozepore was reached on the 1st January 1840, when Lieutenant-General Sir John Keane issued a General Order breaking up the Army of the Indus, and left for Bombay, the corps then marching off to their respective destinations.

The following order was published by Major-General Thackwell on relinquishing his command:

"CAMP NEAR FEROZEPORE, 1st *January* 1840.

"In announcing the order for breaking up of the remainder of the Cavalry Division of the Army of the Indus, Major-General Thackwell, on taking leave, begs Brigadier Persse, commanding the Bengal Brigade, Lieutenant-Colonel Cureton, commanding the 16th Lancers, Lieutenant-Colonel Smyth, commanding the 3rd Light Cavalry, and Captain Alexander, commanding the 4th Local Horse, will accept his thanks for the efficient manner in which they have conducted their several commands during the march from, to their return to, Ferozepore, and he begs they will convey the same to the officers, European and native, non-commissioned officers, and privates of the regiments under their respective orders for their good conduct and steadiness on all occasions.

"He is happy to bear testimony to the discipline and cheerful submission to privations of no ordinary nature of these fine regiments; and if it had been their good fortune to have been opposed to an enemy in the battlefield, the Major-General feels assured that their gallant deeds in arms would have secured for them imperishable renown.

"He begs also that Lieutenant-Colonel Cureton and Captain Bere, Assistant Adjutants-General, and Major Hay, Assistant Quartermaster-General, will accept his thanks for the zealous manner in which they performed their various duties, and he requests to offer the same to the Divisional Staff Medical Officer, Dr. Chapman, and to his aide-de-camp, Cornet Roche, for their ready assistance on all occasions.

"He has also a pleasing duty to perform in acknowledging the good conduct of the remainder of the officers, non-commissioned officers, and soldiers of the Second Column of the Army of the Indus on the march from Kabul to this place, which has been marked by steadiness and correct discipline."

On the 22nd January Major-General Thackwell arrived at Meerut, where he found Mrs. Thackwell awaiting him ; and going on by easy stages *viâ* Delhi and Agra, Cawnpore was reached on the 13th March, by which time General Thackwell had travelled 3,028 miles since leaving that station fourteen months previously.

CHAPTER X

On arrival at Cawnpore Major-General Thackwell appears to have assumed charge of the station under Sir Edmund Williams, commanding the District. The troops in Cawnpore at this time were the 3rd Light Dragoons, some batteries of artillery, both horse and foot, the 31st Regiment, the 3rd and 8th Light Cavalry, and the 7th, 49th, and 66th Regiments of Native Infantry—a force of considerably more than 6,000 men.

For some eighteen months now General Thackwell followed the ordinary routine of garrison life in India—inspections of regiments and the supervision of the cantonment—and continued as of old to keep himself fit and well for any further active service to which he might be called. "I follow," he writes in his diary under date of the 4th October 1840, "the old Afghanistan practice of getting up before daybreak and riding out for about two hours, returning home before the sun becomes powerful."

As the senior officer of cavalry holding a command in the Bengal Presidency, General Thackwell's opinion was naturally often sought on matters connected with his arm of the service; the following extracts from a letter written in August 1840 to the Adjutant-General, reporting on the arms and equipment of the Bengal Light Cavalry, seem to show that, in regard to the views expressed in the first of these extracts, General Thackwell was far in advance of his age, while it may be remarked that the suggestion made in the second has only within the last few months been carried out in its entirety in the British and native armies:

"With regard to the mode in which the Light Cavalry are armed, it occurs to me that instead of fifteen carbines per troop only and a pair of pistols for each man, it would be more advantageous if the regiments were equipped in this particular the same as the Light Dragoons in Her Majesty's Service in this country—every Sepoy (the Sergeants, Band, Trumpeters, and Farriers excepted) being provided with a good carbine and one pistol. . . . In operating upon the open plains of India the necessity for cavalry being thus armed is not so apparent, but when their operations are carried on amongst the jungles, mountain defiles, ravines, broken ground, and in situations where the sword cannot be used, fifteen carbines per troop would be found insufficient to keep in check the matchlocks of the enemy, even if all could be available; but distributed as they might be for the protection of the baggage and in various other ways, but few might be at hand when they were required; and it is manifest there is but little chance of the Sepoys in general becoming good shots with the carbine where the practice must be so limited, and where they are not taught to consider it but as an offensive weapon belonging to a few. At the same time that I urge these considerations, I beg I may not be understood as advocating the fire of cavalry mounted. The sword is their proper weapon, and the other should only be used *when that arm cannot or ought not to reach an enemy.*"

Further on, writing of the large amount of baggage which the native soldiers were permitted to accumulate and carry with them on the march, especially in the way of numerous articles of uniform, General Thackwell writes: "Each of these articles would last longer than the regulated period *if a compensation were given when the article was not required*, and the evil would be greatly diminished."

News from Afghanistan of a disquieting nature was now beginning to filter through, and under date of the 4th October he records that—

"Dost Muhamed by all accounts is now before Bameean with 10,000 men, the garrison 2,000 with only ten days' provisions. Commotions likely to take place among the Afghans and Ghilzais. Major Clibbon's Bombay detachment of 600 infantry, 250 of the Poona Horse, and 4 guns defeated near Kahun by a handful of Beloochees. . . . Two hundred men killed on each side, also four of our officers; all the cattle and guns taken, the latter being spiked. A most mismanaged affair, and the sacrifice will be poor Captain Brown and the unrelieved garrison of Kahun. The chance of a Nepaulese and Sikh war and an actual conflict with China will afford us plenty to do this approaching cold season and next summer."

(Major-General Thackwell's services with the Army of the Indus had been recognised by his nomination and appointment to be a Knight Commander of the Order of the Bath in the

Gazette dated the 20th December 1839, and on the 21st of the following October he enters in his diary: "Received by the mail to-day the Cross and Ribbon of a Knight Commander of the Bath, but the Star and Badge are not come.")

"*24th November*.—On the 5th inst. Kurrak Singh died, and as Nihal Singh was moving with the procession for the funeral obsequies through a gate of the palace, the heavy tread of the elephants caused a beam to fall from the roof which struck him on his skull and caused his death in five hours after. A son of Gulab Singh was killed and others were wounded. If the suspicion that he poisoned his father be well founded, this would seem a judgment of Providence. To-day intelligence has been received that Dost Muhamed has surrendered himself to Shah Shuja after losing a battle in which Dr. Lord, Lieutenants Bradford and Crispin were killed and Fraser and Ponsonby wounded. . . . The account of the action of the 2nd November, in which poor Bradford and Crispin had their heads cut off, states the right troop of the 2nd Cavalry first wavered and that then the two squadrons went about and left their officers to be cut to pieces. They excuse themselves on account of not being able to use their swords. Fraser lost his right arm, it being nearly cut off, and has a wound in the back. Ponsonby has his nose and face cut through and is wounded in two places."

(These officers had of course all been well known to Sir Joseph while under his command in Afghanistan, where Crispin had been his extra aide-de-camp.)

On the 4th January 1841 Sir Edmund Williams was nominated to the officiating command of the

Meerut Division, Major-General Sir Joseph Thackwell succeeding him in the command of the Cawnpore Division—a post which he held until the middle of the following October.

"13*th April.*—The ex-Shah of Afghanistan, Dost Muhamed, arrived here by water on Sunday evening. He remained on the river yesterday, and started at daybreak this morning for Calcutta. He had seven or eight boats, and I saw him when he passed the bridge of boats, as all were upon the roofs of the boats, and I was within forty yards of him—a stout man with a swarthy countenance and bushy beard."

"29*th November.*—Some days ago a rumour arrived that Kabul was in a state of insurrection; that the guns from the Bala Hissar were firing on the town; that Burnes had been murdered; that great fears were entertained for the safety of the 37th and 54th Regiments of Native Infantry, who were left between the Khoord Kabul and Jugduluk Passes; and that Sale's brigade was marching to Jellalabad surrounded by thousands. That Ali Musjid was invested by the Khyberees, and that the position occupied by Captain Ferris was about to be attacked."

"30*th.*—Letters from Kabul of the 9th November state that the Kohistanis have joined the Kassilbash-is, and that the whole country has risen *en masse*; that part of the Forty-fourth and two Sepoy regiments have occupied the Bala Hissar with Shelton; that the guns have caused great destruction in the town; that an attack on the Bala Hissar has been repulsed with great slaughter, and also that a party of the Shah's cavalry has cut to pieces some Afghan horse. That Sir Alexander Burnes and his brother, Captains Swayne and Robinson and Lieutenant Raban of the 44th

Queen's, Captain Maule of the Artillery, Lieutenant and Adjutant Wheeler, Ensigns Gordon and Robinson, 37th N.I.; Broadfoot, European regiment and Second in Command of the Shah's 4th Light Infantry; Ensign E. W. Salusbury of the same; and, report says, Lieutenant Rattray, Mrs. Trevor, and her six children have been murdered. That the envoy and family with other strangers had taken refuge in an entrenched camp. The people had placed the son of Zemaun Shah on the throne, and that several of Shah Shuja's family had joined the insurgents."

"*2nd December.*—Report says that Captain Woodburn, of the Shah's 3rd Infantry, when on the march from Ghuznee to Kabul with 150 men was attacked by a large body of Ghilzais, and that of the whole detachment only one man escaped to tell the sad tale of the slaughter of the rest."

"*3rd.*—A letter from Sir Robert Arbuthnot tells me that he had heard from Clarke, who had received advices from Kabul, that on the 11th an action was fought with the insurgents in which we were completely victorious, recovering the two guns we had lost in the city, and that Sale had defeated the rebels at Jellalabad, but that he expected to be attacked again on the 20th. He felt secure, having placed Jellalabad in a state of defence."

"*7th.*—News to-day that the troops at Kabul were shut up in the Bala Hissar and cantonments with only ten days' provisions. That Macnaghten had written to Sale to move to his relief, whose reply is that he has only twenty days' provisions, and has no money or carriage, and that he cannot move."

"*3rd January* 1842.—A report that the Commander-in-Chief has had advice that a battle took place at Kabul on the 7th December, and that it was a very bloody one, the Afghans having fallen in heaps. Letter from Jellalabad mentions that the cannonading was heard there on the 7th in the line of Kabul from 9 o'clock a.m., until the evening, a distance between the two places of 102½ miles."

"*8th.*—I had a letter from Cureton in which he says, 'Macnaghten writes from Kabul on the 8th December to his brother-in-law at Simla saying their resources were all exhausted and that they must make the best terms they can.' A letter from Jellalabad of the 22nd December says most disastrous news from Kabul, but they have no enemy near them and are not badly off for provisions. The Khyber Pass was closed, but they sent their *dâk* by another but not very secure road. The letters from Ludhiana and Ferozepore say that poor Elphinstone is dead."

"*12th.*—Reports for the last two days that on the 18th December the troops at Kabul gained a complete victory, that the insurgents dispersed in all directions, and that the Chiefs flew to their mountain valleys. A letter from Ferozepore, received to-day by Sir Edmund Williams, says that Macnaghten had written to request that all the troops might be withdrawn from Afghanistan and Dost Muhamed sent up to assume the government. A very different story to that of yesterday, which added that his son Akbar Khan had surrendered to the Envoy!"

When the news of the disaster at Kabul reached India, the first impulse of Lord Auckland's Government was to proclaim—on the 31st January—a determination to expedite "powerful reinforcements to the Afghan frontier, for the maintenance of the honour and interests of the British Government," and it was stated that "the ample military means at disposal will be strenuously applied to these objects." To this end it would appear that Sir Jasper Nicolls, the Commander-in-Chief, had been required to prepare a very much larger force for service in Afghanistan—particularly in cavalry —than was eventually sent or than indeed the military resources of the country could at that moment well afford. In the expectation that it might become necessary to send forward a whole cavalry division, the services of Sir Joseph Thackwell were once more called for, and on the 17th January he received orders to proceed at once by *dâk* to Kurnaul and there join Army Headquarters. He bought his camels and started off them and his horses, while he himself set off on the morning of the 23rd for Kurnaul. On the road he passed his late regiment, the 3rd Light Dragoons, also on their way to the front to join the force under General Pollock, and, not finding the Headquarters at Kurnaul, Sir Joseph pushed on and got into Sir Jasper Nicolls' camp, about 21 miles beyond Umballa, at midday on the 29th.

On the 31st the Commander-in-Chief moved off on his way to Ferozepore, being received *en route* by the Rajah of Pattiala and his troops, who lined the road for more than a mile and a half. General Thackwell describes the Rajah as being "a man of gigantic stature, 6 ft. 4 in. in height with form and features in proportion, and having the reputation of being greatly attached to the English. His health is fast declining, although he is not much more than forty years of age." On the march "there was an official report of the retreat of the two battalions from Ali Musjid, with the loss of 400 men, owing to the want of provisions. Brigadier Wild wounded."

On arrival at Ludhiana Sir Jasper Nicolls held a durbar whereat a number of native officers were presented, and Sir Joseph Thackwell—who did not share the low opinion then far too often held and expressed by officers in the British Service as to the value of native irregular troops—writes: "I had the pleasure of renewing my acquaintance with the native officers of the 4th Irregular Horse —a noble set of fellows!"

Already, no doubt, Sir Joseph must have had misgivings as to whether after all he was likely to be employed, either in "the Avenging Army" moving on Afghanistan, or in an army of observation which there was some idea of forming on the Sutlej to watch the movements of the Sikhs. On the 11th February he brought matters to a head in an interview he had with the Commander-in-Chief—

"and besought him to give me active employment, either the command of the brigade in which the 3rd Light Dragoons will be placed, or of the cavalry, as I was willing to waive my rank and serve under Major-General Pollock. He was highly pleased with my offer, and gave me hopes of being employed."

On the 14th, however, he writes:

"The Commander-in-Chief told me to-day that I might return to Cawnpore, as no cavalry to form a division would be required, no army of observation to be formed, the Sikhs having given proofs of good faith, and although I had volunteered to serve under Pollock in command of the cavalry, it could not be granted."

"16*th.*—Sir Jasper and myself had a good deal of conversation before dinner yesterday and at breakfast this morning, respecting the assembly of troops on the Sutlej; he says that the Council have decided what measures shall be adopted against the Afghans by this time, and that he will be made acquainted with them to-day or to-morrow. He is glad I have not laid my *dâk*, and I hope there is some chance of my being employed."

On the 18th, however, all hope of active employment seeming to be at an end, Sir Joseph took leave of the Commander-in-Chief, who proceeded on his way to Ferozepore, while General Thackwell marched towards Simla for a tour in the hills.

He had intended to have travelled thence over the hills to Dehra, up to Landour, and *viâ* Meerut to Cawnpore, but he found he had not the necessary time, so marched down to Umballa, after spending only a day or two in Simla, and finally rejoined Lady Thackwell at Cawnpore on the 16th March —just in time for the Proclamation Parade, announcing that Lord Ellenborough had assumed the reins of government.

This was an anxious and a sickly season. Pollock's force was fighting its way up the Khyber, and many were the fears that his force was insufficient or that the troops might arrive too late to save the garrison still holding out; news came back but fitfully, delays were exaggerated, and petty reverses magnified. The rainy season, too, set in very late—

no regular rain falling much before the end of August, and cholera of a malignant type broke out at Cawnpore. In April Sir Edmund Williams seems to have been transferred to the command of the Meerut Division, and during the hot weather at least Sir Joseph Thackwell reigned in his stead.

On the 20th June the diary contains the entry:

"Got the General Order for the formation of an Army of Reserve under the Commander-in-Chief. Generals of Division: myself (cavalry), Dennis and Ballin (infantry). Brigadier Graham to command the artillery; Major Smith, Chief Engineer. Brigadiers Orchard, Paul, Moore, and Young. To have an Assistant Adjutant-General and Quartermaster-General to the cavalry, and the same to the infantry divisions."

"*28th July.*—Advices from Captain Somerset, Military Secretary to the Governor-General, that the order had gone up to the Commander-in-Chief on the 25th for the assembly of the Army of Reserve at Ferozepore on the 15th November. Also that the Governor-General and all the suite would leave Allahabad by water on the 20th August for Gurmuktesur Ghat to there meet their tents and land carriage, and march thence by Meerut to Ferozepore."

Handing over the command of the district to General Grey, who had arrived a few days previously, Sir Joseph Thackwell left Cawnpore on the 19th October to join his division in the Army of Reserve, and marching leisurely reached Ferozepore on the 14th November. Here, as senior officer then present, he assumed command of the Army of Reserve or of as much of it as was already assembled.

The Commander-in-Chief, Sir Jasper Nicolls, arrived in camp on the 20th November, and Lord Ellenborough on the 9th December, and Sir Joseph had a fine force of cavalry under his command—the 16th Lancers, the 3rd and 7th Light Cavalry, and the 4th and 6th Irregular Horse, with four troops of horse artillery. The camp was laid out within a mile and a half or two miles of two bridges of boats which had been thrown across the Sutlej, and on the 16th—

"the troops were drawn out in one line on the right of the road leading to the two bridges of boats to go through the form of saluting 'the Illustrious Garrison of Jellalabad' which marched into camp this morning—sleek, plump, and healthy, about 2,000 strong. Twenty-four pieces of horse artillery on the right, eighteen squadrons of cavalry next, then eighteen pieces of artillery, 9-pounders, and on the left twelve battalions of infantry, including four Queen's Regiments. The Jellalabad garrison consisted of the 13th Light Infantry, 720 strong; the 35th, upwards of 800, the Sappers and Miners; No. 6 Field Battery (9-pounders and 24-pound howitzers), about 120; the Mountain Train (2-pounders), about 100; a squadron of the 5th Cavalry, about 100; and a *risala* of Anderson's Horse, about 100 men."

On the following day Pollock's troops marched in, and on the 23rd—

"the Governor-General and the Commander-in-Chief went down to the bridge of boats, at a quarter before 7 a.m., to see General Nott's army and the gates of Somnaut. I accompanied them. The troops looked healthy, in good order, and the horses in pretty good condition. A small advanced guard of cavalry and infantry preceded the column. Then came the 3rd Bombay Cavalry, the 1st Irregular Cavalry, and Christie's Horse—about 1,000 men. The Bombay troop (European) of Horse Artillery, then Anderson's troop of Native. Then a field battery of Europeans, 18 guns; afterwards the infantry—the 40th and 41st Queen's, the 2nd, 16th, 38th, 42nd and 43rd, and the Khelat-i-Ghilzai regiment of Native Infantry—about 6,000. Total between 7,000 and 8,000. Altogether the force cut a very fine appearance."

"*24th.*—About 7 o'clock on the evening of the 23rd, after all the troops and baggage had crossed the river, the great rise of water in it carried away the two bridges of boats; some of these have been carried twenty-five miles down its current, but none have been sunk."

The whole army united under the Commander-in-Chief numbered about 27,000 men, and was composed of 78 guns, 45 squadrons of cavalry, and 28 battalions of infantry.

On the 3rd January the camp began to break up; the Governor-General left for Delhi, and the different corps moved off to their various stations.

General Thackwell obtained leave to visit Mussoorie, and, travelling by easy stages, did not reach Cawnpore again until the 17th February.

During this year a new Commander-in-Chief—and one, too, under whom Sir Joseph Thackwell was to see much service—arrived in India, and on the 24th August the notification of the appointment of Sir Hugh Gough was read to the troops on parade in the different garrisons of the Dependency.

Sir Hugh Gough at once found that both in the north and in the south there were signs that the services of the Indian army might again ere long be required. In September of this year Sher Singh, the Maharaja of Lahore, was assassinated, and the condition of the army of the Sikhs—brave, well trained, and turbulent—gave at once real cause for alarm.

"In the midst of this anxiety a source of trouble, similar if not so formidable, arose in the State of Gwalior. The army of that State was less numerous than that of the Punjab and not quite so highly trained, but the spirit of political dictation was hardly less strong among them."

In November 1817 the Marquis of Hastings had concluded a treaty with Daulat Rao Sindia, under which the latter agreed to abandon the Peshwa, to join the British against the Pindaris, and to surrender for a time two of his fortresses. Ten years later—during the reign of Lord Amherst—Daulat Rao was in declining health, and the affairs of his kingdom were largely in the hands of his favourite wife, Baiza Bai, and her brother Hindu Rao—whose house on the Ridge at Delhi was to become a famous post in the siege. Daulat Rao Sindia himself was childless, and was for some reason unwilling to follow the usual Indian custom and adopt an heir, but wished that this power should rest, after his decease, with Baiza Bai. Before his death, however, in March 1827, he left the decision as to succession absolutely in the hands of the British Resident, and Baiza Bai was

[1] Keene.

permitted to adopt a young and distant relation, who was thereupon placed upon the throne, Baiza Bai retaining the office of Regent. Dynastic disputes, however, at once broke out in the State of Gwalior, and six years after the death of Daulat Rao his widow was compelled to give up the Regency and retire to her private estates in the Deccan, the boy whom she had adopted being recognised by the Government. This youth, however, died some years later, also without issue, but leaving an adopted son who was a minor. It was therefore necessary to appoint a regent or council of regency, and the Gwalior troops refusing to recognise the nominee of the British Government, the Resident, in view of the inimical attitude of the troops of the State, judged it prudent to remove himself in the winter of 1843 to Dholpore, nearer the frontier. The Governor-General now required of the Gwalior authorities the restoration of the Regent and the reduction of the army, whose strength was wholly out of proportion to the requirements of a protected State, and in the meantime Sir Hugh Gough, the Commander-in-Chief, had prepared for all possible eventualities and was ready to take the field with a force of some 12,000 men.

On the 21st October he visited Cawnpore, putting up with General Grey. " I called on him at 11 a.m.," writes Sir Joseph in his diary of this date, "and he was very civil. Told me I was to command all the cavalry for Gwalior."

Later on he duly received his orders and left Cawnpore on the 23rd November for Agra, where he arrived on the 26th and where Army Headquarters was established. Sir Joseph, however, went into camp some 32 miles to the north of the city, where he found already assembled a substantial portion of the division which he was to command in the forthcoming campaign.

CHAPTER XI

THE fort and town of Gwalior were situated about 65 miles almost due south of Agra, being in the most northerly portion of the dominions of Sindia.

"The British districts of Agra and Etawah bounded the Gwalior State on the north-east; the protected States of Dholpore and Rajputana were co-terminous with it on the north-west. Along the whole of these frontiers the boundary line was the River Chumbul; the north-eastern corner of Gwalior extends almost to the point where the united waters of the Chumbul and the Jumna are joined by a smaller tributary known as the Sind, which separated Gwalior from the British districts and protected States of Bundelcund."[1]

In the event of hostilities becoming necessary, Sir Hugh Gough had decided to operate with an army divided into two wings, the right wing to be assembled under his immediate command at Agra—the left, under Major-General Sir John Grey, K.C.B., to concentrate at Jhansi and Koonch in Bundelcund. The Cavalry Division, to be commanded by Sir Joseph Thackwell, comprised four fine brigades containing 32 squadrons, but was

[1] Rait.

equally divided between the two forces—apparently at General Thackwell's own suggestion, for when Sir Hugh visited Cawnpore in November Sir Joseph records in his diary:

"Transacted business with the Chief and fixed on the Assistant Adjutant-General of the Cavalry, Pratt, and Quartermaster-General, Clayton, and he said he had adopted the arrangements I had suggested, and two brigades of cavalry were for Bundelcund."

It is possible that this arrangement was proposed in order to make up in some measure for General Grey's weakness in infantry, he having only one division, while two were with the right wing. With Sir Hugh also was a battering train of 50 pieces, but of these only six 18-pounders and four 8-inch howitzers crossed the Chumbul and arrived at Gwalior.

Sir Hugh Gough has been blamed for this division of the force, the two wings of which were placed on either side of the Gwalior army, which consisted, moreover, of not less than 22,000 veteran troops, trained under European officers and provided with a powerful artillery. There seems no doubt that the Commander-in-Chief, but newly arrived in the country and knowing little or nothing of the armies of native states, had underestimated the fighting powers and resources of his enemy; but he considered that either wing of his army was strong enough to defeat the Gwalior force singly, while the position in which he hoped to force the enemy to accept battle would enable him not only to utterly crush them, but effectually to prevent the dispersion of their force into predatory bodies of armed men.

The cavalry of the right wing of the so-called "Army of Exercise" left its camping-ground on the 16th December for Dholpore, where the Chumbul was to be crossed. Marching by Muttra, Bhurtpore was reached on the 7th, and of the fortress Sir Joseph writes:

"Its celebrated walls are about six miles round, the ramparts of mud, with a deep, dry ditch. . . . The siege by Lord Combermere was on its ESE. face; the two breaches made by the mines are still as they were when the assault took place, except as to remains of mortality. It strikes me the batteries were too distant, as they appear to have produced no great effect, but the breaches were very easy. The town wall opposite to the one to the south ought to have been defended, as it was in a salient round bastion connected with the town by a narrow rampart only. On the south and west sides there is much cultivation, and a lake lies to the west, and at about six miles distant a range of low hills. On this side Lord Lake made his unsuccessful attack in 1805, but for my own part I think it was scarcely to be taken, except by mining, unless the battering train had been very numerous indeed."

Agra was reached on the 9th, and here the whole camp was formed.

"Found in camp," writes General Thackwell, "about 15,000 men, of whom I had the command. The cavalry formed the front of the square together with three troops of Horse Artillery: the infantry, with Nos. 10 and 17 Field Batteries, formed the two flanks."

On the 11th the Governor-General arrived in camp, and next day the 3rd Infantry Brigade, with a field battery and two regiments of native cavalry, were sent towards Dholpore with orders to halt at the second march for further orders. On the 15th and 16th the rest of the cavalry division of the right wing left camp, and was concentrated on the 18th at Dholpore, where the rest of the force soon closed up.

"21*st.*—Marched at half-past 6 o'clock with the 3rd Brigade of Cavalry and the three troops of artillery to Keutri Ghat on the Chumbul River, nearly eight miles—encamped within half a mile of the river."

"22*nd.*—Marched at half-past 7 o'clock to the Keutri ford, about 260 or perhaps 300 yards broad and now but little more than three feet deep, with a hard, firm, sandy bottom. The troops—viz. three troops of artillery, No. 17 Field Battery, the 3rd Brigade of Cavalry, and the 3rd Brigade of Infantry—passed it without accident, and took up a position behind the ravines of the Kohary River, the right resting on Sehoree, and the left towards Hingonah on the same river. Covered the right and left with picquets during the night on a wide, well-cultivated plain. After crossing the ford, the ground for a mile and a half is broken into ravines like the broken billows of a troubled sea; afterwards the country is quite level, and the two roads branching off to Gwalior and Bombay are excellent, though not metalled."

"23*rd.*—The 4th Brigade of Cavalry and the 5th of Infantry, the Governor-General, and the Commander-in-Chief marched into camp to-day. I waited on His Excellency, and saw Smith [Sir Harry], Gough and Grant" [afterwards Sir Patrick].

"Rode in the evening with His Excellency along the right front of the encampment, and he decided on having a wing of the 14th Native Infantry at the village of Sehoree."

"24*th.*—Rode with the Commander-in-Chief along the left front to the village of Chota Hingonah, distant from Sehoree on the Bombay road about five miles. The roads from Keutri ford and Dholpore ferry join before entering the former village and continue on to Gwalior. This road and the one to Bombay are the only practicable routes for carriages, and I believe for horses, across the deep ravines in front of our line, which extends about four miles. My right flank at Sehoree is now well covered, but the left is still exposed. Only by the Gwalior road can the heights beyond the Kohary River be ascended, and they therefore form a very strong position. We got upon them, and in our ascent met with about thirty impudent Mahratta soldiers belonging to Colonel Jacob, and at one time we thought we should have come to blows with them, but a picquet of the 10th Cavalry coming up by accident—as they had been ordered to be discontinued the day before—caused them to change their tone and they were very submissive. It certainly was very fortunate these twenty men came up or I verily believe a collision would have taken place, as the Chief was very irate. . . . Reconnoitred the line of river in front in the afternoon."

"25*th.*—I visited the right picquets and rode to the front beyond the ravines on the Sipree road—a fine plain to the south and east for several miles covered with cultivation."

"27*th.*—I crossed the river at Jetowa and rode beyond the ravines; came home by the Bombay road. Picquets all well posted except the rear one, which is not far enough to the right rear."

"28*th.*—Rode with the Commander-in-Chief this morning across the ravines. A troop of the 4th Cavalry with Lieutenant-Colonel Garden went on the Gwalior road towards Chonda and there met with the advance of the Mahratta army—about 5,000 men. They fired about twelve guns at his party, but no loss occurred—a dog only having been slightly wounded. All commanders of divisions and brigades waited on the Commander-in-Chief at noon to receive instructions about the intended attack to-morrow."

"29*th.*—The army crossed the Kohary Nuddi and advanced through its defiles at 6 o'clock this morning. The right column—consisting of the two troops of the 2nd Brigade of Artillery, the 16th Lancers, the Bodyguard, the 1st Light Cavalry, and two *risalas* of the 4th Irregular Horse (the rear guard to which was one squadron of the 1st Light Cavalry and a *risala* of the 4th Irregulars)—covered the right flank; the centre column consisted of the 3rd Brigade of Infantry under Major-General Valiant and crossed the Kohary at Jetowa and marched by Kaladreh to the left of Mungowleh; and the left column by Chota Hingonah along the main Gwalior road towards Jowra. It consisted of the 4th and 5th Brigades of Infantry, Nos. 10 and 17 Field Batteries, the 4th Native Cavalry and 10th Light Cavalry with the 2nd troop of 3rd Brigade of Artillery. In all about 8,500 men. My column marched along the main Bombay road and turned to the left a little short of Omedghur, leaving the village of Sainte on the left, and arrived on the right of Mungowleh as ordered. It had been delayed a quarter of an hour owing to the horses of a gun falling in the quagmire of the descent to the ford of the Kohary, and on arrival near Mungowleh received an order to proceed towards

Maharajpore and form on the right of the 3rd Brigade of Infantry.

"The Maharaja's army was posted in front of Chonda with a strong body of infantry in advance on the Chota Hingonah road, and another large body of infantry with fourteen guns or more occupied the village of Maharajpore about a mile from the former, and each nearly a mile and a half from Chonda—Maharajpore being on their extreme left on the road to Mungowleh. The Commander-in-Chief sent for me as the troops were coming up, at which time the enemy had opened their fire from both advances, and No. 10 Field Battery and the 2nd troop of artillery were answering them on the left. Colonel Wright's brigade (the 5th) was advancing to attack Maharajpore on its left.[1] After receiving the Commander-in-Chief's orders—who seemed very cool under a sharp cannonade—to turn the village of Maharajpore and attack any troops retiring from it, I made my dispositions—two squadrons of the 16th Lancers on the right and the 1st Light Cavalry forming the first line, and the Bodyguard and two squadrons of the Lancers forming a second line; the guns as they came up were on the left of the first line, supported by the remainder of the 4th Irregulars—not 50 men. The cavalry made a rapid advance and gradually brought its right shoulders forward, during which time the infantry was warmly engaged in carrying the village.

"On arriving on the road leading from Maharajpore to Dampoora, the retreat of the enemy by the road was cut off, and some baggage was taken, and some men were cut down. Near this point Major-General Churchill and Captain Somerset were wounded, and Lieutenant-Colonel Saunders, Under Military Secretary to the Governor-General,

[1] The *enemy's* left of Maharajpore would seem to be meant.

was killed by a musket ball. The cavalry advanced beyond this point, but none of the enemy's infantry were seen retreating; and as a four- or six-gun battery in position among the ravines on the right front of Dampoora opened at about 400 yards on our right flank, and a considerable body of infantry supported the fire of musketry kept up from the ravines before mentioned, and it being also presumed that his flank was protected by a large body of cavalry, I did not think it advisable to continue my advance in the direction of Chonda, but threw my right back and remained fronting Dampoora. During this time the artillery had come up and made good practice against the batteries in the direction of Chonda and Grant's troop against the one near Dampoora, but the enemy had contrived to blow up two of their ammunition wagons by round shot passing through them. Brigadier Scott's two regiments had also been warmly engaged, had taken six guns and had cut down a number of cavalry who had advanced to attack them. This brigade and its troop of horse artillery came unaccountably to the right, round the village of Maharajpore, whereas it ought to have remained on the left flank towards that point, but as Sir Hugh Gough was there I presume they acted by his orders. The 4th Irregular Cavalry were ordered by His Excellency to charge a two-gun battery, which they carried but could not keep, owing to being unsupported when warmly opposed. I had ordered the 1st Light Cavalry to the left to support the two troops of artillery and the attack of Major-General Littler's brigade on Chonda on the left of the ravines, but it does not appear they had an opportunity of doing anything. Major-General Dennis with Brigadier Stacy's brigade—the 4th—came to the rear of the cavalry when Littler advanced, but had orders to halt, otherwise the ravines in front would soon have been cleared, the battery they protected would have been carried, and the cavalry would have passed the ford of the Ahsin and cut up many of the enemy—as it was they left a great deal of baggage behind. Major-General Smith was on their flank, and highly disapproved of the position poor Churchill wanted to place the 3rd Brigade in. He would not allow the cavalry to turn the ravines, or cross the ford to fall on the enemy's rear, therefore as my superior officer I was bound to obey. The position between Jowra, Maharajpore, and Chonda was a fine triangular plain on which cavalry could act in the finest manner."

Elsewhere General Thackwell records:

"The cavalry could not arrive in time for the advance of the infantry, and it was much hurried to get into the second position; it had then to make a rapid movement to get round the 3rd Brigade of Infantry, and advanced to its third and fourth positions under a heavy fire of artillery. The two troops of artillery attached to the 3rd Brigade of Cavalry did not fire a shot at the village of Maharajpore, and did not open fire until it got near the ravines in front of Dampoora."

Of the rest of the force the infantry with Major-General Littler—the 5th Brigade under Colonel Wright—was heavily engaged. Attacking the village of Maharajpore in front while Valiant assailed it, as already described, on the west, they met with a very determined resistance, the

Mahratta army fighting sword in hand, and rallying repeatedly before the village was cleared. Wright then passed on, covered by the artillery, to attack the main position at Chonda, where the enemy made a last very desperate stand, and where the greater number of their guns were captured.

"The battle was over by 12 o'clock; but had the 4th Brigade of Infantry, instead of attacking Chonda, been permitted to cross the river with the cavalry, half the Mahratta army would have been destroyed—or if even a battalion of Stacy's brigade had passed over with the cavalry instead of remaining halted for three-quarters of an hour, a great part of their rear would have been cut off. Thus ended the battle, in which the enemy's guns were well served and they fought with great determination. . . . Encamped on the field of battle, and poor Churchill, whose leg had been amputated, died during the night.

"The Mahratta army was to the amount of about 12,000 men—cavalry probably 3,500 and about 60 guns. Their loss, 56 guns and about 2,000 men. Our total loss was 780 officers and men, of which number the cavalry had 11 men killed and 48 wounded, 78 horses killed and 48 wounded, 4 men and 10 horses missing; 1 officer killed and 5 wounded, of which 1 native officer killed and 4 wounded, of this number three had legs amputated. The 16th Lancers had 2 men and 22 horses killed, 1 sergeant, 5 privates, and 4 horses wounded."

In his despatch on the action Sir Hugh Gough mentions that the cavalry "manœuvred most judiciously on the right and would have got in rear of the position and cut off the retreat of the whole, had they not been prevented by an impassable ravine." He adds:

"To Major-General Sir Joseph Thackwell, K.C.B., to Major-Generals Dennis and Littler, and to Brigadier Gowan, commanding the divisions, my best thanks are due for the manner in which they conducted and led their respective divisions."

General Thackwell mentions in his report Captain Pratt, A.A.G., Captain Clayton, A.Q.M.G., Lieutenant Pattinson, Brigade-Major, Lieutenant Cowell, A.D.C., Captain Harris, Lord Ellenborough's A.D.C., who acted as galloper to Sir Joseph, and Lieutenant Renney of the Engineers, attached to the cavalry division.

"30*th*.—Marched at 11 o'clock to an encampment on the right bank of the Sankh River, eight miles, being about three and a half miles to the southward of Danoila."

"31*st*.—The Queen-Regent came to the durbar of the Governor-General about half-past one, and had a long conference of at least an hour and a half. She was saluted with 21 guns on arriving and the same on going away. The 16th Lancers formed her guard of honour."

"1*st January* 1844.—The Commander-in-Chief assembled the generals of divisions, brigadiers, and commanding officers of regiments at his Durbar tent this morning and publicly declared he was satisfied with the conduct of everybody, as all had nobly done their duty. He was in high spirits and most cordially shook hands. I afterwards called on

Lady Gough, Mrs. Smith, and Mrs. Curtis, and found them not recovered from the shock they received from being present on the field of battle. Poor Major Crommeline of the 1st Cavalry died to-day of the severe wound he received on the 29th ult."

"*3rd.*—The army, or at least the first column, marched this morning at half-past 6 o'clock to an encampment in front and to the right of the Gwalior Residency, about three miles from the old town; no enemy to be seen. Three squadrons of the 16th Lancers formed the advanced guard, followed by two troops of artillery, the fourth squadron of Lancers, the 1st Light Cavalry, the Bodyguard, the 4th Irregular Cavalry, the 3rd Brigade of Infantry with No. 17 Light Field Battery, the 5th Brigade of Infantry with No. 10 Field Battery, the Khelat-i-Ghilzai regiment and the 31st Native Infantry, the Governor-General and the Commander-in-Chief, two companies of Sappers and Miners; the 4th Native Lancers and two guns formed the rear guard."

Major-General Grey's force, fresh from its victory at Punniar, joined the Commander-in-Chief on the 4th, and on the 22nd—the Gwalior army having in the meantime been disarmed and steps taken for the reduction of its numbers from 30,000 to 9,000—"the army assembled in review order at 9 a.m. for the inspection of the Governor-General and the little Maharaja"—a boy about eight years of age—when Sir Joseph led past his fine command of four cavalry brigades and five troops of horse artillery. On the 23rd the Governor-General marched for Allahabad and the force broke up, Major-General Littler remaining at Gwalior with three batteries of artillery, two regiments of native cavalry, and six infantry battalions. Sir Joseph Thackwell himself left the neighbourhood of Gwalior on the 25th on his return to Cawnpore, where he arrived on the 2nd February and resumed command of the station.

On the 14th March the Calcutta Government *Gazette* notified the transfer of Major-General Grey to Meerut and the appointment of Major-General Pollock to command the Cawnpore Division, which necessitated the transfer of Major-General Thackwell, who was senior in rank to Pollock; and consequently a few days later Sir Joseph received a letter from Headquarters notifying his removal to Meerut. He left Cawnpore on the 5th May and arrived at his new station on the 21st, assuming and retaining command of the division until the October following.

Sir Joseph Thackwell had for some time past been anxious to go home to England. The Gwalior campaign and the state of affairs beyond the Sutlej had all combined to keep him in India; but he had now been over seven years away from England, his elder children were growing up, and he had made up his mind to return home. Lady Thackwell and he therefore left Meerut early in the cold weather of 1844, and marching by easy stages the whole way down to Bombay, they arrived there early in February. The journey home was now a much more expeditious business than formerly. When Sir Joseph came out with the 3rd Light Dragoons, the voyage had occupied four months; but now, embarking on the 2nd March, he landed in Southampton on the 11th April—"a raw cold day with some rain." How many returning Anglo-Indians have been welcomed by the same weather on landing once more in their native country!

Sir Joseph had made up his mind to return to India in August, so that his time was greatly occupied—travelling about, visiting relations, and seeing old friends, and the children too from whom he had so long been separated.

"On the 30th anniversary of the Battle of Waterloo dined with the Duke of Wellington, who did me the honour to introduce me to Prince Albert. Met a great number of my acquaintances, who seemed glad to see me, and heard from Lord Fitz-Roy Somerset of poor Valiant's death from cholera at Calcutta. The Duke appeared in good health and spirits, and his voice was firm and strong. About seventy were at dinner."

He attended also a big dinner at the London Tavern, given by the East India Court of Directors, sat for his portrait, and was painted also by Von Orlich, the Prussian officer who had been in India in 1842, in a group of the chiefs of the army assembled at Ferozepore. The latter part of his short stay at home was passed in Ireland, at the home of Lady Thackwell's mother, and it was there that he left his wife and children when on the 5th August he returned to London to prepare for embarkation. He had written late in July to Sir Harry Smith, Adjutant-General of Queen's troops in India, but the reply—dated 24th

September—probably only reached him on arrival in Calcutta. Sir Harry writes:

"I think you may fairly calculate on being appointed to Cawnpore, your old station, but no official notice can be taken of you until the report of your arrival be received. As to the Punjab, affairs are just as they were when you left; but as a thundering army is so cantoned as to admit of a speedy concentration on our north-west frontier, and as you know the Sikhs are a conceited set of rascals, we soldiers must *hope* for the best. Peace is desired, but circumstances, as at Gwalior, may arise when our *interference* creates *expediency*, over which human foresight can have no control; but when you reach Cawnpore you will be so *appuyé* to the course of events, you will be able to take advantage of them and guide yourself accordingly—unless a path be chalked out for you, which I regard as very probable.

"You did not say whether Lady Thackwell is coming; this however we suppose, and Lady S. unites with me in affectionate regards to you both. You know, I hope, old comrade, how ready I am to be of any use in my power."

On the 18th a party of twelve brother officers of the 3rd Light Dragoons entertained Sir Joseph at dinner at the Clarendon Hotel—"a hearty welcome and a pleasant evening,"—and he finally embarked at Southampton on the *Oriental*. He was not alone, for Lady Thackwell's brother, Edmund Roche, his former aide-de-camp, was also on board, with his wife. The *Oriental* was bound for Calcutta, and at Madras Sir Joseph was glad to meet again his old friend and former commander, Sir Edmund

Williams, then commanding the Presidency Division. Calcutta was reached on the 4th October, and on the 18th Sir Joseph set off by *dâk* for Cawnpore, where he arrived on the 30th and assumed charge of the station under Major-General Sir Robert Dick, then in command of the Cawnpore Division.

CHAPTER XII

THERE is perhaps no more stirring and romantic story in the history of the nations, than that of the evolution of a small religious sect into one of the most magnificent military types that the world has seen. It is barely four hundred years ago that Nanak, the wandering devotee, commenced his teachings; at the beginning of the seventeenth century Har Govind caused his followers to lay aside their rosaries and draw the sword in defence of their faith; and it was *his* grandson, Govind Singh, the tenth, the greatest and the last of the Gurus, who founded the Khalsa, the Sikh Commonwealth, converted "a horde of undisciplined peasants into enthusiastic soldiers animated with religious fervour, and laid the corner-stone of that nation, which Ranjit Singh, a hundred years later, raised in the Punjab on the ruins of the Mogul Empire."[1] The story of the Sikhs is full of fascination—their persecuted life, the years of warfare, their ultimate triumph, the days under Hari Singh—

> ". . . who died before Jumrood,
> When the dry and thirsty nullahs ran red with Moslem blood,
> When thro' the Khyber passes the foe rode fast and far,
> And loudly wailed to dark-eyed maids in grovelled Kandahar."

[1] Gordon.

Ranjit Singh had made the stolid yeomen of the Punjab into a standing army, voluntarily enlisted, highly paid, and trained under the eyes of French and Italian officers—Avitabili, Allard, Ventura, Court, and others—who had learnt their trade under Napoleon and had drifted out to India on his fall—"military waifs from war-exhausted Europe." The Punjab had become a powerful State comprising all that country watered by the five rivers, and stretching from Multan to Kashmir and from the northern bank of the Sutlej to the borders of Afghanistan. In 1809—*consule Minto*—a treaty had been negotiated by Charles Metcalfe with Ranjit Singh, under which the Sikhs engaged to restrict their pretensions and operations to the right bank of the Sutlej, and throughout his long reign the Lion of the Punjab loyally adhered to his promises, and cultivated amicable relations with the Government of India. On his death in 1839 "the Punjab provinces, so long kept together by the pressure of his hand, broke, before long, into a tangle of anarchy." [1]

In September 1843 Ranjit's successor, the Maharajah Sher Singh, and his minister had been murdered on the same day, and the condition of the Sikh army became at once a source of alarm both to the Punjab chiefs and to their neighbours, since, as has been said, "the Khalsa, like the Pretorians of Imperial Rome, sold the supreme power, which rested entirely in their hands, to the highest bidder," [2] and there was no knowing when these troops might not be let loose across the

[1] Keene. [2] Hardinge.

frontier of British India. At the time of the Gwalior campaign the unrest beyond the Sutlej had given an additional anxiety to the Governor-General and the Commander-in-Chief, then preparing for operations on the Chumbul; a prolonged or indecisive campaign in this quarter might have brought down the Sikhs to co-operate with the Mahrattas, and Sir Hugh Gough had arranged for the formation of a force under Sir Robert Dick, to watch the line of the Sutlej, while he also strengthened our outposts at Ferozepore and Ludhiana. The Gwalior campaign terminated so quickly that the Sikhs had no time to act, and war, then so imminent, was postponed for two years.

In the meantime the infant Dhuleep Singh had been placed upon the throne, with his mother—the Jezebel of the Punjab—as Queen-Regent, but the State was divided into three factions; that of the Rajput Chief of Jammu, whose nephew Hira Singh was the minister; that of the Sikh nobles; and finally that of the army *punchayats* or committees. The nominal rulers of the Punjab had no feeling of hostility towards the British, but the Queen-Regent was disinclined for peace, the treasury was empty, the rulers had little authority, and it was only too apparent that matters were drifting rapidly towards war. Sir Henry Hardinge, who had in July 1844 succeeded Lord Ellenborough as Governor-General, had early concerted measures of precaution with Sir Hugh Gough; and as a result of these measures the British force at and above Umballa was augmented by December 1844 to 32,500 men and 68 guns, while 13,000 more men with 30 additional guns stood between Umballa and Meerut, and the train of boats, already provided in anticipation of requirements by Lord Ellenborough before quitting India, was brought up from the Indus to Ferozepore.

At this time the army of the Khalsa was returned at about 87,000 men, with nearly 500 guns; the Sikh troops were politically insubordinate, "but their military discipline was high and their spirit that of a dominant class." [1]

The hot weather of 1845—the summer which Sir Joseph Thackwell spent in England—passed quietly, but the Governor-General had been warned by the Jummu Raja, Gulab Singh, that there would be war after the fast of the Dasahra—the day on which Rama set forth on his great campaign.

On the 22nd November there is an entry in Sir Joseph's diary—"a rumour came from Ferozepore that the Sikhs were preparing to cross the Sutlej in force"; and on the following day—"Dr. Graham writes to Nuttall from Ferozepore that the Sikhs are assembling 60,000 men opposite the ghauts to cross the Sutlej, that they have 20,000 men at Kussoor and 20,000 more within 20 miles of that place." By this time Sir Joseph had probably had a hint that it might be as well to be ready, for on the 26th we find that he "went camel-hunting," and on the 16th December there is an entry—"got a letter last night from Colonel Gough informing me that I had a chance of being posted to the command of the cavalry."

In the meantime rumour had as usual been busy.

[1] Keene.

"Got letters," he writes, "from Colonels Scott and Cureton, and Major Balders; the two first certain that we must have war, and the last that the Governor-General will not cross the Sutlej if he can possibly help it. The Governor-General left for Patiala, as said by the latter, on the 6th, and the Commander-in-Chief was to follow in a day or two."

Intelligence came that the army had made a forward movement from Umballa towards Sirhind, and that the troops were closing up from and upon Meerut; and finally on the 17th we read: "Received the order to join the Headquarters as soon as possible."

On the 20th December he wrote to Lady Thackwell from Cawnpore:

"Plots have ripened since I last wrote, and the *dénouement* you will possibly get by the next mail, if they should not break forth by this. The reports on the morning of the 18th were that the Sikh army had crossed the Sutlej and meant to attack Ferozepore, into the miserable fort of which place all the ladies had taken refuge. Sir Henry Hardinge had been unwilling to believe that the Sikhs were in earnest, and has in consequence delayed to form and brigade the army until the enemy were over the river, and he has incurred every chance of defeat in detail, and will probably not be able to bring 10,000 men into the field against very likely six times that number. On the 17th I was dining with Dick, when at nearly 11 at night we both got an order to join Headquarters as expeditiously as possible, and I laid my *dâk* that night for 10 o'clock this morning and shall hope to reach Umballa by the morning of the 25th—400 miles—and must no doubt afterwards ride to reach Headquarters, if I am able to do so without an escort. Sir Henry has issued a manifesto declaring all the Sikh States on this side of the Sutlej to have lapsed to the Company, and we have a report that the Ferozepore troops have driven the Sikhs over the Sutlej; but this wants better authority."

General Thackwell set out by *dâk* on the 20th, and while on his journey, on the 29th at Busseean, he—

"heard to-day positively that the Sikh army had been defeated on the 21st, 22nd, and 23rd of December with great loss, and had recrossed the Sutlej, leaving behind them 130 guns. Our troops are said to have suffered severely in men and horses."

(This was of course the battle of Ferozeshah.)

Writing home on the 30th December from Wudnee, where he had overtaken a column under Colonel Campbell of the 9th Lancers marching to join the army, Sir Joseph says:

"I arrived at this place this morning, about fifty miles from Ferozepore, which place I hope to reach in two marches, or to reach Headquarters if it be towards the Hurriki Ghat on the Sutlej. With all our haste from Cawnpore, Roche and I have been most unlucky, for a fight took place on the 21st, 22nd, and 23rd of December, and the Sikh army has recrossed the Sutlej with great loss, leaving behind them 130 pieces of artillery or more. As I told you in my letter from Cawnpore, the Sikh army of more than 60,000 men would have to be fought by ours not numbering more than ten or twelve thousand, and our loss in consequence has been severe. It is said the Sikhs have made overtures for peace, but nothing will now satisfy the Governor-General but the surrender of Lahore and the disbanding of the Sikh army, I imagine; and therefore we shall have another fight on more equal terms, as nearly 10,000 men are now closing up, and I shall have a fine body of cavalry. We have had no despatches from Headquarters until this morning, as everything was stopped by the Sikh garrison in the fort here, which surrendered at discretion two hours ago."

"*1st January* 1846.—Marched at half-past 9, and in two miles' distance came upon the ground on which the battle with the enemy's advance guard took place on the 18th December, in which they were driven back, and 15 of their guns taken—mostly field-guns. In about two miles more came to the village of Lohana, where there had been some fighting—and in three miles more came to the village of Ferozeshah, where the main battle with the Sikhs took place on the 21st and 22nd. The Sikh army was formed in an oblong round the village, and consisted, it is said, of about 60,000 men, and they had more than 120 pieces of cannon, nearly the whole of which was of large calibre—12 and 18-pounders with a few 24-pounders. Continued my march on to Sultankhanwala from Moodkee—12½ miles—and on to the Headquarters camp near Malawala, 13 miles from Ferozepore up the left bank of the Sutlej, and about 10 miles from the last-named stage—total this day, 22½ miles. Waited upon Sir Hugh Gough, the Commander-in-Chief, who seemed delighted to see me and said the 3rd Light Dragoons had behaved nobly, that he had never seen such gallant fellows."

He wrote home on arrival at Headquarters Camp, 13 miles from Ferozepore up the Sutlej :

"*2nd January* 1846.—I reached this place yesterday, and surprised everybody with the rapidity of my journey from Cawnpore through latterly a disaffected country. Sir Hugh Gough received me in the most gracious manner, and was much pleased at my joining so soon. He told me the noble bravery of the Third surpassed everything he had witnessed, in charging heavy batteries and cavalry, and everybody says they gained the battle by their noble daring; but their loss has been severe. We have large reinforcements coming up. I hope everything will go on well if there is any more fighting; but most people think the Sikhs have had enough of it, and will not attempt to come over the river again, and I fear we have not the means of attacking Lahore this season."

On the 6th January the 9th Lancers, 3rd Light Cavalry, and 4th Irregular Cavalry, with three troops of horse artillery, marched into camp, when the mounted force under command of Sir Joseph Thackwell was thus composed :

1st Brigade.	*2nd Brigade.*
Brigadier T. Scott, C.B.	Brigadier A. M. Campbell, C.B.K.H.
3rd Light Dragoons.	9th Lancers.
4th Light Cavalry.	11th Light Cavalry.
5th Light Cavalry.	2nd Irregular Cavalry.
9th Irregular Cavalry.	8th Irregular Cavalry.

3rd Brigade.	*4th Brigade.*
Brigadier C. R. Cureton, C.B.	Brigadier D. Harriott.[1]
16th Lancers.	1st Light Cavalry.
Gov.-Gen.'s Bodyguard.	8th Light Cavalry.
3rd Light Cavalry.	3rd Irregular Cavalry.
4th Irregular Cavalry.	

There were also eleven troops of Horse Artillery.

Lieutenant Edmund Roche (3rd Light Dragoons) had again joined the staff of the commander of the Cavalry Division as aide-de-camp, but had almost at once been appointed to officiate as D.A.Q.M.-General *vice* Captain C. F. Havelock, (9th Foot), wounded at Ferozeshah, Lieutenant Roche's place as A.D.C. being taken by Lieutenant T. J. Francis (9th Lancers). The D.A.A.-General, Cavalry Division, was Captain J. Tritton (3rd Light Dragoons), who had already had a horse killed under him at Ferozeshah.

To the north-east of Ferozepore the Sutlej makes a turn to the west, and in this bend there were five places where the river might be crossed: (1) at Ferozepore itself, where the main road passing the stream leads on to Kussoor and Lahore; (2) a ford at Nuggar; (3) another at Tilleewala; (4) opposite Sobraon, where the Sikhs had thrown across a bridge of boats covered by a *tête-de-pont*; and (5) a ford at Hurriki in the angle where the

[1] This arrangement of the brigades appears under date of the 1st January in General Thackwell's diary, but Brigadier Harriott had been wounded at Ferozeshah, and the 4th Brigade seems to have ceased to exist as a unit, though its components still formed part of the army.

Sutlej, having again turned southward, is joined by the waters of the Beas.

On the 12th January the whole force shifted its ground to the right in order more closely to observe the movements of the Sikh force posted on the right bank in the vicinity of Sobraon. The Headquarters camp shifted to Bootewala, the headquarters of the Cavalry Division with Brigadier Scott's brigade moving to Malawala in rear; Sir Harry Smith's Division, with Cureton's cavalry brigade, was placed on the extreme right opposite Hurriki; Major-General Gilbert was in the centre with Sir Robert Dick's division and Campbell's cavalry on his immediate left—the whole opposite the bridge of boats at Sobraon; Sir John Grey, posted at Attaree, watched the Nuggar ford; and Sir John Littler's troops occupied the cantonment and entrenchment at Ferozepore.

" *Wednesday, 14th.*—The Sikhs came over in force, having established a kind of *tête-de-pont* for the protection of their bridge, defended by their heavy guns on the other side of the river. The Commander-in-Chief had two divisions of infantry and some cavalry out; they fired a good deal at us with light guns, but all their shots fell short, and we only replied with two 8-inch howitzers, but intended to reply with a 24-pounder, which however burst, but without injury to any save one man. But little if any loss was sustained on either side, and soon after 4 p.m. the Sikhs withdrew their guns to the other side of the river."

On the 15th January Sir Joseph wrote to Lady Thackwell:

"The Sikh army is on the other side of the Sutlej, and we are watching the fords on it and their bridge, which is about four miles in our front. They defend it with such heavy batteries on the other side of the river, that it is rather more difficult for us to get to them than for their army to attack us. If they should do so they will suffer for it, but I do not think they will venture on such a step. We are glad to gain time for our heavy cannon, reinforcements, and commissariat supplies and carriages to come up, and I believe we shall get troops from Bombay and Madras to supply the stations where they have been withdrawn, and of course fresh regiments will be sent out from England to supply their places. Our Court of Directors' eyes will be now opened to a just estimate of the formidable power of the Sikhs, and to the necessity of at least 10,000 Europeans in addition to our present numbers, for the safety of this Empire depends on the dispersion of the Sikh army, whether it can be done this season or the next. If our means are adequate we have a hot summer campaign before us, but perhaps not hotter than that Edmund and myself weathered in Afghanistan."

In the meantime it had been reported that a force of Sikhs of all arms had crossed the upper waters of the Sutlej at Phillour, with the intention not only of threatening Ludhiana—then held by three battalions of native infantry under Brigadier Godby—but of striking at the British line of communications near Busseean. Sir Harry Smith's division, with Cureton's cavalry brigade, was detached from the main army on the 18th to subdue a small fort held by Sikh mercenaries at Dhurmkote, which was halfway between Ferozepore and Ludhiana, and he was now directed to march on the latter place—a movement which led to the victorious action of Aliwal on the 28th, where the cavalry,[1] and in particular the Sixteenth Lancers, greatly distinguished themselves.

On the 31st January Cureton wrote to Sir Joseph:

"I wrote an 'official' to Tucker to let you know that we had an action on the 20th, and I then fully intended to have written a private letter to you, but every moment of my time has been occupied with public duty. Pattinson, who knew my anxiety, promised he would write you all particulars, and I hope he has done so. It will, I know, be most gratifying to you to know that the cavalry behaved extremely well, the 16th Lancers and 3rd Light Cavalry suffered severely, but fully effected their object; but the despatch which has, I hear, been furnished to Headquarters will, I have no doubt, give full particulars. I do not think the Sikhs will be much inclined to face the British in the open again—whenever they do the result will be the same. Smyth of the Sixteenth is shot through the thigh, but doing well; Fyler has a similar wound but more serious; Orme is wounded in two or three places and considered in danger; Bere, Pattle, and Morris were all wounded in the face, but not seriously. The 50th Foot have suffered a good deal, and had eight officers killed and wounded; the loss, however, on the whole is not heavy, considering the force engaged. The river at this place is shallow, and easily crossed by the villages, but the fords appear very intricate, and I have no doubt great numbers of the enemy

[1] Cureton had under him at Aliwal the 1st and the 5th Light Cavalry and the 8th Irregular Cavalry, besides his own—the 3rd Brigade.

were drowned in crossing, as they had to rush headlong in without reference to a ford. We have taken a beautiful park of artillery from them, mostly heavy metal, but you will probably have seen the return of captured ordnance before you get this. I suppose the guns will be sent to Ludhiana; I fancy old Ranjit's guns will soon be exhausted if we take fifty or sixty in every battle. Poor Swettenham's loss is much felt in the regiment; he was a universal favourite. I do not think you had ever seen Williams, another officer who fell; the latter came out with me in July."

The immediate result of this successful action was the evacuation by the Sikhs of all the forts they had hitherto held on the left bank of the Sutlej, with the exception of the entrenchment at Sobraon; and Sir Hugh Gough now only waited to attack the enemy in this position until the troops under Sir Harry Smith should have rejoined the main army, and the siege train and ammunition should have arrived from Delhi.

"23*rd.*—A good deal of firing at the outposts to-day—the enemy showing a good many sowars—and my red jacket seemed very attractive to them, as I could not show myself without having lots of bullets whizzing near me. Had some talk with Prince Waldemar of Prussia."

"24*th.*—The 5th Cavalry marched to Dhurmkote, and Brigadier Scott was directed to bring his head-quarters to this camp. The Sikh sowars made a forward rush upon the picquets at Sobraon, and it was imagined from the reports of the orderlies that it was an advance by the Sikh army—however, it turned out to be only a few sowars, who were driven back by the picquets of the 2nd Irregular Cavalry."

"*27th.*—The Sikhs occupied the watch tower at Sobraon last night, and the cavalry picquets of course fell back. There has been a good deal of skirmishing to-day, but no results of consequence. Yesterday a sowar of the 9th Irregulars was killed and another was wounded. Waited on the Commander-in-Chief and got his orders for the attack to-morrow morning, but it was countermanded at 10 o'clock p.m."

"*28th.*—Soon after 10 a.m. I heard a heavy cannonade near Ludhiana, supposed to be a fight between Sir Harry Smith and Ranjur Singh, and soon after 3 the picquet at Aleewala was driven in by the Sikh cavalry and infantry, on hearing which I sent a squadron of the 3rd Light Dragoons to its support and soon after marched with the remainder of the Third, followed by a troop of Horse Artillery and two battalions of infantry; but at Asyah heard that Captain Beecher had recovered the post, killing some sixteen or twenty of the enemy, being supported by the squadron of the Third. I halted all my troops where the order found them.

"Examined all the ground to the river, and saw about a thousand Sikh sowars, and perhaps many more and infantry were concealed in the jungle. When most of these had retired within their lines, I sent the troops to camp except the squadron of the 3rd Light Dragoons at Aleewala—three-quarters of a mile in front of which were still about 400 sowars with some camel swivels, but when these retired the squadron was ordered back to camp."

"*29th.*—At 7 a.m. I walked to the Sobraon picquets, and when I returned found the line turned out to fire a salute of 21 guns, with three cheers by the men, for Sir Harry Smith's victory yesterday near Ludhiana. The substance of which was that Runjur Singh and the Sikh army had advanced to attack him, when, after an action of one hour and twenty-five minutes, the Sikh army was totally defeated with the loss of all their artillery (50 guns) and baggage, and had been driven across the Sutlej. Soon on horseback, and went down the line with the Governor-General and the Commander-in-Chief."

"30*th*.—Visited the front picquets soon after 7 a.m., and from the clouds of dust, noise of carriage wheels, and swivel detonations, and the opinions of the outposts came to the conclusion that the main body of the Sikh army was retiring, but I am not certain that this conjecture is correct. The Commander-in-Chief imagined we should be attacked to-day, but I never thought so."

"31*st*.—The Sikhs quiet all day. The sowars of the Sobraon picquet and the Sikh sowars in friendly conversation at the posts of the vedettes."

"6*th February*.—Major-General Smith's force came into camp on the right, where the 3rd Brigade of Cavalry, with the exception of the 4th Irregulars, is posted. The 4th and 5th Light Cavalry joined, and Lieutenant-Colonel Lane's troop of Horse Artillery joined Scott's brigade, and the 2nd Brigade of Cavalry with Major Campbell's troop of artillery are posted to the left front of Scott's brigade. The 8th and 9th Irregulars are near Kameelwala as before, and the 11th Light Cavalry have not yet joined this brigade. The Commander-in-Chief and many of the Head-quarter camp went out to meet the Aliwal Division."

On the afternoon of the 9th February all Generals of Division, Brigadiers, and Heads of Departments were summoned to the tent of the Commander-in-Chief to hear the proposed arrangements for the next day's battle. Sir Hugh Gough explained that the strongly fortified position of the Sikhs was to be attacked at daylight, and gave the following detailed orders: the main attack was to be made on the left, by the Third Division under Sir Robert Dick, supported by the fire of thirteen heavy guns; this division was to be drawn up in two lines with a reserve, and, having three brigades, it was stronger than the First and Second Divisions, which contained only two brigades each, while the 7th Brigade, which was to head the attack, was to be strengthened by the attachment of the 53rd Foot from General Grey's division. In rear of Dick's division was the 1st Cavalry Brigade (with which was the Cavalry Divisional Headquarters) with three regiments of native infantry—one of these also detached from Grey. The Second Division under Major-General Gilbert was to be in the centre, "deployed for support or attack," with a battery of heavy guns on either flank. Sir Harry Smith with the First Division—much weakened by its losses at Aliwal—was to attack on the right, supported by the 2nd Brigade of Cavalry. The 3rd Cavalry Brigade under Brigadier Cureton was directed to move towards the ford at Hurriki, and, by feigned attempts to cross, to divert the enemy's attention, and occupy the Sikh cavalry posted opposite Hurriki on the right bank. Major-General Sir John Grey, with the 8th Light Cavalry and three regiments of native infantry, watched the Nuggar ford opposite Attaree.

Sir Joseph Thackwell's diary gives the following

account of the action of the cavalry under his immediate command:

"*Thursday,* 19*th February.*—A lovely day at half-past 4. The troops turned out and got into position at daylight, the heavy guns began to vomit fire and continued to do so until after 10 a.m., when the infantry, light guns, and cavalry moved to attack the strong position of the Sikh army at Sobraon; and in less than an hour after this the entire camp was in our possession and the remains of the Sikh army were all on the right bank of the river, leaving in their entrenched camp all their guns and nearly 10,000 men killed, wounded, and drowned. The cavalry were under a heavy fire in supporting the infantry, and I led two squadrons of the 3rd Light Dragoons, the 4th and 5th Light Cavalry into the enemy's camp in single file, and several charges on the enemy were led by me. Many of the Sikhs are very brave and defended themselves manfully. At the time that Scott's brigade took ground to the right to support the Second Division I was in front and about to move part of the cavalry to support the Third Division, which had carried the entrenchments on our left; I therefore left the 9th Irregular Cavalry—the 8th being already on the left—for that object. On passing at a trot beyond the watch-tower I met the Governor-General, who said, 'When you get into the entrenchment don't spare them.' The Sirmoor battalion on the left of the Second Division was retiring, but when they saw the Third ride to the trenches they rallied and passed between the horses into the entrenchments. When the Third got about 120 yards from the Sikh right, I rode forward and discovered a place where we could get into the entrenchments in single file—about sixty yards from the Sikh right and 150 from the flanking battery—and as I brought the Third up, the enemy began gradually to give way on their right. I was the first cavalryman who entered over the ditch and up the parapet in rear of the right regiment; and as soon as the first squadron was formed, I led the charge over very broken ground to near the ford. It was obliged to give way before the mass of retreating Sikhs, but it rallied, and when the second squadron was formed, another charge was made, but the same reasons obliged the squadrons to give way. At this time the 4th and 5th Cavalry, and, I believe, the other two squadrons of the 3rd Light Dragoons—under Brigadier Scott, who had entered the entrenchments further to the Sikh right—came up and several charges took place, and the ford was filled with the Sikhs, many of whom were drowned or slaughtered by our artillery and musketry fire. Owing to these cavalry attacks the First and Second Divisions were enabled to enter the lines with but little loss. The first squadron of the Third suffered from our own artillery fire until Smith discovered we were not Sikhs. The 9th and 8th Irregulars were left in support, the latter regiment being in support of the left from the beginning of the action. The two left squadrons of the Third were to support the regiment ordered into the entrenchments."

In a private letter home describing the battle Sir Joseph writes:

"It is due to the Sikhs to say that they fought bravely, for though defeated and broken, they never ran, but fought with their tulwars to the last, and I witnessed several acts of great bravery in some of their Sirdars and men. It was a most beautiful sight to see the advance of the Third Division of Infantry, or rather of two of its brigades (six regiments), after the heavy guns had been at work for two hours, supported by the fire of 18 field guns and the 1st Brigade of Cavalry. Everybody moved as at a field day, though numbers were falling from the enemy's round shot, grape, camel swivels, and musketry; no dust obscured the view over the plain, and the wounded men were seen dotting it in great numbers as they went to the rear. The left brigade carried the entrenchments first, then the second brigade, and these made but slow progress, when the Second Division was ordered to attack, but was repulsed. At this time the regiment on our right, clearing the ramparts, was brought to a stand by the determined bravery of the Sikhs, when I led two squadrons of the 3rd Light Dragoons down to the ditch of the rampart and along the glacis under a heavy fire of grape from 3 guns in a flanking battery at 150 yards—most of the shot going over us—to within about 60 yards of the Sikh right, which began to give way and enabled the battalion near us in the entrenchments to make progress. I then found a *bund* in the ditch, over which I led the first squadron in single file and over the parapet into the entrenchment and camp. When the squadron was formed, it made a gallant charge nearly to the ford, but was obliged to give way by the pressure of the mass of the retreating Sikhs, and to ascend the high bank it had before with some difficulty descended. Luckily for me my horse was not hit, or I should not have lived to have told the tale. By this time the second squadron had come up, and, both advancing, made a successful charge down to the ford, the whole breadth of which—and I may say the river half a mile over—was filled with the discomfited enemy. The heavy masses of the retreating Sikh left obliged the two squadrons to retire, but our diversion enabled the First Division to carry the left entrenchments, and the light cavalry having at length come up, charged the retreating masses, and the last of the Sikh army was driven into the river and cannonaded by our light guns."

In the records of the 3rd Light Dragoons we read that—

"in the heat of the infantry attack on the enemy's right, and at a moment when a partial check had taken place at this point, Major-General Sir Joseph Thackwell, commanding the Cavalry Division, advanced the 3rd Light Dragoons to the foot of the defences in support of the infantry. The 'King's Own' were here halted under a heavy fire of round shot and grape, whilst an opening was being made by our Sappers, through which the Dragoons passed in single file, and on the enemy's side of these works the squadrons were re-formed. The infantry bugles now sounded the 'cease firing,' and the 3rd Light Dragoons, led by their veteran commanders Sir Joseph Thackwell and Lieutenant-Colonel Michael White, dashed headlong into the thickest of the enemy's masses, and were for a time in the very centre of upwards of 30,000 resolute warriors who were determined to die rather than yield."

In the account of the battle by the son of the Governor-General, the second Viscount Hardinge writes:

"At this crisis"—the temporary check on both flanks of the infantry divisions assaulting the entrenchments—"the cavalry were ordered up under Sir Joseph Thackwell, who rode with his squadrons in single file through an opening in the entrenchment. It seemed as if they were doomed to destruction. Many fell in the ranks when within the camp. But the 3rd Light Dragoons, showing the same invincible bravery as on the night of the 21st of December, quickly re-formed and charged the serried ranks of the Sikh infantry."

In Sir Hugh Gough's despatch he says:

"The Sikhs, even when at particular points their entrenchments were mastered with the bayonet, strove to regain them by the fiercest conflict—sword in hand. Nor was it until the cavalry of the left, under Major-General Sir Joseph Thackwell, had moved forward and ridden through the openings in the entrenchments made by our Sappers, in single file, and re-formed as they passed them; and the 3rd Dragoons, whom no obstacle usually held formidable by horse appears to check, had on this day, as at Ferozeshah, galloped over and cut down the obstinate defenders of batteries and field works, and until the full weight of three divisions of infantry, with every field artillery gun which could be sent to their aid, had been cast into the scale, that victory finally declared for the British. Major-General Sir Joseph Thackwell has established a claim on this day to the rare commendation of having achieved much with a cavalry force, where the duty to be done consisted of an attack on field works, usually supposed to be the particular province of infantry and artillery. His vigilance and activity throughout our operations, and the superior manner in which our outpost duties have been carried on under his superintendence, demand my warmest acknowledgments."

Writing home on the 26th March, by which date Sir Hugh Gough's despatch on the battle of Sobraon had been published, Sir Joseph says:

"On the 15th inst. Sir John Grey left Lahore for his command at Meerut. He is disappointed at the Governor-General not having mentioned his name as commanding the troops which first crossed the Sutlej, and I feel nearly as much, as he has not particularised me, who did more than Smith or Gilbert, for it was my attack which enabled those two divisions to get into the Sikh entrenchments from which Gilbert's division, as well as Smith's, had been driven back twice with great loss; and lest there should be a garbled account in the papers as emanating from the Commander-in-Chief, I will transcribe what he has announced to the Governor-General, and published to the army respecting myself. But it seems he was not aware that I was the person, under a heavy fire within 60 yards of the Sikh right, and a flanking battery at 150 yards firing grape shot, who found a dam across the ditch which enabled me first, and two squadrons of the 3rd Light Dragoons in single file afterwards, to enter the camp and make several successful charges. When I led the cavalry down to the trenches, the left battalion of Gilbert's was retiring, and the right regiment of ours within the trenches was brought to a standstill; but as I rode along the outside of the ditch the Sikh right began to give way, and that battalion was enabled to make progress, and the Sirmoor battalion of Gilbert's was able to enter the camp under our protection. It was a miracle we were not properly riddled, but from the constant fire the trails of the guns had so sunk in the sand that the gunners could not depress the muzzles sufficiently, and therefore most of the grape went over our heads." (Then follows an extract from Sir Hugh's despatch.) "The Commander-in-Chief in taking leave of the 3rd Light Dragoons paid me some high eulogiums. There were no Sappers whatever employed to make a way, but the Sikhs had left dams at intervals across the ditch for their own convenience."

And finally, here is an extract from a letter written some few years later to Sir Joseph by Sir Harry Smith, then commanding at the Cape:

"I think I see you at this moment, with your one arm, riding into the trenches at Sobraon"; and again, "shoving in your unwinged shoulder into the gap at Sobraon—the most gallant 'go' of you and the 3rd Dragoons I ever witnessed."

There can be no doubt that Major-General Sir Joseph Thackwell greatly felt—not so much for himself personally, but as commander of the Cavalry Division which under him had done so much to secure and complete the victory of Sobraon—the total omission of all thanks to himself in Lord Hardinge's General Order of the 14th February. Each one of the three general officers commanding infantry divisions were specially named therein, and their services, and those of their commands, particularly eulogised; but there was no special recognition of the importance of the action of the cavalry, or of the unusual circumstances of its most opportune employment under the personal leadership of its commander. Sir Joseph addressed Lord Gough on the

19th March, asking that the omission might be made good. On this letter being forwarded to the Governor-General, Lord Hardinge replied as follows in a memorandum dated 23rd April, from which it is very evident that His Excellency had missed the whole point of Sir Joseph Thackwell's appeal—which was not for a recognition of his own *personal* acts, greatly as they had contributed to the successful issue of the battle, but that the Cavalry Division, through its commander, should receive, equally with the Major-Generals commanding divisions of the other arms, the recognition and approbation of the first authority in India.

"I have received Major-General Sir Joseph Thackwell's letter of the 18th March, addressed to the Adjutant-General of the Army, which I only saw a few days ago, representing that in the Governor-General's order of the 14th February no mention is made of the charges of Cavalry in the Sikh entrenchment camp, except the praise bestowed on Her Majesty's 3rd Light Dragoons.

"The Governor-General does not see the Divisional reports made by the General Officers to the Commander-in-Chief.

"On referring to the Commander-in-Chief's despatch, it is stated that the Cavalry of the left, under Major-General Sir J. Thackwell, moved forward and rode through the openings in the entrenchments and that the 3rd Light Dragoons had on that day, as at Ferozeshah, galloped over and cut down the obstinate defenders of Batteries and Field Works.

"I saw the bodies of the soldiers of the 3rd Light Dragoons who had been killed within the

Enemy's camp. I saw no other cavalry soldier killed; and in reference to the lists of the killed and wounded it appears that no other cavalry soldier under the immediate command of the Major-General was killed.

"Her Majesty's 3rd Light Dragoons lost 5 men killed, 4 officers wounded and 22 men wounded.

"In the whole of the other regiments on our left, not a single officer or man was killed, 15 men were wounded.

"In the Commander-in-Chief's despatch the only regiment of cavalry named was the 3rd Light Dragoons.

"I think His Excellency was quite right. I only mentioned the 3rd Light Dragoons, as His Excellency had done.

"It nowhere appears in the Commander-in-Chief's despatch that the two squadrons of that noble regiment, the 3rd Light Dragoons, were led to the charge by the Major-General Commanding the whole of the Cavalry of the Army.

"I have no doubt of the fact, since I have perused the Major-General's letter, but how was I to be aware of it?

"I knew the Commanding Officer of that Regiment was in the Field and commanding his own Regiment, and I naturally inferred that *he* and not the *Major-General* did lead the two squadrons, and I conceived, when a Regiment is named by the Commander-in-Chief as having distinguished itself, that the merit is due to the Commanding Officer, when no other Officer is named by the Commander-in-Chief.

"Unless any disorder should arise, or reluctance be shown, requiring a General Officer's interference, surely it is fair to give to the Commanding Officer the credit of having commanded his own regiment.

"I praised the only Regiment of Cavalry named by the Commander-in-Chief; and according to the custom of the Service, I mentioned the Lieutenant-Colonel in command of the Regiment, and not the Major-General. In the previous General Order by the Governor-General of the 2nd February, after the battle of Aliwal, I specially thanked the Cavalry who, from the nature of the ground, took a larger share in that battle, and I stated that the 'Native Cavalry had on every occasion proved its superior powers, whether in the General Actions which have been fought, or in the various skirmishes at the Outposts, such as that in which Captain Beecher was gallantly engaged with a small party of the Irregular Cavalry at Aleewala, on the morning of the 27th January.'

"As regards the Cavalry generally, I am not aware of any omission which requires me to issue, towards the end of April, a supplementary order to inform the Army why I had praised the 3rd Light Dragoons and neglected to praise the Regiments or individuals of that arm.

"My own opinion is, that the practice of loading the General Orders with praises of every Department and everybody is an erroneous system. The thanks of the high authorities empowered by the Sovereign and the East India Company to convey these thanks to the Army, by publishing them for general information, must lose their value if they are given with profusion, or without discrimination.

"It is still more undesirable that Military Men should constitute themselves to be Judges of the degree in which they may be entitled to be praised.

"An officer possessing a reputation so well earned as Major-General Sir J. Thackwell's would be the last person to encourage an inconvenient system; and as he must know the estimation in which he is so deservedly held, I am confident on reflection he will come to the conclusion that no supplementary order is required from me, and at any rate that it is not my intention to issue one."

It is open to question, however, whether history has ever done sufficient justice to Sir Joseph Thackwell's timely intervention in the fight. Our artillery, having expended all their ammunition, had ceased firing; our infantry had already twice been repulsed on either flank; when the quick eye and ready hand of the cavalry commander enabled him at the right moment to throw a new weapon into the scale and restore the balance of the hard-fought battle.

CHAPTER XIII

At 3 o'clock on the morning after the battle, Sir Joseph Thackwell, with the 1st and part of the 2nd Cavalry Brigade, moved down to Attaree, where a bridge of boats was to be thrown over the Sutlej, preparatory to the army crossing the river, and on the 12th Sir Joseph wrote home:

"We cross the Sutlej either to-night or to-morrow morning, and in three or four days' time we hope to be at Lahore. The natives say that everything is over, as many of the Khalsa troops have gone to their homes. My poor old friend Sir Robert Dick got his death-wound in storming the right batteries, and died in the evening."

It was not, however, until the morning of the 13th that the cavalry division was able to cross over to the right bank and move on to Kussoor, where a position was taken up on the Lahore side of the town.

"We arrived here," writes Sir Joseph to Lady Thackwell on the 26th from Lahore, "without any opposition, on the 20th inst. The Sikh army had suffered so much in the battle of the 19th that they became quite dispirited, and it was pretty clear that no more fighting would take place. They said if they could not defend themselves in their entrenched camp, it would be in vain to try again when nearly all their guns had been taken, and they had lost many of their Sirdars and considerably more than 10,000 of their men in the last encounter. Gulab Singh had sent *vakeels* to sue for peace before that event, but said he was not strong enough to contend with the Khalsa army and that no treaty could be effected as long as their supremacy lasted. That was decided on the 10th in the battlefield, and it was clear the disbanding of the Sikh army would be the consequence. On the 13th two brigades of cavalry, Headquarters, and the First Division of Infantry reached Kussoor, the fort there having previously surrendered; and on the morrow, or rather the day after, Gulab Singh came into camp and the terms of peace offered by the Governor-General were gladly accepted by him, who had expected nothing less than the appropriation of the Punjab to ourselves. On the 17th he again came to Kussoor, and the army marched to Lullianee on the 18th, at which place Dhuleep Singh, the Maharaja, came to the Governor-General's durbar and remained in his camp until the 20th—until we had taken up our position before Lahore; and at 4 p.m. a troop of artillery and four regiments of cavalry escorted him to his palace in the fort. . . . We expect, if nothing unforeseen should occur, to break up about the 10th March, and I shall hope to be at Umballa before the end of the month and at Cawnpore by the 10th April, which is, I believe, to be my divisional command. If there should be a brevet, taking Grey in, I shall probably obtain a permanent command—if the brevet does not include him, Smith will most likely get Dick's, and I may be sent to another Presidency if I am to be employed. But if there is to be no brevet this year and a Major-General is sent out to replace Dick, I shall procure leave of absence and start for Bombay in November or the beginning of December next."

The remains of the Sikh army were still, however, in the field and a source of some anxiety, for on the 28th February there is an entry in Sir Joseph's diary:

"Major Lawrence was alarmed last night by the approach of the Sikh army with their two-and-twenty guns to within four coss of Lahore, and in consequence the picquets were placed on the alert, and a squadron per regiment was ordered to saddle in addition to the inlying picquets."

But it was not long before the Sikh battalions were marched one by one into Lahore and there disbanded, no men enlisted since the time of Kurruk Singh being permitted to enter the new army, which moreover had to serve at the same rate of pay as in the days of Ranjit Singh, instead of the double rates which had lately been demanded from and paid by the Lahore Government.

On the 4th March Sir Charles Napier, who had been moving up with a force from Scinde to co-operate with the army under the Commander-in-Chief, arrived in camp; on the next day there was a grand review of the army; and on the 9th a great durbar was held in the state tent of the Governor-General, for the ratification and promulgation of the treaty which had been signed on the previous day. On the 10th there was another review, when Sir Joseph Thackwell was in command of nearly 3,500 horsemen, and at which he records that "Raja Lal Singh was in a complete suit of plate armour." After the review was over he attended the Governor-General at 3 p.m., and with the Commander-in-chief, Prince Waldemar of Prussia, Sir Charles Napier, and many others went to the durbar of the Maharaja held in the palace in the citadel.

"... The Koh-i-nor diamond was shown—said to be the largest known in the world; many longing eyes rested upon it and I fear coveted the bauble. The Sikh soldiers looked well, both horse and foot, and perhaps about 1,500 might have been drawn up. They presented arms very well and their bands played 'God save the Queen' in good style. The artillery fired the salute very regularly."

Writing home under date of the 13th General Thackwell says:

"We have had lots of durbars and dinners and the usual complimentary toasts and speeches, but, thank God, they are now over and the Governor-General left yesterday for Umritsar with the troops going to remain in the Doab. . . . As you might have expected, I am appointed to the Cawnpore Division in the place of poor Dick, and have pocketed twelve months' *batta* to the tune of £4,752—not quite as good as the Koh-i-nor which we saw at the durbar, but I fancy as good as the sacking of Govindghur would have been, as I suspect there is but little treasure there."

On the 22nd the camp broke up, the greater part of the army returning to garrisons further south, while Major-General Sir John Littler remained until the end of the year in Lahore with a force amounting to some 9,000 men. By a special clause of the treaty these troops were to remain in Lahore "for the protection of the Maharaja and his Government," while the reduction of the Sikh army—now restricted to 12,000 cavalry and twenty-five battalions of infantry—was being carried out. British brigades, organised as movable columns ready to take the field, were also placed at Jullundur and Ferozepore, while altogether some 50,000 men with 60 guns were cantoned at or above Meerut, so that a large force might be immediately available in the event of further trouble in the north-west. Sir Joseph Thackwell himself left the neighbourhood of Lahore on the 23rd with the bulk of the army, and, marching by Umballa and Meerut, reached Cawnpore on the 21st April and assumed command of the division.

He relates that on the 27th March on his way towards Ludhiana he—

"went over our old camping-ground, and I paid a visit to the field of Sobraon and witnessed a horrid sight. The river line of the Sikh camp is about 2,750 yards, and the line of the entrenchments is about 3,655 or perhaps more, and along its extent were some Europeans still unburied, and between 2,000 and 3,000 Sikhs in a state of fearful decomposition. . . . A poor Sikh horse with a broken leg was still alive, a skeleton, but I caused the *coup-de-grâce* to be given him from a pistol."

In the meantime people at home had not been unmindful of the great services rendered by the army in India. On the evening of the 2nd April the thanks of both Houses of Parliament were offered to, among other officers—

"Major-General Sir Joseph Thackwell, Knight Commander of the most Honourable Order of the Bath, for the distinguished services rendered in the eminently successful operations at the Battle of Sobraon."

On the same day a special General Court of the Proprietors of Stock was held at the East India House, where a vote of thanks was proposed and unanimously adopted to the troops engaged in the war of the Sutlej. At a Special Court of the Court of Common Council of the City of London, the thanks of the Court were awarded to the officers, European and native—"for the intrepidity, perseverance, and discipline evinced by them." And finally the Mayor and Council of the Borough of Liverpool passed a vote of thanks expressing their sense of the services rendered by the army of the Sutlej during the late operations, and recording the name of Major-General Sir Joseph Thackwell amongst those of other General Officers, for that—

"in the opinion of the Council the annals of British valour, glorious as they are, have rarely furnished more signal instances of true personal bravery and devoted heroism than were so generally manifested both by officers and men on these memorable occasions."

In forwarding this last under cover of a letter from himself the Governor-General wrote:

"It gives me the greatest satisfaction to convey to you the approbation which is felt by our fellow-subjects in England for your eminent services and personal conduct during the late short but decisive campaign."

Sir Joseph Thackwell, who still, it will be remembered, was borne on the strength of the 3rd Light Dragoons, was naturally very proud of their gallant behaviour in the Sutlej campaign; he had written to Lord Charles Manners, the Colonel of the regiment, giving him an account of the operations in which the 3rd had taken so conspicuous a part, and in his reply Lord Charles says:

"Your communication has filled me with pride and exultation, but at the same time, I must assure you, with sorrow and regret. It is indeed a proud circumstance to be connected with a regiment composed of such sterling stuff as the 3rd Dragoons, but most deeply do I mourn and bewail the loss of so many fine fellows."

Major-General Thackwell's hopes of obtaining a divisional command were for the time to be disappointed, for Sir Harry Smith was almost at once given the charge of the Cawnpore Division in place of Sir Robert Dick killed in action, and Sir Joseph therefore reverted again to the command only of the station. He had, however, great hopes of soon obtaining promotion, his claims having been urged at the Horse Guards by Lord Hardinge, but he was anxious to get home for a few months. Sir Harry Smith too was in bad health and was torn between a desire to go home to

England and a wish, for financial reasons, to retain a lien on his command. Some correspondence passed between him and Sir Joseph Thackwell, and it seems to have been ultimately arranged by Sir Harry Smith with the authorities, that Sir Joseph should officiate for him in command of the Cawnpore Division during his absence in England, vacating it in favour of Sir Harry should Sir Joseph not in the meantime have been promoted a substantive Major-General on the Indian establishment. Sir Joseph Thackwell, however, started on his way to Bombay in January 1847, having been fortunate in first seeing something of his eldest son Edward, who had come out to join the 3rd Light Dragoons, and travelling by easy stages he arrived in due course at Bombay. Here, however, he would appear to have received advices that his promotion was imminent, while Sir Harry Smith had also in the interim started for England. Sir Joseph then cancelled his passage home and went round by steamer to Calcutta, where he arrived on the 6th May. Leaving that city on the 2nd June for Allahabad on a flat towed by a steamer, he found himself back again in Cawnpore on the 27th of the same month.

By a letter from Lord Fitzroy Somerset, dated the 23rd July 1847, Sir Joseph Thackwell was "placed upon the staff of the army serving in the East Indies during the absence on sick leave" of Sir Harry Smith from the 1st May, and later—on the 23rd September—the appointment was confirmed, Sir Harry Smith having in the meantime been appointed Governor and Commander-in-Chief at the Cape.

For some time past there had been intolerable maladministration in the Kingdom of Oudh, and Lord Hardinge had arranged to proceed to Lucknow in the autumn of this year to administer a personal rebuke to the King, and to inform him that, if oppression and misrule should continue to prevail, the kingdom would be brought under British rule—a threat already made in 1831 by Lord William Bentinck and which was finally carried out by Lord Hardinge's immediate successor. The Governor-General arrived at Cawnpore on the 6th November, and on the 10th the King of Oudh, gorgeously apparelled—(" the value of the King's dress," writes Sir Joseph in his diary, " was stated to be nine lakhs of rupees ")—arrived to pay his respects. On the 12th the King returned to Lucknow, and on the following day the Governor-General marched thither accompanied by Sir Joseph Thackwell, who had been invited to attend. The vicinity of Lucknow was reached on the 17th and the camp of the Governor-General was pitched in the Dilkusha Bagh—the site of the present cantonment—

" and all the Governor-General's guests in full dress mounted on elephants at 6.30 a.m. to visit and breakfast with the King of Oudh. His Majesty met the Governor-General at the bridge over the Goomtee, or a branch of that river,[1] a mile and a half from the camp, and we continued —a *cortège* of fifty or sixty elephants—along the

[1] Probably the Nasir-u-Din Haidar Canal.

suburbs of the city, which are mean and wretched, and, making a detour, entered the town and arrived at the Motee Mahal soon after 9 a.m.—about 7½ miles from the camp—and immediately went to breakfast.

"After this the King presented each lady and gentleman with the usual tinsel neck ornament, and we took our departure to a second breakfast at the Resident's—a fine large building with a detached large Banqueting House, having a fine saloon and a dining-room more than a hundred feet long. The Governor-General and many of the guests occupied the house; I and most of the single men occupied tents in the compound, which together with grounds and gardens are very pretty."

"18*th*.—I went to the cantonments, about 2¼ miles from the iron bridge over the Goomtee, which is deep and about 70 yards wide. It lies to the north-east of this bridge and is well placed on a dry, sandy soil, but the exercise ground is too limited. At 9 a.m. the King returned the Governor-General's visit, and the breakfast was excellent—all the officers and ladies in the station assisted at it, and about thirty of the Oudh Court, most of whom ate voraciously. A rich necklace and a gold and diamond ring were given to the King, and the usual present of a silver tinsel necklace was placed round his neck and those of the young Princes and all his Sirdars. When this ceremony was over the King took his departure under the usual salute of 21 guns, but his Minister was left behind with Mr. Elliot and Colonel Richmond to arrange, I fancy, something respecting the treaty which had been decided upon. Went to the Dilkusha Bagh, a pretty palace of the King's near Constantia the celebrated, built by General Claude Martin, a Frenchman in the service of the Nawab of Oudh."

"19*th*.—At half-past 3, went with the Governor-

General and a crowd in the King's carriages to see the aviary and menagerie. The tigers were tremendous beasts; many fine sheep, antelopes, hog-deer, birds of all sorts, and monkeys were exhibited. Some fights of sheep, antelopes, hog-deer, and a pair of partridges were amusing. After this drove round the cantonment, which is for three sepoy regiments and a company of Gholandaze. Dined in the Banqueting House, and about 140 guests sat down to dinner."

"20*th.*—At 8 a.m. attended the Governor-General, in all the pomp of elephants, guards, and guns, to the Motee Mahal, the palace before mentioned, where the display of wild-beast fights took place. They were tame and uninteresting; the rams did not butt; the elephants did not push except in play; the hog-deer and antelopes fought well. The display of horsemanship, and the art of defence with sword, sword in two hands, with club and mace was dexterous and amusing."

Sir Joseph Thackwell left Lucknow on the same day, and was back in Cawnpore by the 22nd, preparatory to a move to Meerut, whither he had been transferred, and where he arrived on the 29th December, having halted *en route* at Agra and Alighur to carry out inspections at these stations. At this time the Meerut Division was one of the most important in Upper India, and the station itself contained a very large garrison: there were three troops of horse artillery; two field batteries; two companies of foot artillery; the 9th Lancers; the 5th and 10th Light Cavalry; the 32nd and 98th Foot, and several regiments of native infantry, of which the Ludhiana Regiment of Sikhs seems greatly to have impressed Sir Joseph Thackwell.

"Their manœuvring," he says, "was admirable. In the Grenadier Company was no man under 5 ft. 11 in., and the right-hand man (6 ft. 6 in.) was a giant. One of the finest companies I ever saw. A sergeant of the Bundelcund Legion was also a noble-looking fellow of nearly equal height."

Those who have seen the 15th Sikhs will admit that the regiment is now much the same in physique and appearance as when inspected in January 1848 by Major-General Sir Joseph Thackwell.

Captain Campbell, who had been aide-de-camp to Sir Joseph in Cawnpore, had now left, his place being taken by Lieutenant Edward Thackwell, 3rd Light Dragoons, the General's eldest son: and on the 21st February Lady Thackwell arrived from England, accompanied by her eldest daughter Elizabeth.

Ere this Lord Hardinge had left India for England, and had been succeeded as Governor-General by Lord Dalhousie, but, prior to the departure of the former, his conviction that the late campaign had resulted in a permanent peace led him to take the hazardous step of effecting large reductions in the strength of the army in India; and there seems some reason to believe that his desire for economy in the military expenditure—then urgently needed—may have coloured his sanguine forecast as to the maintenance of peace. Since the year 1837—the last year of peace —the strength of the army in India had been gradually increased by 120,000 men, and with the close of the campaign on the Sutlej Lord Hardinge made up his mind to effect large reductions in the native army, while at the same time redistributing the entire force so that the North-west Frontier and the Punjab might, as far as possible, be secured against any possible contingency. He decided to reduce the rank and file of the native army by 50,000 men by offering a bonus to every man willing to take his discharge, and by anticipating the invalid retirements by twelve months; and he was thus enabled rapidly to bring down the establishments of all regiments of native cavalry to 420 sabres, and those of native infantry regiments to 800 bayonets. In the artillery the number of horses per battery was reduced from 130 to 90. The actual number of regiments of native cavalry was at the same time increased by eight. In acknowledging the receipt of the memorandum embodying these proposals, the Commander-in-Chief wrote:

"I deeply regret the financial difficulties, and the consequent reductions rendered indispensable thereby, particularly until time shall have tested the feelings which the late arrangements with the Lahore Durbar may produce."

He pointed out at the same time that the proposed reduction in the establishment of native infantry regiments was really larger than was apparent, since the majority of the corps on the frontier were considerably over strength; he begged for a small increase in the establishments of British Cavalry, and asked that the Company's European infantry might be made up to 950—the same as those of the Queen's Regiments; but while the Governor-General acceded to some of the minor suggestions made by the Commander-in-Chief, he declined to make any modification in the actual reductions proposed, and which were therefore carried into effect.

In defence of these measures it has been claimed that, since the actual number of officers in the reduced regiments remained the same, the ranks could rapidly be expanded on the approach of war, and that efficiency was not reduced in the same ratio as the expense; that the total financial relief to the strain on the Indian Exchequer amounted to close upon two million pounds; and that a better strategical distribution of troops would more than counterbalance any actual reduction in mere numbers, and would ensure that no part of our most vulnerable frontier was left unguarded, or at least unwatched.

It may be that Lord Hardinge's measures might have answered his purpose had trouble arisen where alone it seems to have been considered possible, and in anticipation of which the great scheme of defence and reinforcement had been devised; but—

> "contrary to all expectations, it was not in the Sikh army or the Sikh community, or even on account of Sikh interests, that the disturbance arose which led to the final Sikh war; but from a comparatively insignificant trouble on an outlying Mahomedan frontier."[1]

[1] Gough and Innes.

CHAPTER XIV

ALTHOUGH it is a fortress which has enjoyed from time immemorial the reputation of great strength, Multan has on several occasions been successfully besieged. It was taken by the Mahomedans under Mahomed Ben Kasim at the close of the eighth century; Mahmood of Ghuznee captured it at the commencement of the eleventh; while Tamerlane took it at the close of the fourteenth. After several fruitless attempts, extending over several years, the fortress was in 1818 captured by Ranjit Singh, but not until 19,000 of his followers had been slain during the siege operations, and this Afghan outpost was then added to the Punjab.

When the close of the first Sikh war seemed to promise a protracted period of tranquillity, many of the districts of the Punjab—besides those in the Jullundur Doab which had been annexed to the British dominions in India under the terms of the treaty of Lahore—had been placed under the charge of British officers. The district of Multan was, however, still under native administration, the fortress and the province being held by the Dewan Mulraj, the infamous son of one of Ranjit Singh's best and most trusted officials. Mulraj was not contented under the new conditions of government, and towards the end of 1847 he intimated his wish to resign his office, but was asked to reconsider his application. Early in the new year he repeated his request to be relieved, and Sirdar Khan Singh was accordingly detailed to take his place, and proceeded to Multan at the end of March, accompanied by Mr. Agnew of the Civil Service and Lieutenant Anderson of the Bombay Army. There can be no doubt that in all that followed, Mulraj was in league with the Rani, and that the attention of the British was designedly diverted to Multan, while in the north the train was laid for a general revolt of the Sikhs.

Mulraj countenanced the murder of the two Englishmen, to avenge whom Herbert Edwardes hurried down from Dera Futteh Khan with his levies, calling for help from Van Cortlandt at Dera Ismail Khan and from the chief of Bahawalpore, of whose friendship we have before heard in the passage through his dominions of the Army of the Indus.

There is no intention of here describing the course of the Multan campaign, except in so far as it affected the operations in the Punjab to which it was the prelude. It will probably be enough to say that on hearing of what had transpired at Multan, Sir Frederick Currie—who early in the year had succeeded Henry Lawrence as Resident at Lahore—though aware that few, if any, of the Sikh troops could safely be employed against Mulraj for the reduction of the fortress, determined to use the most trustworthy of them for the purpose of taking possession of the districts attached to the Multan Government. He therefore arranged for the employment of five columns—one under Edwardes and Van Cortlandt, another under the Khan of Bahawalpore, and the remaining three composed of Sikhs under Sikh Sirdars of whom the chief was Sher Singh.

Later Currie sent a British force from Lahore and Ferozepore to move on Multan, and to this Lord Gough—who throughout had been opposed to undertaking operations at this season of the year and until all arrangements for the assembly of an adequate force were fully completed—at once added another infantry brigade, two troops of horse artillery and a siege train, not caring that an isolated British brigade should be sent into the field with allies ill-trained or notoriously disaffected. These reinforcements, under General Whish, reached the neighbourhood of Multan in the third week in August, and in the middle of the following month the Sikh commander and his following deserted to Mulraj. The siege of Multan, which was by then in progress, had to be raised, and the small British force was itself in turn invested in its camp on the Lower Chenab. On October 9th Sher Singh marched northwards again from Multan, calling to arms, as he went, the disbanded warriors of the Khalsa and raising the standard of revolt throughout the Sikh districts.

By the beginning of November Multan was still holding out in the south, while in the western Punjab proper the old Khalsa soldiers were flocking to Sher Singh, whose father, Chutter Singh, had enlisted Mahomedan support by offering to hand over Peshawar to Dost Muhamed, the Ameer of Kabul.

Thus the action of the Sikh Sirdars had rendered imperative the conquest and annexation of their country.

Lord Gough had early foreseen the possibility of the whole of the Punjab flaming into rebellion; he was ever strongly opposed to the employment of small isolated forces; while he was fully cognisant of the climatic risks attendant upon a summer campaign. He was, therefore, altogether in favour of delaying operations until the advent of the cold weather; but none the less he persistently advocated the concerting of measures for the gradual assembly and equipment of a force with which the invasion and conquest of the Punjab might successfully be undertaken. For such a purpose he estimated that an army of 24,000 men with 78 guns would be necessary, and he especially urged that all the regiments of native infantry and cavalry might at once be recruited up to the higher establishments at which they had stood at the close of the first Sikh war—pointing out that a very large number of our disbanded sepoys were immediately available for re-enlistment in the native army, and that if we failed to obtain their services "those opposed to us will assuredly exert every nerve to get them."

Lord Dalhousie did not accept all the propositions of the Commander-in-Chief: he declined to reverse the policy of his predecessor by sanctioning the increased establishments of native regiments; the request for immediate preparations was refused; and even in the estimate of the strength of the force required for hostilities in the Punjab, the Governor-General differed from the expert opinion of the Commander-in-Chief. It was not, therefore, until the 24th August that sanction was definitely accorded to Lord Gough to take the necessary steps for meeting any outbreak on the frontier, and, accordingly, in the beginning of September the force at Ferozepore was doubled in strength by moving up the troops detailed for the ordinary relief of the force at Lahore; but it was not before the last day of September that Lord Dalhousie finally directed that Lord Gough should carry into effect the proposals made by the Commander-in-Chief in the previous May. It was, however, by this time impossible that the Army of the Punjab could be assembled and ready to take the field before the middle of November.

When it seemed certain that a second Sikh war must be entered upon, Lord Gough had forces immediately available—though not sufficiently numerous for the work in hand—at Lahore, Jullundur, and Ferozepore. Late in October he directed General Wheeler, at Jullundur, to move forward his brigade across the Beas, and he shortly afterwards increased the force on the further bank of the Ravi by a cavalry brigade under Cureton, and a brigade of infantry under Godby, followed by a second under Brigadier Eckford. These two latter had moved forward about the 3rd November, and by the 16th the Commander-in-Chief had ordered on from Ferozepore six regiments of infantry, a light field battery, and six siege guns, with a reserve company of artillery and a pontoon train. In advance he already had (exclusive of Wheeler's brigade in the Bari Doab) 8 regiments of cavalry (of which 3 were British), 5 troops of horse artillery, 2 light field batteries, and 6 battalions of infantry. Following in rear were an infantry brigade, 2 regiments of irregular cavalry, and a siege train ; whilst on the frontier—not counting the troops holding Ferozepore, Ludhiana, and Umballa—there were another brigade of infantry, and one of irregular cavalry.

By great exertions, and in a comparatively brief space of time, Lord Gough had been enabled to collect under his own immediate command, an army of 20,000 men with nearly 100 guns, while the arrival of the contingent of Bombay troops before Multan, had increased the strength of the forces there under General Whish to 17,000 men with 64 heavy guns. Without perhaps unduly anticipating events, it may here conveniently be mentioned, that the surrender of the fortress of Multan on the 22nd January 1849 permitted the troops which had been investing that stronghold to march north to join Lord Gough. The Bengal portion of the Multan force set out on the 27th January, and the Bombay column on the 31st, reaching Headquarters in time to take a share in the crowning victory of Gujerat.

There is but little, if anything, in General Thackwell's diary of the earlier half of the year 1848 regarding the progress of the events above briefly described; but in September he chronicles the continual movements of troops towards the frontier, and was himself no doubt anxiously looking forward for orders to take the field.

"*Tuesday*, 19*th September*.—The 5th Cavalry marched from this on the 14th, and the 69th Native Infantry on the 16th for Ferozepore, and the native troop of artillery on the 18th for Muttra to relieve the native troop which has marched to the former place. No. 5 Light Field Battery and the 4th Native Infantry have also marched in the direction of the Sutlej."

"29*th*.—Early this morning marched to Ferozepore H.M.'s 9th Lancers, 1st and 2nd Troops of Artillery, and No. 6 Light Field Battery towards Umballa."

"30*th*.—A letter from the Commander-in-Chief warning me to be ready for service."

Shortly after this Sir Joseph must have received definite orders, for on the night of the 7th October he sent off his baggage, and following himself on the 24th, he reached Ferozepore on the 5th November, the Commander-in-Chief arriving in camp the next day. In a letter to Lady Thackwell dated the 6th, Sir Joseph says:

"The Commander-in-Chief received me very kindly, and said he had intended to have given me a separate command, but circumstances had at present changed, but hoped an opportunity would occur, as he had considered me unlucky in not having got *honours*"—(in allusion, no doubt, to the fact that Sir Joseph Thackwell had received no recognition of his services in the Gwalior and Sutlej campaigns)—"implying that his recommendations had not been attended to—however, *n'importe.*"

The Army of the Punjab was constituted as follows: There was a cavalry division of two brigades—in all, seven regiments—under Brigadier Cureton. This officer was considered too junior for the command of an infantry brigade without displacing other officers—such as Colin Campbell—whom it was intended to employ. Cureton had been holding the appointment on the staff of the army of Adjutant-General of Queen's Troops; he was an experienced officer, and it was natural that the Commander-in-Chief should wish to avail himself of his services, but his employment in command of the cavalry division rendered it impossible to appoint at least two cavalry brigadiers of equal experience and senior in service to Cureton.

There were three infantry divisions, of which the First was before Multan; the Second, comprising two brigades, was commanded by Major-General Sir Walter Gilbert; while the Third, under Sir Joseph Thackwell, was thus constituted:

1st Brigade.	*2nd Brigade.*
Brigadier Pennycuick.	Brigadier Hoggan.
H.M.'s 24th Foot.	H.M.'s 61st Foot.
24th Native Infantry.	36th Native Infantry.
45th Native Infantry.	46th Native Infantry.

3rd Brigade.

Brigadier Penny.
15th Native Infantry.
20th Native Infantry.
69th Native Infantry.

There were six troops of horse artillery, three light field batteries, and two heavy batteries. Hoggan's brigade was at this time at Saharun, on the Chenab, with Cureton and Colin Campbell.

On the 8th November Major-General Thackwell left Ferozepore with his first brigade and Captain Kinleside's light field battery (No. 5), forming the advance of the Commander-in-Chief's column.

After four marches Noewala was reached, where he halted, and the force with Lord Gough closed up.

"*22nd.*—Marched to Ramnuggar, eleven miles. We marched in battle array, Gilbert's Division on the right, the Third Division on the road, and Pope's brigade of cavalry and the Reserve Artillery in column on the left. Brigadier-Generals Campbell and Cureton had marched with the Commander-in-Chief at 3 a.m., to try and surprise the Sikhs said to be encamped near Ramnuggar. They found them near the river, under cover of their whole army on the other side. Some injudicious charges were made, which resulted in the death of poor Cureton and, it is supposed, of Lieutenant-Colonel Havelock of the 14th Light Dragoons. We lost one gun, stuck in the sand, and altogether what was expected was not done. The heavy batteries on the Sikh side of the river caused the

16

greater part of the loss. Poor Cureton was shot through the heart by a matchlock ball."[1]

Sir Joseph had been instructed to follow the Commander-in-Chief at 7 a.m.; the march was hastened on hearing the guns, but the action was over before Thackwell's force could come up, and the army encamped in front of Ramnuggar, and within a mile of the left bank of the Chenab.

On the 24th Lord Gough appears to have sent for Sir Joseph Thackwell, and directed him to take over command of the Cavalry Division, explaining to him at the same time the reasons which had influenced him in giving the command to Cureton in the first instance. He also published a general order appointing Sir Joseph Thackwell second in command of the army, and specifying the particular duties which that officer would be required to perform, which included the provision and superintendence of all measures for the security and sanitary control of the camp. The command of the Third Division was bestowed upon Brigadier Colin Campbell.

It seemed to Lord Gough that it was a matter of urgency that the Sikhs should be driven from their position on the right bank of the Chenab opposite Ramnuggar. Their remaining there, so close to the capital of the Punjab, not only

[1] Cureton's career had been a remarkable one. He enlisted as Charles Roberts in the 14th Light Dragoons, and served with them all through the Peninsula War. He obtained a commission, commanded the 16th Lancers in the latter part of the first Afghan war, and a cavalry brigade in the Sutlej campaign; was Adjutant-General of Queen's troops in India, and when commanding a Cavalry Division at Ramnuggar fell with his old corps, the 14th Light Dragoons.

constituted something of a menace, but encouraged those—and they were not few—who were disaffected, and who were in the rear of the British force; while the longer the Sikhs were permitted to retain their position, the more time was allowed for the arrival of the reinforcements which would be set free by the daily expected fall of Attock. It seemed impracticable to force a passage at Ramnuggar, where the river is very broad and where the enemy was strongly entrenched, and Lord Gough accordingly decided to remain there himself, holding the Sikh army in front of his position, while he threw a strong force across the Chenab by one of the fords higher up the stream. It was known that at Wazirabad there was a good and practicable ford, and that there was moreover a ferry, but between Ramnuggar and Wazirabad there were three other fords, and it was hoped that by one of these the detached force might be able to cross to the right bank. Of these three fords, that at Ghurri-ki-Patten, seven miles from Lord Gough's camp, was known to be practicable, but was believed to be too closely guarded for a crossing to be there safely attempted; at Ranni-ki-Patten, twelve or thirteen miles from Ramnuggar, there was another ford which was known to be difficult; and finally, about a mile nearer to Wazirabad, there was the ford of Ali-Sher-ke-Chuk—a practicable but rather dangerous crossing. The depth of water at any of these fords was known to be not less than four feet.

Major-General Sir Joseph Thackwell was selected for the command of this detached force, which consisted of White's Cavalry Brigade, the 3rd Irregular Cavalry, three troops of horse artillery, two light field batteries, and the Third Infantry Division (three brigades).[1] The pontoon train and two days' supplies also accompanied this force. Brigadier Colin Campbell, who commanded the infantry, has stated that—

"the movement was, in my view, and in that of the General, a hazardous one—the placing a force under 7,000 in a position in which they could not be supported, and where they might be opposed by 30,000."

But I have been unable to find in Sir Joseph's diary or correspondence anything to show that he entertained the forebodings thus attributed to him.

It had been intended that this detached force should rendezvous on the right of the camp of the 3rd Light Dragoons on the Wazirabad road, at 1 a.m. on the 1st December: but the night was very dark, and only the cavalry and artillery reached the place of assembly at the hour appointed, and Sir Joseph moved on with these two arms, leaving the infantry to follow. The General's diary for this date reads as follows: "Marched at 1 a.m. up the left bank of the Chenab *to the ford of Ali-Sher-ke-Chuk, a little above Ranni-khan-ki-Patten.*"

Sir Colin states in his diary that the division under his command marched at 2.30 a.m., but gives no particular reason for his troops failing

[1] The 12th Irregular Cavalry were also temporarily attached.

to be up to time, beyond that "we had great difficulty in finding the place of rendezvous. Two brigades of infantry went astray." But as the Deputy Assistant Quartermaster-General of the Third Division was a very young and inexperienced officer, it is perhaps not unfair to assume that faulty staff work was largely responsible for the very late start of the infantry portion of Sir Joseph Thackwell's command. Prior to leaving the camp at Ramnuggar, Sir Joseph received the following note from Lieutenant-Colonel Grant, Adjutant-General of the Army:

"8 p.m., *30th November* 1848.

"My dear Sir Joseph,

"The Commander-in-Chief hopes you understand distinctly that unless you get across the river in time to rest and breakfast your troops, so as to admit of your marching them to the left of the enemy's position by 1 p.m. to-morrow, it is his wish that you should make a second day of it.

"The distance to the ford is now reckoned to be thirteen miles, and to cross the river, even with a good ford, with the force you will have, must be a work of time. . . . Night work is to be avoided at all times; if, therefore, you cannot bring all your troops fresh and with ample daylight before them, it is much better, allow me to suggest to you, that the attack should be deferred one day. God bless you.

"Yours very sincerely,

"PAT. GRANT."

The force seems to have arrived in the vicinity of the fords at Ranni-ki-Patten and Ali-Sher-ke-Chuk about 11 a.m., after a somewhat harassing march, delayed moreover by the difficulty in dragging the heavy pontoons over the sandy track followed by the troops. Arrived here, Sir Joseph caused both the fords to be carefully examined. Of the first at Ranni-ki-Patten it has been recorded that—

"the Chenab here consisted of four branches or channels, one beyond the other. The sandbanks bordering and dividing these streams are insecure; moreover, the river here is notorious for its numerous quicksands, the danger of which is proverbial. A large party of the enemy were now descried posted for the protection of the ford. . . . Our artillery would not have been able to cover the passage of the British infantry, for the bank on the other side of the river was out of range of sight and shot."[1]

Lieutenant Paton, Deputy Assistant Quartermaster-General of the Army, who had been detailed by Headquarters to accompany the force, and who, it is fair to assume, was one of the officers of the Quartermaster-General's Department who are stated in Professor Rait's "Life of Lord Gough" to have made recent inspections of the neighbouring fords of the Chenab, spent some three hours in endeavouring to test the capabilities of the upper of the two fords—that at Ali-Sher-ke-Chuk. Sir Colin Campbell's diary of this date contains the following statement:

"After a march of fourteen or fifteen miles, we arrived at a part of the river which was guarded on the opposite bank by the enemy. It was the

[1] Thackwell.

ford of Ranni-ki-Patten; *and above this, about a mile higher up, was the ford at which it was intended the force should pass.* Lieutenant Paton, Assistant Quartermaster-General, was sent to examine it and the approaches to it. His report was that it was of this shape  and breast-high in some places; that the sand through which the guns must pass was very deep and heavy, and that the bullocks with the pontoon train would certainly not be able to drag the pontoons through it. . . . *The enemy, moreover, were on the opposite side, ready to oppose our passage, and in such cover as to make it difficult for our guns to drive them from it.*"

In the two books[1] recently published on the Sikh wars, the writers seem to have fallen into an error as to the ford by which Sir Joseph Thackwell was directed to cross, and also as to the reasons which decided him to use his discretion about moving on to, and crossing the river at, Wazirabad. It appears to be quite clear, both from Sir Joseph's own diary, and from that of Colin Campbell, his next in command, that the force started from Ramnuggar under orders to cross, if practicable, at the upper of the three fords between Ramnuggar and Wazirabad. Some anonymous contemporary writers have stated that the force was intended, and ought, to have crossed at Ghurri-ki-Patten, but this contention is untenable, and the writer of "The Life and Campaigns of Lord Gough" has definitely settled this point by admitting that Ghurri-ki-Patten "was known to be too well guarded."

[1] "Life of Lord Gough," "The Sikhs and the Sikh Wars."

The authors, however, of this book and of "The Sikhs and the Sikh Wars," have all spoken of "the ford at Ranniki" as the one by which Sir Joseph should have effected his passage of the Chenab; and this error is to some extent excused by the slipshod method employed by contemporary writers in speaking—as does also the author of "The Narrative of the Second Seikh War"—of the "ford" or "fords" of Ranni-ki-Patten. As a matter of fact there were *two* fords at or in the immediate vicinity of this spot—that at Ranni-ki-Patten itself, and the one at Ali-Sher-ke-Chuk, only a mile at most further up the river. Both presented considerable natural difficulties; the further bank at both places was held by the enemy; while at neither could the passage be covered by the fire of the British guns.

Sir Joseph had been detached to the right to effect, if he could, an unopposed crossing, because "it was impossible to force a passage at Ramnuggar." Would he have been justified in attempting to force one where the natural difficulties were even greater, with an unsupported and infinitely smaller body of troops, and in the face of an opposition which he knew to exist, but the strength of which he had no means of gauging? The author of "The Life of Lord Gough" has formed the conclusion, from the wording of a portion of "The Narrative of the Second Seikh War," that Sir Joseph decided against the practicability of a crossing near Ranni-ki-Patten—"not in the view of the presence of bodies of the enemy on the other side, but because of the nature of the ford."

The author of "The Life of Lord Gough" is not only in error as to the ford by which General Thackwell was ordered to pass the Chenab, but, in proof that the crossing at Ranni-ki-Patten was practicable, he states that on the following day Brigadier Hearsey passed over it some irregular cavalry, and that later on Brigadier Markham crossed with the whole of his brigade. Apart from the fact that in neither of these subsequent operations was there any likelihood whatever of the crossing being opposed, it may, with regard to Hearsey, be pointed out that the passage of a river by a few mounted men is a very different operation from putting across, in the face of an enemy, a force composed of the three arms; while of Markham and his brigade, Lord Gough states that they crossed at Kanokee[1] *with the help of no less than forty-seven boats*, from which one may reasonably surmise that this body was not under the necessity of trying whether the *ford* was practicable or not.

It is true that Sir Joseph in his despatch does not actually state in so many words that the farther bank at Ali-Sher-ke-Chuk was held by the enemy, but he gives as a reason for not there attempting the passage, that he could not so place his guns as to cover the crossing; while the presence of the enemy at both fords is mentioned by Sir Colin Campbell in his diary, and by Sir Joseph

[1] I cannot find Kanokee on any contemporary or modern map, but Durand (*Calcutta Review*) says that Markham crossed at Ghurri-ki-Patten.

himself in a letter dated 19th May 1849. It seems therefore quite clear that Sir Joseph Thackwell reconnoitred both fords in the neighbourhood of Ranni-ki-Patten, that he perceived at both he would have to force a passage under natural difficulties and military disadvantages, and that it was the presence at both places of strong hostile bodies, and not the nature of the fords only, which induced Sir Joseph Thackwell to move on to Wazirabad—an alternative permitted him by the terms of his instructions.

The length of the march and the difficulty of supply induced Sir Colin to suggest that the failure to cross by the fords near Ranni-ki-Patten necessitated a return to camp at Ramnuggar, but Sir Joseph resolved to push on, and sent forward Lieutenant John Nicholson, assistant to the Resident at Lahore, and who was acting as chief civil authority with Thackwell's force, and this officer managed, not only to collect fifteen large boats, but to stake out two of the fords over the three branches into which the Chenab is there divided, before the troops arrived at the ford and ferry, greatly fatigued, about 5.30 or 6 p.m. Here there was no appearance of an enemy, and the 24th Foot, with two native regiments and two guns, under Brigadier Pennycuick, were at once ferried across to secure the passage. The 3rd Irregular Cavalry crossed by the ford, and another infantry brigade, endeavouring to wade across, was overtaken by the darkness and halted and bivouacked on a sandbank between the second and third branches of the stream. The night was bitterly cold, no supplies could be passed over to the advanced troops, and once at least during the night an alarm was raised on the right bank; but the night passed quietly, and as soon as it was light the passage of the rest of the force was energetically proceeded with, and by noon the whole of the troops were across the river. The pontoon train, two light field guns, and two 18-pounders were sent back to Ramnuggar, escorted by the 12th Irregular Cavalry and two companies of native infantry, and with them went one of the Commander-in-Chief's aides-de-camp—who had thus far accompanied Sir Joseph—bearing to Lord Gough the news that the detached force had successfully negotiated the passage of the Chenab.

This chapter may perhaps fittingly be closed with some extracts from a report dated 7th December, on the crossing of the Chenab, from Lieutenant Baird Smith[1] of the Royal Engineers, who was Senior Assistant Field Engineer with the right column. After detailing the composition of the force, and the hour at which the march commenced, he goes on to say:

"The movement of the column was directed in the first instance on Ali-ke-Chuk, at which place a ford practicable for artillery and infantry was reported to exist. On arriving, however, at the point it was found that the information was incorrect, and that the access to the ford was so difficult that to move the artillery upon it was impracticable. The river being divided into three streams would have caused indefinite delay, had the passage been attempted by means of the pontoon train.

[1] Afterwards Chief Engineer at the siege of Delhi.

"Under the circumstances the selection of another point at which to effect the passage of the river was inevitable, and for the following reasons the Major-General Commanding resolved to move at once upon Wazirabad.

"(1) The distance from Ali-ke-Chuk to Wazirabad did not exceed six miles, and the troops were quite equal to the additional march.

"(2) At Wazirabad a ford practicable for infantry and cavalry was known to exist.

"(3) For the transport of the artillery fifteen boats, of which twelve were first-class river craft, and three of a smaller kind, were in possession of a party of Afghan Horse attached to Captain Nicholson, Assistant to the Resident at Lahore, and were available for immediate use.

"(4) There were the best grounds for believing that the enemy was not in force on the opposite or right bank of the river at Wazirabad, and that consequently the passage would be undisputed.

"My professional opinion having been required, I gave it decidedly in favour of the movement on Wazirabad: and requesting permission to precede the column with the view of making such arrangements as time would permit, I proceeded with Lieutenant H. Yule of the Engineers and Captain Nicholson, and reached the ford to the north-west and about two miles distant from the town about 3.30 p.m.

"At the ford the river was found to be divided into three streams—and in no part of the ford itself did the water exceed 3 ft. 10 in. in depth. Two of the streams only were staked out on the night of the 1st, the third was unmarked until daybreak on the 2nd, in consequence of the pressure of work and want of materials. The ferry was about three-quarters of a mile above the ford.

"The passage of the troops and guns at the ferry

continued throughout the night in a regular and systematic manner, without confusion or accident of any kind. By about sunrise on the 2nd, 6 regiments of infantry with 18 guns, under command of Brigadier-General Campbell, were in possession of the right bank of the river, occupying a strong position on the high ground bounding the valley of the Chenab. . . . The passage . . . was completed exactly at noon on the 2nd. Considering the operations to have commenced at 6 p.m. on the 1st, and to have been completed at 12 noon on the 2nd December, the passage of the force, consisting of 28 guns, 4 regiments of cavalry, 7 regiments and 2 companies of infantry, with the baggage and commissariat, the transit occupied 18 hours, and this will not, I trust, be considered long, when it is remembered that two-thirds of the work was done during the night, and with only a few hours' previous preparation."

CHAPTER XV

At the suggestion of Colin Campbell, the march down the right bank of the Chenab was not commenced until two hours after the transfer across the stream of the troops and impedimenta was effected, in order that the native soldiers, composing the major portion of the column, might have their usual one meal before fresh exertions were demanded of them, and a start was consequently not made until 2 p.m.

The force marched in order of battle: three brigades in brigade column of companies, at half distance—left in front, at deploying intervals; the brigade of cavalry under Brigadier White, with strong flanking parties and rear-guard, covered the right—the more exposed flank; while the 3rd Irregular Cavalry marched on the left of the infantry with orders to patrol right down to the river and clear the right bank aided by infantry, if necessary. Each infantry corps was covered by its own skirmishers. After a march of some ten miles the village of Dowrawala was reached about dusk—the movement had been entirely unopposed, no sign of the enemy having been met with. Some time during the evening (" before evening," says the author of the "Narrative of the Second Seikh War"—"during the night," says Lawrence Archer) a shuter-sowar, or camel despatch rider, arrived in camp, bearing the following letter from the Commander-in-Chief:

CAMP, *2nd December*.

"MY DEAR THACKWELL,

"I congratulate you at having crossed. I shall make as great a *tomasha* as I can here to-day, in order to keep them here—or rather, to keep their guns here; and if the bridge of boats reach me in time I shall throw a body across, so as to join your advance; therefore do not think a dust approaching you from the river is the enemy. They are entrenching themselves to protect this ford, but I hope to get across by the bridge of boats, or by some other manœuvre.

"They are in a great fright, and well they may, for they cannot take off their guns without fearful loss. Two of their regiments have promised to join me. God bless you.

"GOUGH.

"Do not hurry; bring your force well up in hand.

"G."

At 6 a.m. the next morning, Sunday, 3rd December, Sir Joseph marched off again, hoping to be able to attack the left and rear of the Sikh position in front of Ramnuggar at 11 o'clock. The force moved on, clearing away the enemy from the river bank, and had already gone about two miles south of the ford at Ghurri-ki-Patten when another communication was received from Lord Gough:

CAMP, *2nd December*.

"MY DEAR THACKWELL,

"I congratulate you at having crossed. I shall make as great a fuss as possible here to-morrow by cannonade, to keep their guns here, and if the boats reach me in time I hope to throw a body to co-operate with your left. Do not hurry your men: bring them and your guns well up in hand, and we are sure of success. I have two regiments in the enemy's camp ready to come over to me. I shall push across whenever I can, when you come into action. Keep me acquainted with your movements. I take it they will fall back from the fords when the river is your advance. They are in a great fright. With God's blessing, before to-morrow night we, united, will give them reason to be so.

"Yours very sincerely,

"GOUGH."

At the same time that this letter was handed to Sir Joseph Thackwell, or very shortly afterwards, two others were received—the one from Lord Gough, and the other, of very much later date, from one of the Headquarters Staff:

CAMP, *2nd December*.

"MY DEAR THACKWELL,

"This is the third note I have written you this day. I find they are retiring from my immediate front, merely keeping the fords well protected. I am, therefore, going to open two batteries upon them this day. Do not hurry on; keep your men fresh and well in hand. I hope to reinforce you to-morrow morning; don't think every dust an enemy. With God's blessing, by to-morrow night, we shall have all their guns. Two of their regiments are ready to come over to me. God bless you all.

"GOUGH.

"This man will bring an answer."

(What had happened was that the Sikhs had divined the object for which Thackwell had been detached, and, leaving a body to hold the Commander-in-Chief in front of Ramnuggar, had marched up the right bank in force to oppose his second in command.)

"TO SIR JOSEPH THACKWELL.

"When General Thackwell has taken possession of Ghurri-ki-Patten, Lord Gough desires that he will not move his force on to the attack till reinforced from Ramnuggar by a brigade of infantry and cavalry, which are prepared to move at a moment's notice.

"By Order,

"J. B. TREMENHERE.

"3rd December, 8¼ a.m."

It was probably at this time that the force was halted for breakfast and to replenish the pouch ammunition up to sixty rounds per man. It was 11 a.m. when the column again moved on to some villages, nearly three miles to the right front of the Ghurri-ki-Patten ford, where the troops were halted, and whence Sir Joseph sent a wing of the 56th Native Infantry and part of the 3rd Irregular Cavalry to hold the ford, to which 600 of the enemy's cavalry had been seen approaching. Colin Campbell mentions in his diary that on halting near these villages "Sir Joseph Thackwell received a letter from the Commander-in-Chief to inform him that he had sent a reinforcement of the 9th Lancers and 14th Light Dragoons to join by the ford of Ghurri-ki-Patten"; the letter to which allusion is thus made is no doubt the following:

HEADQUARTER CAMP,
3rd December, 10 *o'clock a.m.*

"MY DEAR SIR JOSEPH,

"We have heard this morning of your approach to the Ghurri-ki-Patten Ghaut, and I am desired to acquaint you that the Commander-in-Chief is despatching the 14th Light Dragoons, a detachment of the 9th Irregulars, and a Brigade of Infantry (the 2nd European Regiment, Dett. 45th, and the 70th N.I.), to join you at the Ghaut.

"Our heavy guns are all in position on the river, close to the ford near the clump of trees, and ready to aid you as well they can (*sic*) in your attack on the enemy's position.

"Bagot takes this to you.

"Yours very sincerely,
"PAT. GRANT."

The troops having halted, Nicholson's Pathan Horse were sent out to the front, and three of the four villages of Langwala, Tarwala, Rutta, and Kamookhan were each occupied by a company of infantry, the column itself remaining some little distance in rear, in front of the village of Sadulapore. General Thackwell rode off to the left to see that the troops guarding the ford had been placed in the right position, and then returning to his troops proceeded to ride along the line of villages. Nicholson's troopers had early reported that some 200 Sikh horsemen had been noticed, and from some correspondence later between Campbell and Thackwell, it would appear that, during the absence of the latter, reports had reached Campbell that the Sikhs in front were becoming more numerous; but this information was apparently not conveyed to Sir Joseph, whose first intimation of the enemy being in force was from a patrol of the 5th Light Cavalry, in front of the right, being cannonaded, at the moment when Sir Joseph was passing the village of Rutta. There not appearing to be time—owing to the rapid advance of the Sikhs—to move the British line up to the villages, which were moreover surrounded by fields of tall and thick sugar-cane, Sir Joseph Thackwell directed Brigadier-General Campbell to retire some 200 yards and deploy immediately in front of the village of Sadulapore—the companies falling back from the advanced villages, which were at once occupied by the Sikhs, who showed in great strength.

The short retrograde movement was interpreted as a retreat and the Sikh attack rapidly developed. Sher Singh was himself here in command; he brought some twenty guns into action, massed a numerous cavalry with horse artillery on the left, while the fields of sugar-cane seemed alive with matchlockmen.

The Sikhs clung to their cover and opened a heavy fire from their guns, threatening both flanks at the same time with their cavalry with a view of getting to the rear and cutting off the baggage. The fire was so heavy that Sir Joseph ordered the infantry to lie down, while the 3rd Dragoons,

8th Light Cavalry, and Major Christie's troop of horse artillery checked the advance of the enemy on the right flank, and Captain Warner's troop was moved to the left of the infantry, and, supported by the 5th Light Cavalry and the remainder of the 3rd Irregulars, drove back the Sikhs. Sir Colin Campbell's diary states that he twice asked Thackwell's leave to advance the infantry to the attack, and in the "Life of Sir Charles Napier" a letter is quoted wherein that officer declares that Campbell three times begged for leave to advance and take the guns. Sir Joseph, however, always denied that such application was made, and his contradiction appears in the second edition of the "Life." In a letter also to Dr. Macgregor, written on the 19th May 1849, Sir Joseph says:

"Anonymous writers have presumed to say that some persons urged me to attack the enemy's position. This assertion is unfounded. I have no doubt that all were ready to obey orders, but not one suggested that the enemy should be attacked."

While in a letter written to Sir Colin Campbell on the 8th August of the same year, in reference to the question of attack, Sir Joseph Thackwell reminds Campbell that "you were most cautious in not giving an opinion," and in a further communication, written to Sir Colin on the 24th May 1850, he recalls to his recollection that "I gave you for perusal Lord Gough's note prohibiting any further advance."

For some two hours the cannonade continued and, the fire of the enemy then slackening, the cavalry on the right were directed to charge and capture the Sikh guns if possible, Sir Joseph intending to support them by moving the brigades in echelon from the right; no opportunity offering, however, for the action of the cavalry, General Thackwell deemed it advisable to remain in his position and not to attempt to drive back an enemy so strongly posted on the right and centre, with the probability of having afterwards to attack them in the entrenched position to which they would inevitably fall back.

As soon as Lord Gough had heard of the successful crossing of the Chenab on the morning of the 2nd December, he had directed a heavy cannonade to commence upon the Sikh batteries and encampment at Ramnuggar, and this was maintained during the 3rd so as to hold the main body of the Sikh army to this point and thus to delude the enemy into the belief that it was intended there to force the passage of the river. Although, however, the heavy firing at Sadulapore on the afternoon of the 3rd must have been plainly audible at Ramnuggar, and it should have been thereby apparent that a large portion of the Sikh force which had been in Lord Gough's front had gone off to oppose Sir Joseph Thackwell, the Commander-in-Chief seems to have made no real effort to "push across" as promised in his second letter of the 2nd December, but remained inactive, save for a distant and ineffective bombardment of the half-empty trenches on the further bank.

The Sikhs replied to this bombardment with only six guns, but they were so cleverly concealed that the British guns were unable to silence them; and leaving these six guns to reply to Mowatt's and Shakespear's batteries, the enemy gradually withdrew the remainder and finally fell back on the night of the 3rd, leaving only empty emplacements on the river bank.

Sir Joseph invited the opinions of the next senior officers as to the expediency of an advance, but, with the exception of Brigadier Pennycuick, all were agreed that it would be best—the afternoon being now well advanced—to remain where they were: the enemy were still in strength in front; the line of villages formed a strong position; a deep nullah ran along the left front; and it was reported that the ground about the villages, and in rear of them up to Sher Singh's position at Ramnuggar, was thick with fields of sugar-cane.

As the Sikh fire began to diminish in intensity, another letter was received from Lord Gough by the hand of Lieutenant Tytler, 9th Irregular Cavalry, and in this communication Sir Joseph was at last left free to act as he considered best, whether the promised reinforcements had reached him or not.

The time of receipt of this letter is noted by Sir Joseph Thackwell as 3.30 p.m.

The casualties in the action of Sadulapore amounted to 73, of whom 33 were killed.

At about midnight on the same date the force in front of Thackwell also began quickly to fall back from the position it had taken up, and the whole Sikh army then, forsaking the richly cultivated lands on the banks of the Chenab, proceeded to take up and entrench a fresh position in the thick jungle bordering the Jhelum River.

At dawn on the 4th Lord Gough, finding the position on the river bank evacuated, sent the 9th Lancers and 14th Light Dragoons in pursuit, and at the same hour Godby's brigade—the promised reinforcements—was still effecting its passage of the ford of Ghurri-ki-Patten. The crossing had been found to be exceedingly difficult; it was too deep to admit of the troops wading through it; the pontoons were of no service—probably owing to the number and breadth of the several channels; and finally boats—which Sir Joseph had been directed to bring down from Wazirabad—had to be made use of. It was not until 5 p.m. on the 3rd that, just as the action at Sadulapore was closing, the work of embarkation commenced, and by 8 p.m. only one complete regiment, the 2nd Europeans, had been disembarked on the further bank. At 7 o'clock on the morning of the 4th, Brigadier Godby joined Sir Joseph, bringing with him the 2nd Europeans and the 70th Native Infantry, and it was 9 a.m. before the remainder of Godby's troops left the western bank and marched to join Thackwell—close upon 24 hours since these reinforcements had left Ramnuggar, which was only some 10 miles distant.

In Lord Gough's despatch dealing with the operations connected with the passage of the Chenab, and which is dated the 5th December—

before he had received Sir Joseph's report written on the 6th—but which Lord Gough himself states he did not receive until the night of the 9th, there are several points which, in justice to Sir Joseph Thackwell, seem to call for comment.

In paragraph 4 the following sentence occurs:

> "This officer" (*i.e.* Sir Joseph) "moved upon a ford which I had every reason to consider very practicable (and which I have since ascertained was so), but which the Major-General deemed so difficult and dangerous, that he proceeded (as he was instructed should such turn out to be the case) to Wazirabad."

It appears that about the middle of November Lord Gough had directed reports to be prepared on the fords of the Chenab, that William Hodson had been employed on this duty, and that the records of some of these inquiries have been preserved. While encamped at Ramnuggar the Commander-in-Chief had instructed the Field Engineer, Major Tremenhere, to ascertain the practicability of the fords on both flanks, in co-operation with the officers of the Quartermaster-General's department, and in "The Life of Lord Gough" the nature of the four fords upstream from Ramnuggar has been thus epitomised:

> "*Ghurri-ki-Patten.*—Practicable, but too well guarded.
>
> "*Ranni-ki-Patten.*—Practicable, but objectionable on account of the steep bank, the strength of the stream, and the passage not being straight.

"*Ali-Sher-ke-Chuk.*—[Where, according to the journals of both Sir Joseph and Sir Colin, the force was intended to cross,] is described as 'dangerous.'

"*Wazirabad.*—The bed of the stream hard and level, the current slack, and the passage of sufficient breadth for crossing without risk of any sort."

From the above it may reasonably be inferred that of the four crossings the last, where the passage was actually effected, was the easiest, and that Ghurri-ki-Patten came next in order of practicability—that its only objectionable feature was the fact that it was guarded; we have seen, however, what difficulties were experienced by Godby's single brigade in crossing, unopposed, at this place. The practicability of any ford from a military point of view cannot, however, be considered apart from the matter of the presence or absence of the enemy on the further bank. Lord Gough's despatch on the 5th December describes the fords of the Chenab as being "most strictly watched by a numerous and vigilant enemy and presenting more difficulties than most rivers." The evidence of Thackwell and Campbell is unanimous that at both Ranni-ki-Patten and Ali-Sher-ke-Chuk the further bank was held by the enemy; and under the circumstances it seems that Sir Joseph was justified in availing himself of the discretionary power allowed him and in moving on to Wazirabad, instead of endeavouring to force a passage at one of these fords, known to the Quartermaster-General's department of the army as "objectionable" or "dangerous," and where a crossing would be strongly opposed.

At the end of paragraph 5 of the same despatch there occurs the following sentence:

> "I was enabled to detach another brigade of infantry, under Brigadier Godby, at daylight on the 3rd, which effected the passage, with the aid of pontoon train, six miles up the river, and got into communication with Major-General Sir Joseph Thackwell."

That the passage of Godby's brigade was effected quite independently of the pontoons and with boats brought thither by Sir Joseph, is a small error due to the preparation of the despatch on insufficient data, but what is more important is that the wording of the latter part of the sentence leaves it to be inferred that Godby was in communication with Sir Joseph *prior* to the action of Sadulapore, or indeed that he had actually joined forces and so strengthened Thackwell sufficiently to attack the Sikh army—which, as has been shown, was not the case.

At the beginning of paragraph 6 of the same despatch we read:

> "Having communicated to Sir Joseph my views and intentions, and although giving discretionary powers to attack any portion of the Sikh force sent to oppose him, I expressed a wish that, when he covered the crossing of Brigadier Godby's brigade, he should await their junction, except the enemy attempted to retreat; this induced him to halt

within about three or four miles of the left of their position."

The instructions, conveyed to Sir Joseph before leaving Ramnuggar by the Adjutant-General of the army in the letter already quoted, are very distinct: unless he can be in position ready to attack the left of the Sikh entrenchment not later than 1 o'clock in the day, the attack is to be deferred; "night work is to be avoided at all times;" provision is to be made for "ample daylight" before an attack is initiated. (The importance of not commencing operations late in the day has, it may be remarked, been proved over and over again in Indian frontier warfare, and disregard of this rule has repeatedly led to disaster.) The letter from Major Tremenhere positively directs General Thackwell not to attack until reinforced by a brigade of infantry and cavalry, which however had not started at 8.15 on the morning of the 3rd. Neither here nor elsewhere is there any reservation about the enemy's retreating, while it was surely natural that the receipt of such explicit orders should have induced Thackwell to halt—sanction to advance, with or without reinforcements, not being received until the action had been already some considerable time in progress.

When Lord Gough sent away the force under Sir Joseph Thackwell, the Commander-in-Chief did not apparently count upon securing more than the mere crossing of the Chenab; he hoped, indeed, that the Sikhs would remain quietly in their trenches opposite Ramnuggar until attacked in flank by

Thackwell, and in front by the force under his immediate command, but the possibility of the Sikh commander holding *him* in front while an overwhelming force was directed against Thackwell's advance, does not appear to have entered into the Commander-in-Chief's calculations. A letter quoted in the "Life of Lord Gough," written to his son, and dated the day Thackwell's force left Ramnuggar, seems not only to show that the Commander-in-Chief was not very sanguine as to the prospect of success in what Colin Campbell calls "a hazardous movement," but that on that date at least he was certainly doubtful as to whether the river could be crossed where Thackwell had been ordered to make the attempt.

"I attack to-morrow," he wrote; "if my flank will be successful, that is, if it can get across, everything will go well; if, on the contrary, they find the ford impracticable, they will have to go on to Wazirabad and force a passage. If my friends opposite me move in any numbers, one way or the other, I shall punish them; but with a treacherous river in my front, I cannot prevent their running away. They have all the boats, and, by keeping eight or ten guns at the only ford here, and sending off the rest a couple of days' march ahead, they may say, 'Catch me who can.'"

Sir Joseph Thackwell had carried out a difficult operation, had executed to the letter the somewhat contradictory orders received, and only the flight of the Sikhs prevented the combined attack by him and Lord Gough upon the entrenched position at Ramnuggar. But the menace of Thackwell's advance was mainly responsible for the hurried retirement of the enemy to the barren country of the Jhelum, the abandonment of six guns, and the destruction of a large quantity of gunpowder.

For the discrepancies, to which allusion has been made, between some of the statements in Lord Gough's despatch of the 5th December and the documentary evidence now here published, there appears to be a very simple explanation, which is, *that no copies were kept by the Headquarter Staff of the instructions sent to Major-General Sir Joseph Thackwell*, so that in the preparation of his despatch the Commander-in-Chief had to rely, for any record of his orders to his Second-in-Command, upon the memory of his chief staff officer. On the 6th December at Helah, some thirteen miles as the crow flies to the west of Ramnuggar, Sir Joseph received a letter from the Adjutant-General, which is dated "Ramnuggar, 5th December 1848, 7½ a.m.," and of which the concluding sentence runs as follows:

> "*Will you do me the favour to let me have copies of the several notes, this one included, I have addressed to you by the Commander-in-Chief's direction since you left this ground—there was no time to copy them before they were sent off.*"

It is to be regretted that, even if the Commander-in-Chief could not have withheld his despatch to the Governor-General until receipt of Sir Joseph Thackwell's report on his operations, Lord Gough did not wait for the arrival of the copies of the above-mentioned "notes" to verify the correctness of his recollection of his own instructions.

CHAPTER XVI

On the morning of the 4th December Major-General Thackwell—leaving Brigadier Campbell to follow with the infantry portion of the force—pushed on with the cavalry in pursuit of the Sikhs, with the intention of harassing their retreat by the Dinghi, Jullalpore, and Pind Dadun Khan roads. He was not, however, able to overtake any of their guns, nor indeed was anything seen of the enemy, in spite of the fact that he pursued them direct and does not appear to have received — until too late to comply—orders for him to move first towards the vacated entrenchment opposite Ramnuggar. Sir Joseph took up a position for the night with his whole force on the Jullalpore road, about two and a half miles from the village of Helah, and twelve in front of Ramnuggar, and here the 14th Light Dragoons joined the column in the evening.

The following letters were received from Headquarters during the day. The first is from Lord Gough, evidently written before he knew of the general retirement of the enemy.

RAMNUGGAR, 2 a.m
4th December.

"MY DEAR THACKWELL,

"Not having heard from you, I hardly know what to say. I hope Brigadier Godby has joined you with reinforcements both with cavalry and guns, but this fearful river and want of boats embarrasses us much. Pray communicate with me by an officer, who can cross at our pontoon bridge, which I will ride over to this morning if I can be spared here. I should like to see Nicholson at it, at this side, with some intelligent officer from you—say at 7 o'clock—that is, if the enemy have not retired during the night.

"Sincerely yours,
"GOUGH.

"I hope you have got your ammunition. What is the enemy's position—how far from you?"

The next letter is from Colonel Grant, dated the same day as the last, but at 8.15 a.m.

"MY DEAR SIR JOSEPH,

"The enemy has blown up his magazines and abandoned his position directly in front of the clump of trees. The Commander-in-Chief requests you will feel your way up in this direction, and he will send the 9th Lancers across the now open ford here to reinforce you and enable you to pursue with fresh cavalry the enemy supposed to be retreating, and perhaps admit of your cutting off a portion of his guns before he gets away to any great distance.

"Provisions for your men and horses will be crossed here and be in readiness for you by the time you reach the enemy's abandoned position.

"We have not had a line from you subsequent to your note of yesterday intimating that you had got possession of the Ghurri-ki-Patten ford—not a word about your action in the afternoon has reached us, and the Chief is intensely anxious for intelligence of your proceedings.

"Yours sincerely,

"PAT. GRANT."

The date of the following is indecipherable, the paper being much worn and rubbed, but from the contents it would appear to have been written shortly after the foregoing letter from Grant:

"MY DEAR THACKWELL,

"Four Jhansie guns have left the ford at 4 o'clock this morning drawn by bullocks. The 9th Lancers are just going to cross to reinforce you. Feel to your left with infantry to this ford and to your front with cavalry.

"I leave everything to you, but don't endanger your men; capture the guns if you can.

"Yours very sincerely,

"GOUGH."

On Tuesday the 5th Sir Joseph's diary records:

"Marched at 7 a.m. to Helah. The Third Division encamped in rear of Helah. The 14th Dragoons and 8th Light Cavalry pushed on to Dinghi to return to-night. The 9th Lancers (which had joined during the night) and 5th Light Cavalry, with Huish's troop of artillery, moved on eight miles in front" (on the Jullalpore road) "to return to-night. I have the 3rd Light Dragoons, two troops of artillery and the 3rd Irregular Cavalry. The troops detailed returned in the night. A Sikh rear-guard is at Pind Dadun Khan within two *coss* of Jullalpore."

None of the enemy's troops or guns were, however, overtaken.

In a note from the Commander-in-Chief dated 6 a.m. the 5th December, Lord Gough apparently contemplated his advance troops moving on to Dinghi, for he impresses on General Thackwell the importance of all pursuing troops returning to camp at Dinghi for the night.

"White's brigade," he continues, "had better halt to-day with the Horse Artillery and get all their tents and baggage up with their provisions. I shall cross over and encamp with the Cavalry Brigade under White or the infantry under Campbell, if not too far in advance. Let me know by an officer where each will be this evening encamped and the distance. Concentrate as much as you can, but do not fatigue and overwork your men."

Later, however, a note, written an hour and a half after that above quoted from Lord Gough, was received from Colonel Grant saying that as the Commander-in-Chief had since learnt that the enemy had gone off by three routes towards the Jhelum, Thackwell was now to collect his entire force and form a camp at Helah, which was to be his "fixed point," although he was at liberty to send his cavalry out in any direction he might think desirable, but it was again impressed upon him that they were invariably to be recalled by dusk.

On the 6th Lord Gough paid a flying visit to Helah, returning to Ramnuggar on the following day. "He inspected the cavalry," writes Sir Joseph, "and was much pleased with their appearance and the position I had taken up. He also seemed pleased at the operations which brought me here."

18

On this day Sir Joseph Thackwell wrote his report on the operations covered by the period since he left Ramnuggar on the night of the 30th November, but there is no explanation forthcoming as to why it was not seen by the Commander-in-Chief until the evening of the 9th December. But the probable reason is that while Sir Joseph's despatch was written in the rough on the 5th, the belated arrival of the baggage from the rear had prevented its being before reduced to the proper official form. On the 9th Lord Gough, in the post-script to a note, says, "For God's sake send me in your report; the press will be open-mouthed at its long delay." The despatch was then no doubt on its way to Ramnuggar and cannot have been long retained by the Commander-in-Chief before transmission to the Governor-General, then at Umballa, since Lord Gough's covering letter is dater the 10th. It seems that it was Lord Dalhousie who was responsible for the delay in publication, since he took exception to the insertion in the report of the name of a particular individual, and appears to have returned the document to the Commander-in-Chief in order that the name might be expunged.

Sir Joseph had attached to his staff a young fellow in whom he took a kindly interest, of the name of Angelo. This young gentleman was seeking to establish a claim to a cadetship in the Company's Service—which he was subsequently granted—and having volunteered for service in the campaign, was, with the sanction of the Head-quarter Staff, allowed to join Sir Joseph in the capacity of extra aide-de-camp. That the appointment had official sanction there can be no doubt, for in General Thackwell's diary, under the date of the 7th November, appears the entry: "I called on Grant and settled about young Angelo," and in the Sadulapore despatch Sir Joseph had mentioned Angelo's services, alluding to him as "extra aide-de-camp." There was excellent precedent for such a procedure, for in Lord Gough's despatch dated 19th December 1845, on the battle of Moodkee, he includes among the casualties in the "Cavalry Division Staff" the name of "Volunteer Mr. A. Alexander, A.D.C. to Brigadier Gough," as severely wounded.

To Lord Dalhousie it did not, however, seem fitting that such a title should be accorded to one who held no military rank, and the despatch was returned to the Commander-in-Chief by the Governor-General with the suggestion that the paragraph to which exception had been taken should be expunged. Writing to Sir Joseph on the 24th December, Lord Gough says that he had received the enclosure "last night from the G.G." and that "he would strongly advise" General Thackwell "to do as the Governor-General suggests." On the same day Sir Joseph replied that, "agreeable to the suggestion of the Governor-General, and to your Lordship's advice, I beg that the name of Mr. Volunteer Angelo may be erased from my despatch."

The report was again returned to the Governor-General, and was finally published on the 31st January at Ferozepore—more than a month later. Contemporary writers have generally attributed to the Commander-in-Chief the blame for the belated publication of the Sadulapore despatch, but it seems clear that it was Lord Dalhousie's somewhat unnecessary insistence upon a matter of trivial punctilio, which was responsible for the narrative of the crossing of the Chenab being given to the world, when at least local interest had greatly subsided, and when the minds of men in India and at home were occupied with later events, and oppressed with the desperate fighting at Chillianwala.

In his covering letter to Sir Joseph Thackwell's report, Lord Gough expresses "warm approval of the conduct of the Major-General," and although he thereby does a tardy and none too ample justice to his second in command, the Commander-in-Chief still says nothing to correct any false impression which might have been conveyed by the terms of his despatch of the 5th December, describing the operations of the force detached from Ramnuggar—a despatch, moreover, which was written before the whole facts were before him. The evidence which has been given in Chapter XIV makes it clear that the ford at Ali-Sher-ke-Chuk was the one by which Sir Joseph was to endeavour to cross; and a perusal of General Thackwell's despatch shows that very particular attention was given to its examination, and that the decision to proceed to Wazirabad was not come to until the possibility of crossing either at Ranni-ki-Patten or at Ali-Sher-ke-Chuk had been considered from every point of view. It would be interesting to know the grounds upon which the statement is made on page 196 of "The Life of Lord Gough," that "Thackwell never found the real ford."

General Thackwell's loyal determination to leave no stone unturned to carry out his orders, is shown by the resolve—contrary to the advice of Sir Colin Campbell—to try the distant, but—as reported—easy ford of Wazirabad. Sir Joseph's despatch proves that even here the difficulties of the crossing had been much underestimated; although Nicholson had staked out two-thirds of the ford, three sowars of the 3rd Irregular Cavalry were drowned while attempting the passage after darkness had begun to fall, while it was necessary to pass almost the whole of the infantry over in boats.

It will be seen from Thackwell's report that on the morning of the 3rd December he was actually on the march to attack the left of the Sikh position, and had arrived within four miles of it, when he received orders to wait for reinforcements crossing at Ghurri-ki-Patten. At that time—it would have been about 9.30 a.m., since he hoped to attack at 11—he had already marched several miles beyond the ford, and had now to halt, send to secure the passage, and await the promised reinforcements; but Sir Joseph must have already plainly realised that there would be no time to march on the Sikh position and attack by daylight, if he was to wait until Godby joined him.

At 2 p.m. he was himself attacked; and after the action had been some time in progress,—at 3.30 p.m., as recorded by General Thackwell,—the prohibition against attacking single-handed was removed, but it was 4 p.m. before the British guns had obtained any real superiority over the Sikh artillery, or sufficient to warrant an advance against an enemy in the strong position from which it would have been necessary to dislodge them, before moving on to a fresh attack upon the flank of the entrenchment opposite Ramnuggar.

To the relief which the Commander-in-Chief must have felt at the successful accomplishment of the passage of the Chenab and the consequent retirement of the Sikhs, there succeeded, as was perhaps natural, a feeling that more might have been done, since these results had been attained so expeditiously and with such little loss of life; and that perhaps an opportunity had been missed of punishing the enemy during their retreat to the Jhelum. A perusal of the Governor-General's order shows that Lord Dalhousie, with a finer sense of proportion, better appreciated the advantage gained and the nature of the difficult and hazardous operations by which success had been assured.

Other and less charitable reasons for the delay in the publication of Sir Joseph's report have been put forward by some writers—notably by Thorburn and Malleson. The former says that after Lord Gough's "grandiloquent despatch" of the 5th December, the immediate publication of "Thackwell's plain unvarnished tale" would have tended to diminish the value of the remarks by the Commander-in-Chief, who had hastened to announce to the Governor-General results which actual events hardly bore out.

To return to Sir Joseph's diary.

"*7th December.*—All quiet, but it seems certain that the Sikhs are entrenching themselves at Moong, having been encouraged by the arrival of 4 battalions and 12 guns from Peshawar, with a regiment of sowars. Four battalions and 12 guns more are expected in a few days."

"*8th.*—Brigadier Campbell and the infantry moved into line to-day and occupied the centre of my position."

"*9th.*—Brigadier Brooke came in last evening with 3 troops of Reserve (Horse) Artillery."

"*10th.*—Went out with a squadron of the cavalry, and I made a reconnaissance towards Moong, the position of the Sikh army, to near the village of Chuk; found the road practicable, but from the continued jungle for nearly six miles, it is by no means desirable to advance by such a route with the chance of opposition, if another can be found less difficult."

Lord Gough had ere this received very explicit instructions from the Governor-General, not to move forward until the fall of Multan should set free the column under General Whish.

"*12th.*—Lord Gough made a reconnaissance to Phallia, and I accompanied him. He had a guard of two squadrons of the 3rd Light Dragoons, one of the 8th and one of the 5th Light Cavalry. Phallia is on the road from Ramnuggar to Pind Dadun Khan."

Lord Gough now resolved to concentrate his force at Wazirabad, holding Gujerat as an advanced post, and on the 14th he wrote to Thackwell saying he could make no movement until the 26th, when he expected the arrival of "an immense convoy" with the 20th Native Infantry, and would then send a brigade to Wazirabad and commence his own move thither. Writing again on Saturday the 16th the Commander-in-Chief says:

"I have this day sent Pope's brigade to Wazirabad, and we shall all probably make a movement on Monday. The position I propose to take up is with my right (or rather your right) rather in rear of or resting on Bagran with the left thrown back in the direction of Kienwallee. This covers both Gujerat and Ramnuggar, and brings me so near to Dinghi that if the Sikhs occupy it I may be enabled, *if I can get leave,* to attack them. At present I am under the ban of the G.G. I shall with Headquarters cross the river on Monday, please God, and will join your camp with the Headquarters either on that or the following day. Keep all this, however, to yourself, but be prepared to move at a moment's notice. You will take care not to let the Sikhs occupy Gujerat; you must be before them."

The various reconnaissances which had been made of the approaches to the enemy's position at Moong, and a report transmitted by Thackwell late on the 16th that it was rumoured that Sher Singh was likely to occupy Dinghi with the Peshawar troops, induced Lord Gough to change his intention of proceeding to Wazirabad; and Sir Joseph was directed to stand fast at Helah, since it seemed to the Commander-in-Chief that, with Sher Singh at Dinghi, the Sikhs might cross the Chenab below Ramnuggar and strike at the

British communications. Headquarters crossed the river on the 18th, but remained on the right bank.

"*19th.*—At nearly 3 p.m. the enemy fired a salute of between 60 and 70 guns—I counted 61—but the reason is only conjectured. Some say it is on account of the treaty between Chutter Singh and Dost Mohamed; others that four regiments from Multan and a large body of Gulab Singh's troops were then within fifteen miles of their camp and about to join them."

On the 29th December the Headquarters closed up to Janukee.

"*2nd January* 1849.—It is said the Sikhs are throwing a bridge over the Jhelum at Tupai about a mile below Lassoorie and that Sher Singh is three miles from Moong."

"*8th.*—At 1 p.m. I made a reconnaissance to a hill about seven miles from camp and three from the Sikh post at Tronwala. Had a beautiful view over the country and the Sikh camp at Russool, all quite clear, and distant from Tronwala about 6 miles."

On the 9th Lord Gough moved to Loah Tibba, three miles nearer Dinghi, from which the Sikhs had now retired—the prohibition against an advance having apparently been definitely removed on the 7th—and on the following day he reached Lussoorie, where he was joined by the force under Sir Joseph Thackwell and by Brigadier Penny's brigade, which latter had up to this remained at Ramnuggar. On the 12th the whole army marched to Dinghi and encamped nearly a mile to the westward of the town.

When Alexander the Great had conquered Persia he passed on to India, ever the goal of every conqueror, who had pointed out to those whom he led that its wealth would repay them for all their toils and sufferings. He passed through the Khyber, crossed over the Indus at Attock in April 327 B.C., and—

"found the Punjab divided into a number of Hindu states, jealous of each other to a degree, which kept them from uniting to oppose his progress. On the banks of the Jhelum or Hydaspes, he was encountered by a considerable native army under a local raja, whom the Greeks knew by the name of Porus. But the tactics of the Macedonian were too much for the multitudinous army of the unpractised Asiatic. Leaving his camp in its original position, Alexander moved a strong division, unperceived, a few miles up the river, to a ford. Here he crossed early in the morning and fell upon the Indian host while they were engaged in opposing the portion of the Macedonian force that had been left in their front."[1]

It was within a stone's throw of this historic battlefield that Sher Singh, with the Sikh Sirdars, had taken up his ground to oppose the advance of the British.

The army under Sher Singh, estimated by Lord Gough at 30,000 to 40,000 men, with 62 guns, was daily expecting to be reinforced by the troops under Chutter Singh set free by the fall of Attock. To oppose the enemy already present in his front, the British Commander-in-Chief could dispose of from 11,000 to 12,000 men, with 60 guns. These

[1] Keene.

are Lord Gough's own figures as given to the Governor-General in a letter dated 11th January; but in an account of the "Battles of Chillianwala and Gujerat," contributed to the *Journal of the Royal United Service Institution* in March 1895 by General Sir Charles Gough, V.C., K.C.B., it is stated that "the force at Lord Gough's disposal amounted to about 24,000 men all told and 66 guns." These are approximately the numbers given in Appendix B of Lawrence Archer's "Commentaries on the Punjab Campaign," and which are taken from the returns dated 1st January 1849; but it would seem that these figures cannot altogether be relied upon, for in many cases the strengths of the corps are shown as hardly differing from those given on the 1st of the October previous, while in one or two instances regiments are taken as being even stronger at Chillianwala than at the commencement of the campaign.

It is possible that this might be accounted for in the case of native cavalry and infantry corps by the return of the men from furlough, but it may probably be taken for granted that European regiments at any rate took the field at their full available strength, and that their numbers are far more likely to have been reduced than augmented during the progress of the campaign. The authors, moreover, of "The Sikhs and the Sikh Wars" put the whole British force at about 14,000 men, with 66 guns, placing the effectives of the units of cavalry and infantry at not less than 200 per corps fewer than are shown in the returns quoted by Lawrence Archer. If, then, from this total are subtracted the necessary baggage guard—which on the occasion of Chillianwala consisted of three guns of a field battery, two regiments of Irregular Horse, and a battalion of Native Infantry—it would seem that Lord Gough was not far wrong in his estimate of the numbers of men and guns which he expected to be able to bring into action, as stated in his letter to Lord Dalhousie written two days prior to the battle.

Once again, as at Sobraon, had the Sikhs taken up a position with their backs to a wide river. Their left rested on the heights of Russool—a rugged range of low hills sloping gently towards the plain—while on the side towards the Jhelum, from which it is separated by a sandy plain, there were innumerable ravines and precipitous cliffs. The village of Russool itself was perched upon a high bluff surrounded by deep ravines, and was only approachable by a narrow causeway some eight feet in width. Sir Joseph, who, some days after the battle, went over the position, says that "the Sikhs had three lines of batteries to the east of it."

From Russool the line, passing by Lullianee, Kot Baloch, Fatehshah-ke-Chuk, and Moong, ended on the right at Luckneewala. Sher Singh himself, with the chief portion of his immediate followers, was about Moong and Kot Baloch, the Bunnu troops were on the right at Luckneewala, the Peshawar force was at Fatehshah-ke-Chuk, while the line of the Jhelum and the strongly entrenched and defended position about Russool—in rear of which latter boats had been collected, and preparations made for the construction of a bridge

—were held by large numbers of Sikh irregular troops. The whole country along the front of the position taken up by the Sikhs, for several miles to the south and east of the village of Chillianwala, was covered by dense thorny jungle, attaining in some places to the height of seven or eight feet; it was, however, rather less thick about Dinghi, whence, too, a frequented road led towards Russool. Near Moojeawala, and thence to Chillianwala, and for about half a mile beyond this village, the ground was fairly open.

On the evening of the 12th January Lord Gough called together his generals and brigadiers to receive his instructions for the morrow.

These officers were furnished with a rough plan of the enemy's position, put together from the various reports of the reconnaissances carried out by the cavalry while at Helah, and also with a statement of the strength of the Sikh forces. This latter paper, which appears to be in Colonel Grant's handwriting, contains the following information:

"*Memo. of the enemy's strength*, 12*th January.*

"At Luckneewala (the right of their position): the Bunnu troops under Ram Singh, one regiment of cavalry, four of infantry, and eleven guns.

"Fatehshah-ke-Chuk: Sirdar Atar Singh and Lal Singh, with two regiments of cavalry, six old and four new corps of infantry, and seventeen guns.

"At Lulliawala: Sher Singh, with one regiment of cavalry, five old and four new regiments of infantry, and twenty guns, with main body of Ghoorchurrahs, about 4,000.

"At Russool (the extreme left) are two new infantry corps, and seven guns.

"At Moong: Soorut Singh, with three guns in" (illegible).

At 7 a.m. on the 31st the army advanced towards Chillianwala, whence the Commander-in-Chief proposed to effect a reconnaissance, and elaborate his plans for attacking the Sikh centre, breaking through their entrenched line and separating the regular from the irregular troops.

"The army," says Sir Joseph, "advanced in brigade column of cavalry, artillery, and infantry towards the Sikh position. Each column formed its own advanced guard, and the heads of columns were to be 100 yards from each other. The right wing moved in column, left in front, and the left wing right in front, directed by the heavy battery, in front of which was the Commander-in-Chief."

A considerable detour was made to the right, partly in order to deceive the enemy, and partly to avoid the jungle; and on approaching the village of Chillianwala it was found that a low hill on the right of and overlooking the village was occupied by a Sikh outpost, composed of both cavalry and infantry, which was quickly dislodged.

The whole army advanced up to Chillianwala, and from the top of a house in the village, Lord Gough was enabled to obtain a fair view of the jungle-covered ground for some three miles to his front. It is true that he was prevented from discovering precisely the strength and limits of the enemy's position, or of accurately locating their guns; but he could see that Sher Singh had moved to the front out of his entrenchments. It was by now, however, close upon 2 o'clock, the jungle was very thick, and Lord Gough had resolved to postpone the attack until the next day, encamping for the night in the vicinity of Chillianwala, where there was abundant water. Preparations for laying out the camp were already in progress, when the Sikhs advanced some horse artillery guns, and opened fire upon the British outposts. To silence these, the heavy guns under Major Horsford were immediately sent forward to the front of the village of Chillianwala, whereupon the Sikh commander replied at once with nearly the whole of his artillery, thus exposing the extent and situation of his position, which hitherto the jungle had partially concealed. For half an hour a comparatively ineffectual fire was maintained by either side. It was by now 3 p.m.; to encamp was impossible, retreat was out of the question, and Lord Gough resolved upon an immediate attack.

CHAPTER XVII

In order properly to follow the course of a battle fought in thick jungle and of which the incidents are necessarily somewhat confused, it is as well that the arrangement of the divisions and brigades, and the position at the outset of the several arms should be clearly understood.

On the right of the British line, which was about a mile and a half in length, was the Second Cavalry Brigade under Brigadier Pope, C.B., of the 5th Light Cavalry—the 9th Lancers, the 14th Light Dragoons, the 1st and 6th Light Cavalry. With this brigade were the three troops of Horse Artillery commanded by Lane, Christie, and Huish—the whole under Colonel Grant.

On the left of the Second Cavalry Brigade was Gilbert's (the Second) Infantry Division, with Godby's brigade on the right and that of Mountain on the left, while between the two brigades was No. 17 Field Battery, under Dawes.

In the centre of the line and between the two infantry divisions were Horsford's two heavy batteries, each composed of four 18-pounders and two 8-inch howitzers.

Then came Colin Campbell's division, Pennycuick's brigade being on the right next to the heavy guns, while Hoggan was on the left. No. 5 Field Battery, under Mowatt, and half of No. 10 Field Battery, under Robertson, were attached to the Third Division, and Campbell accordingly placed Mowatt's guns between his brigades, and those of Robertson on the left of Hoggan.

On the left of the army were three troops of Horse Artillery, commanded respectively by Warner, Duncan, and Fordyce—the whole under Colonel Brind[1]—flanked by the First Cavalry Brigade, commanded by Brigadier White of the 3rd Light Dragoons, with whom was Sir Joseph Thackwell. The brigade was composed of the 3rd Light Dragoons and the 5th and 8th Light Cavalry.

The Reserve was formed of two native infantry regiments of Brigadier Penny's brigade, his remaining battalion, with two regiments of Irregular Horse and the remaining three guns of Robertson's battery, being, as has already been stated, placed in charge of the baggage, under Brigadier Hearsey.

The troops, having deployed, were ordered to lie down, whilst the heavy guns, supported by the field batteries with the infantry divisions, opened fire upon the Sikh centre, where the bulk of the enemy's artillery appeared to be posted, and after about an hour's cannonading—"about 2 o'clock," states Colin Campbell, while Sir Charles Gough says "3 p.m. or a little later"—the Third Division was directed to advance, and very shortly afterwards the same orders were conveyed to Sir

[1] Afterwards General Sir James Brind, G.C.B.

Walter Gilbert, the cavalry under Pope being posted to protect the right flank and support the movement.

The author of "The Narrative of the Second Seikh War" states that the Staff Officer who brought to Campbell the order to advance, directed him to "carry the guns in his front without delay at the point of the bayonet." Campbell himself makes no mention of this in his journal, but the officer who succeeded to the command of the Twenty-fourth, on the colonel being killed, has left it on record that Campbell particularly enjoined the regiment to advance without firing; while Lawrence-Archer, who was attached to the 24th Regiment, states that the Brigadier-General, addressing the men, said, "there must be no firing, the bayonet must do the work"— before riding off to the left to accompany the advance of Hoggan's brigade. In his report on the action, too, Campbell states that "the batteries were carried without a shot being fired by the regiment or a musquet taken from the shoulder," which Lord Gough describes as "an act of madness." Before the advance commenced, Campbell had made up his mind that the nature of the ground made it impossible that he could personally direct the attack of both the brigades of his division, and he accordingly decided to remain with his left brigade as being the one more liable to be outflanked. If this abrogation of his proper duties of a divisional commander can be justified—which may well be questioned from the fact that Gilbert, advancing through even thicker country, retained throughout his hold on both his brigades—it seems unfortunate that of the two, Campbell did not elect to move with Pennycuick's brigade, in front of which the ground seems to have been even more wooded than it was on the left; while the right brigade contained, moreover, a regiment which, although far stronger[1] than any other corps on the ground, was full of young, ardent soldiers just out from home, led by officers and commanded by a colonel to whom very many of them were complete strangers.

The three troops of horse artillery on the left moved forward under Colonel Brind and engaged the Sikh batteries directly in front. At the commencement of the advance White's cavalry brigade was much thrown back—this flank being thus refused—so that the further the forward movement of the general line was continued, the greater must the interval become between White's cavalry and the left of the Third Division. Sir Joseph Thackwell was fully alive to this danger, and had early despatched Lieutenant Tucker, his D.A.Q.M.G., to see that the distance was not permitted to become excessive. Brind's guns were firing on the Sikh batteries for some three-quarters of an hour, and had twice silenced them, the enemy returning to their guns directly the British fire showed any sign of slackening. On the extreme south of the enemy's position the Sikhs had assembled a large body of cavalry, and these horse-

[1] The Twenty-fourth—*pace* Macpherson—went into action exactly 1,100 strong, including officers.

men became more daring, and manœuvred with the intention of turning the British left; "but the eagle eye of Sir Joseph Thackwell," to quote the records of the 3rd Dragoons, "instantly detected their object," and he ordered a squadron of the 3rd Light Dragoons and the 5th Light Cavalry to charge, drive back the Sikh horsemen, and then endeavour to take the guns in flank, whilst a part of the remaining cavalry of the First Brigade attacked in front. The 5th Light Cavalry came upon a large mass of Sikhs, and were received with a heavy musketry fire, and, being much broken up by the thick jungle through which they had passed, were repulsed, but were quickly rallied on the 8th Light Cavalry.

The 4th (the grey) squadron of the 3rd Light Dragoons, under Captain Walter Unett, coming possibly on a somewhat smaller body of the enemy, broke through them. The following is Unett's account of this dashing charge:

"On returning from reporting myself to the officer commanding the 5th Light Cavalry, I saw the enemy's line, who just commenced to open fire at a distance of about seventy or eighty yards, at the edge of a low, thorny jungle. We were on low ground, but open to the enemy, had just passed through jungle with stunted trees, in line with the 5th Light Cavalry, the same as one regiment. Their bugle sounded the 'charge,' and we instantly started and drove through the centre of the enemy's *gole* at the utmost speed as the formation of the enemy and nature of the ground where they stood would admit of. They closed in on our flank, and it was *pell-mell*. In one minute

I had received three blows from different men on my left when engaged on my right front, viz. a sword-cut from the right side of the neck to the left shoulder, a spear-wound on the right side, and a blow with a matchlock across the loins. We then got pretty clear of the enemy, and I killed a Goorchurrah at least 900 yards (upwards of half a mile) in rear of their *gole*. I then found myself entirely alone, and the first man I saw was Private Galloway. We rallied a few more, as we were dreadfully broken up, and instantly charged back through the enemy to our old ground. They did not offer so much opposition, but opened out, and abused us as we passed. I mention this little circumstance to show that we were (or many of us) considerably in rear of the enemy's *gole*—I have always thought a mile. We could not, from the casualties, confusion, and jungle, all meet and come back at the same time, but we were pretty near the three officers with parties. I think this caused greater panic and dismay to the enemy, for they could not tell at what point we were coming through them, and on our re-forming they retired off the field. As to the 5th Light Cavalry, we never saw anything of them, and by Captain Wheatley's official report they were 'repulsed and driven back.' Our loss in killed and wounded will pretty well show where we charged: out of my squadron of 106 men, including 3 officers, we had 23 killed and 17 wounded—total 40,[1] with 18 horses killed and 8 wounded (including one officer's horse)—total 26, and on re-forming we had only 48 men in their saddles."

[1] Captain Unett does not mention that of the three officers present two were wounded. The other subaltern was Lieutenant Stisted, son of the Colonel Stisted with whom Sir Joseph had exchanged to the 3rd Light Dragoons in 1837.

The account of Lieutenant MacQueen, who commanded the right troop of the squadron in the charge, differs slightly in some particulars. He says:

"Soon after the enemy's guns had opened on our brigade, Colonel Yerbury received an order to send a squadron to join the 5th Light Cavalry to assist in driving back a body of the enemy's horse who were at that time threatening our flank. . . . At this time I could only see the tops of the spears which some of the Goorchurrahs carried; but after crossing the low ground and ascending the short slope on the opposite side, we came in full view of the enemy at about 350 yards to our front. They appeared to be about 800 strong, and they stood in line, but were formed into two bodies by a small interval which separated them. As soon as we came in full view of them the trumpet sounded the 'gallop,' and we went on as steadily as the jungle through which we were passing would admit of, until we came to within 60 or 70 yards of them, when they opened a heavy fire of musketry on us. At this moment the order to charge was given, and as the left body of the Goorchurrahs stood exactly perpendicular to our front, the charge of our squadron was made right through the centre of that body.

"The squadron was rallied again at about 300 yards to the rear of where the enemy stood, and then, but not till then, I perceived that the Fifth had not charged with us, and that we were alone. At the same time I also observed that the Goorchurrahs had re-formed their line on nearly the same ground on which we had just charged through them. They seemed merely to have countermarched, and were apparently again waiting to receive our attack. We therefore again advanced towards them, but this time, instead of receiving our charge as before, after firing a few shots they opened out and got behind the trees and bushes in the jungle, and as we passed through them they wheeled their horses round, and came on pell-mell with us, nearly to the ground where we first joined the 5th Cavalry and commenced our first advance. Here we again rallied the squadron, when the Goorchurrahs who had followed us turned and fled. The squadron went to join the regiment, which had gone off nearly a mile to the right."

It is now time to see what had necessitated the movement to the right by the Cavalry Brigade immediately under the orders of Sir Joseph Thackwell.

Colin Campbell states in his journal that the Staff Officer who directed him to advance, told him he would be supported by Brind's guns on the left, but Colonel Brind does not appear to have considered that he was in any way at Campbell's disposal, for in a letter to Sir Joseph he mentions having afterwards heard that some assertion of this nature had been made, and pertinently asks, "If my guns were under his orders, why did he not exercise some control ?"

Campbell's two brigades appear to have advanced simultaneously, but became almost at once separated, and never saw one another after the order to advance was given. Mowatt, with the battery between the two brigades, was to cover the advance of Pennycuick, and moved forward in line with the skirmishers of that brigade; but the advance was pushed so fast by Pennycuick that the front of Mowatt's guns would appear to have been quickly masked. Lawrence-Archer, who was with the 24th Regiment in the centre of this brigade, says that ground was more than once taken to the right, a procedure which, in such thick jungle as covered the front, naturally tended to remove the brigade farther from the infantry on the left, and from the supporting guns.

"So close was the undergrowth, so numerous the thorny trees and bushes, that not only was touch lost between brigades, but even between the regiments of each brigade: the maintenance of connection was hardly possible. By degrees the line of infantry, except on our left, where obstacles were fewer, resolved themselves into companies and groups of scrambling soldiers, none able to see clearly twenty yards in any direction, all pressing forward as fast as their weary frames and nervous tension permitted, toward the Sikh batteries in their front. Trees and scrub were full of the enemy's sharpshooters."[1]

And throughout this rapid movement the men under Pennycuick, and especially the 24th Regiment, were pounded by the heavy guns of the enemy; but, rushing on without firing, the Twenty-fourth arrived breathless at the guns, in confused bodies, and captured them at the point of the bayonet. Their loss, however, had been heavy. Three field officers and five company commanders, besides other officers, had been killed or wounded, and the casualties among the other ranks were many; so that when, unsupported, exhausted by their

[1] Thorburn.

exertions and the rapid advance, they were received by an overwhelming fire from the Sikh infantry in rear of the guns, the remnant of the Twenty-fourth fell back upon the village whence it had advanced, the native infantry on either flank giving way at the same time, and the whole being followed up by the Sikh cavalry, until the advance of the brigades on the left and right caused the exultant horsemen to withdraw.

In the meantime, Hoggan's brigade, admirably led by Campbell in person, was greatly assisted in its advance by Mowatt's guns—when no longer able to cover Pennycuick's movements—by the Horse Artillery, under Brind, and by Robertson's three guns, which engaged and silenced the heavy battery of the Sikhs, which otherwise would have enfiladed the left brigade during its advance. Neither in his journal nor in his report of the action does Sir Colin give any especial credit to the artillery on his either flank for the success which his advance achieved and merited, and which he thus describes:

"Although the jungle through which the 7th Brigade passed was close and thick, causing frequent breaks to be made in the line, yet by regulating the pace so as to make allowance for these obstructions, the left brigade, after an advance of half a mile, reached a comparatively open tract of country in a tolerably connected line. On this open tract we found formed in our front a large body of cavalry, and regular Sikh infantry which had played upon us during our advance.

"H.M.'s 61st Regiment charged this cavalry,

and put it to an immediate and disorderly flight, while the 36th Native Infantry on the right made an attack upon their infantry, which, however, was not successful, and in consequence they came down, accompanied by two guns, upon the 36th Regiment, obliging it to retreat in rear of H.M.'s Sixty-first.

"The two right companies of the Sixty-first were instantly made to change front to the right, and while the remainder of the regiment was ordered to form rapidly in the same direction, the two right companies charged the two guns, and captured them. The fire of these two companies upon the enemy who were in pursuit of the Thirty-sixth compelled them to desist and retreat.

"While the remainder of H.M.'s Sixty-first was forming on those two companies, the enemy brought forward two more guns and fresh infantry, upon which those who had desisted from pursuit of the Thirty-sixth again formed, and the whole opened a heavy fire; this force was likewise charged by H.M.'s 61st Regiment, put to the rout, and the guns captured. At the same time the 45th Native Infantry, in its movement to form on the left of H.M.'s Sixty-first, was attacked by a large body of the enemy's cavalry, which it gallantly repulsed.

"The formation of the brigade on the flank of the enemy's line now being completed, it moved forward, driving everything before it, capturing in all thirteen guns, until it met with Brigadier Mountain's brigade advancing from the opposite direction. The enemy retreated upon their guns, which were in position along their line in twos and threes, which they defended to the last moment in succession, and were only obtained possession of by us after a sharp contest, such as I have described in the capture of the first two guns, and they were all charged and taken by H.M.'s 61st Regiment. During these operations we were on two or three occasions threatened by the enemy's cavalry on our flank and rear, and were obliged to face about and drive them off. The guns were all spiked, but having no means with the force to remove them, and it being too small to admit of any portion being withdrawn for their protection, they were, with the exception of the last three taken, unavoidably left upon the field."

In a letter written by General Mountain some years afterwards to Sir Joseph Thackwell, speaking of this period of the battle, Mountain says:

"We cleared the jungle, took all the guns in our front, and silenced all opposition, and, having done this, were then sent to our left and rear to the assistance of Sir Colin Campbell, who was threatened by the enemy, and had asked for a reinforcement."

On finding how sensibly the interval was increasing between himself and the Third Division, owing to Campbell's movement to the right, Sir Joseph Thackwell had sent Warner's troop of horse artillery, escorted by a squadron of the 8th Light Cavalry, to join and support Hoggan's brigade, and the rest of the cavalry brigade was finally moved bodily to the right, and formed up near Campbell and Mountain, opposite the centre of the Sikh position.

Gilbert's division, its front covered by skirmishers, had advanced steadily through the jungle in the order already previously described, but its forward movement was checked and endangered by the action of Pope's cavalry. The Commander of the Second Cavalry Brigade, seeing a large body of the enemy's horsemen on the slopes of the hills near Russool, detached, to protect his right flank, a force under Colonel Lane of the Horse Artillery, consisting of a wing each of the 9th Lancers, the 1st and 6th Light Cavalry, Lane's own guns and two of Christie's. There is but little to chronicle in regard to Lane's further share in the events of the day, for he appears to have become completely separated from the remainder of the cavalry; he was given no further orders by Pope throughout the action, he was unable to afford any assistance to his Brigadier, and, although his guns did open fire and check some of the enemy threatening his own flank, he practically took but little further part in the battle, and appears to have remained in comparative ignorance of all that transpired on his immediate left.

In regard to what occurred under Pope, it has in many accounts been stated that the Brigadier, seeing a body of Sikh Horse in front of his line, ordered the charge, but the evidence of the 9th Lancers and the 14th Light Dragoons—of Hope Grant, Thompson, and Chetwynd—does not agree with this. A few of the enemy's cavalry had certainly been noticed in front, and the guns were pushed forward—Pope, while the artillery were coming into action, bringing on his brigade in one long line at a trot. There were no scouts or skirmishers in front, and no support of any sort or kind was formed in rear. The line advanced in the following order: the two squadrons of the 9th Lancers were on the extreme right, on their left were three troops of the 1st Light Cavalry, then three troops of the 6th Light Cavalry, while on the left of the line were the four squadrons of the 14th Light Dragoons; beyond these again were the guns. The Brigadier led the line in front of the native cavalry, by the centre of which the four regiments were dressing and regulating their pace. The cavalry had already passed beyond the line of the guns, and had to some extent covered their front, thus masking their fire, when—according to the Commander of the second squadron of the 14th—the pace dropped to a walk, and then the whole line came to a halt at the same moment that some Sikh horsemen became visible in the jungle immediately in front.

Hope Grant, who commanded the two squadrons of the 9th Lancers, stated in his report that his men—

"were proceeding on steadily and changing their direction a little to their left, when the native cavalry began to cheer and charge. I confess at the time I could see no enemy except a party of about fifty horsemen a good deal to our right flank, which . . . I took to be some of our own Irregular Horse."

Several explanations have been put forward to account for the rearward movement. Durand, in the *Calcutta Review*, mentions that some of the sowars of the native cavalry afterwards told Brigadier Godby, that their British officers were fifty or sixty yards in front of the line when the enemy's horse were sighted, and were then ordered back to their squadrons. That they came galloping back at the moment when the cry of "Threes About" was raised, and so unwittingly helped to strengthen the impression that the order was given by authority. But it seems probable that what then really occurred was that the Brigadier (of whom it is asserted that he was not wounded in the advance, as several writers have stated, but during the retirement), recognising that his forward movement had masked the fire of his own guns and wishing to repair the error, gave, or intended to give, the command "Threes *Right*," so as to clear the front; but at this moment a close body of some forty or fifty Sikh horsemen charged hotly into the halted squadrons of native cavalry, when the repeated word of command, as then passed on, became "Threes *About*." That considerable confusion and a certain amount of individual panic ensued is admitted by Hope Grant and other leaders; some of the guns which had been left in rear of the cavalry were ridden over, and the Sikh sowars, following up, cut down several of the gunners and captured four guns, two waggons, and many horses. It has in some quarters been represented that the whole cavalry brigade fell victims to an unreasoning panic and that there was a general reckless stampede to the rear, but it would seem that this was very far from being the case. The testimony of some of the artillery officers—the corps which suffered most—describes the retreat as more of the nature of the ordinary temporary retirement of a parade movement, that it was "deliberate," that "the men rode slowly to the rear"; while at any rate *some* of the squadrons retired in good order and at no faster pace than a slow trot. The fatal mistake was that there was no support in rear of the line—nothing upon which the cavalry could re-form: there was no common halting-point.

The story of this unhappy occurrence has been handed down to posterity as one in which British and native cavalry, actually committed to a charge, suddenly went about and fled panic-stricken from the field. It seems clear, however, that no charge was either ordered or made, neither the "Gallop" nor the "Charge" was sounded, nor was there any present need for either; but the going about at an unfortunate moment in obedience to an imaginary order, induced among individuals of the four regiments a panic, easily communicable in retirement, and to which the broken, wooded nature of the ground—causing separation rather than assisting re-formation—easily lent itself.

Sir Joseph Thackwell greatly regretted that he had placed himself on the left; he had, as mentioned in his report on the action, intended to join the right column of cavalry, but it being stated that the enemy seemed to be making a movement to the left rear of the British, the Commander of the cavalry division concluded that at least an equally large body of the enemy's horse would be on the left as was present on the right. Sir Joseph was President of a Court of Inquiry convened to inquire into the conduct of the cavalry of the Second Brigade, and the verdict of the veteran cavalry commander, who had probably at that time seen more service with the mounted arm than any dragoon then serving in any army, may well be accepted as final.

"I feel assured," he said, "from what I have heard of Brigadier Pope and the conduct of the cavalry of the right, that their retrograde movement originated more from mistake, than a fear of encountering an insignificant enemy."

The untoward incident above described occurred just as Gilbert was leading his division to the attack, and he, finding his right uncovered by the retirement of the cavalry, was obliged to refuse his right brigade (Godby's), to some extent, so as to protect his flank, while he advanced on the village between Kot Baloch and Tupai. During the whole of this advance Dawes' field battery was of the very greatest service. Covered by the fire of these guns, the brigade under Mountain advanced with great steadiness upon a strong battery of the enemy, charged and captured the guns, while Godby's brigade also carried the Sikh guns in their front. The flank of this brigade was now turned by the enemy and it was attacked on three sides, but the onslaught was splendidly met, and Dawes, moving to the right, poured in a heavy fire and assisted to disperse the enemy. The Reserve, under Penny, had been ordered up by Lord Gough and, coming into action, was at once hotly engaged on Godby's right; and Campbell's advance being now visible on the left, Hoggan's and Mountain's brigades wheeled inwards and the position was at last taken, and the Sikhs driven from the field. They fled towards Russool and across Lane's front, and he, now opening a heavy fire of grape, completely dispersed them.

As darkness fell upon the field the British were in possession of the whole of the Sikh line and the enemy had been driven back upon the Jhelum, taking with them the four guns captured on the right, but leaving thirteen of their own in our hands and several more spiked and useless on the ground.

It was eight o'clock before the majority of the wounded had been brought in, and then, there being no water on the field, the British withdrew to the village of Chillianwala for the night—an unfortunate necessity, since, under cover of the darkness, parties of Sikhs returned to the deserted battlefield and carried off nearly all the guns which had not been brought into camp that evening.

The total loss of the British force was 22 officers, 16 native officers and 659 men killed or missing, while the wounded included 67 British officers, 27 native officers and 1,547 rank and file, or a total killed and wounded of 2,338—roughly, 17 per cent. of the whole force engaged. The loss of the enemy was never accurately estimated, but there is no doubt that they too suffered very severely.

Durand wrote in the *Calcutta Review* of General Thackwell at Chillianwala that—

"he did very great service on that memorable day by maintaining an imposing front, working Brind's

20

guns to advantage, and showing, by the gallant Unett's daring charge, that Atar Singh's advance from his ground without the support of his batteries of position would meet with no respect from those ready horsemen, and that, once in motion, the Sikh chief might look for rough handling from the 3rd Dragoons and their native comrades. Thackwell acted wisely, cautiously, and firmly."

CHAPTER XVIII

As was perhaps only to be expected, the generalship of Lord Gough on the 13th and the published details of this attack upon a difficult position over broken ground, induced a storm of criticism in the Press, not only in England but in India, and there were few soldiers of any eminence, who had taken part in the campaign, who did not find themselves the mark of anonymous scribblers. It has well been said that—

"in the days of the Lawrences, and even later, Government servants who had the gift were openly Press correspondents or leader-writers, and in their literary efforts not only criticised the proceedings of individual officers, but even of departments and the Government itself. In some cases they edited newspapers, and owed their advancement in the service to the freedom and ability with which they used their pens."[1]

Among those who were attacked was Colin Campbell, and to him the criticism with which he was assailed and the mistakes which were attributed to him in the conduct of his division at Chillianwala, seem to have given more annoyance than similar irresponsible vapourings gave to his contemporaries,

[1] Thorburn.

or to Lord Gough himself. The strategy of the campaign and the tactics of Chillianwala were discussed by nobody with greater freedom than by the editor of the *Mofussilite*, a newspaper published at Agra. This writer was John Lang, a barrister and a journalistic freelance, better known perhaps for his celebrated and successful defence of Jotee Pershad, the notorious army contractor.

Most soldiers followed the example of Lord Gough and took no notice of newspaper comment, but Colin Campbell seems to have been especially susceptible to criticism of this kind, and on the subject of the inadequate use made of his guns and the small artillery support given to his division at Chillianwala—in regard to which he seems to have been blamed in the columns of the *Mofussilite*—he had some correspondence with Sir Joseph Thackwell, from which extracts will now be given.

On the 26th July Campbell wrote to Sir Joseph:

"I take the liberty of enclosing for your perusal a letter I have received from Lieutenant Robertson of the artillery, who was posted with one half—three guns—of No. 10 Field Battery on the left of the Third Division of infantry when formed in line on the left of the village of Chillianwala on the 13th of January, on its being ordered by the Commander-in-Chief to advance in that order to the attack of the enemy in position on the other side of the jungle immediately in its front. Lieutenant Robertson states that he accompanied the Third Division for a short distance, when he was ordered by a staff officer, whom he cannot recollect, to move with his three guns to his left to assist the horse artillery, the three troops of which arm in that direction were attached to the cavalry on the left of the line under your immediate personal command on that day.

"Will you do me the favour to inform me if the order received by Lieutenant Robertson to move to his left with his three guns to assist the horse artillery was given by you? And if not, if you are aware by whom the order was given to Lieutenant Robertson?

"It could not have been given by an officer of the Headquarter Staff, because Lord Gough, previous to the advance of the division, sent two officers—Lieutenant Bagot, A.D.C., and Lieutenant Johnstone, Deputy Judge Advocate—to inform me that he had desired these officers to convey his wishes to the horse artillery on the left to accompany and assist my attack."

In reply to the above Sir Joseph, writing on the 8th August, says:

"I have read the whole of Lieutenant Robertson's letter, and he has given me the first intimation I ever received of his three guns having acted on the flank of the enemy's battery opposed to Lieutenant-Colonel Brind's guns. This half battery and No. 5 Field Battery were attached to your division, and Lieutenant Robertson should not have received any orders except through you. Now, I have much pleasure in answering your question whether the order for Lieutenant Robertson to move to the left to assist Colonel Brind was given by me; to this I say no such order was given or even contemplated by me. I had my own guns and you had yours, nor did I ever hear of such an order having been given by anybody. Nor, as I before observed, was I aware that any guns fired upon the enemy opposed to Lieutenant-Colonel Brind, except his own, until the arrival of Lieutenant Robertson's letter. In my advance the order to Lieutenant-Colonel Brind was to keep his battery within about a hundred yards of your left. He, however, got to a greater distance, owing to the jungle and the enemy's right opening their guns upon him; but I do not think he was more than four hundred yards from your left at that time. Of course the gap became great when your left was obliged to close to the right. Neither Lieutenant Bagot nor Lieutenant Johnstone ever came to me to give any orders from the Commander-in-Chief respecting Brind's guns; and if he had received such an order from either of these officers, I think he would have reported it to me, as I was with the cavalry on his left. When the 5th Brigade failed, Brigadier Brooke[1] came to say the left wing was to close to the right, and this was the first intimation that anything was wrong which came to my knowledge, as I contemplated turning the enemy's right flank. Warner's troop of horse artillery joined your left and Moone's squadron of the 8th Light Cavalry. Whilst I was with this party, White took his brigade towards the centre, by, I believe, the Chief's orders, and your flank was greatly exposed."

It would seem that Sir Joseph must have written a further letter to Sir Colin telling him that he was making inquiries among the different members of his staff about the order given to Robertson, for, writing again on the 15th November, Campbell says:

"As you told me in your former letter that the order referred to by Lieutenant Robertson did not emanate from you, I am sorry you troubled yourself with writing to your staff. Of the fact, however, that Lieutenant Robertson did get an order to leave his place in his own division and proceed

[1] Commanding Horse Artillery.

to the assistance of Colonel Brind, who with his eighteen horse artillery guns was opposed to some eight or ten of the enemy, and that I was consequently deprived of his aid when I most needed him on my own flank, there is no doubt whatever. There is no mystery on that head. Some irresponsible person took it upon himself to order Lieutenant Robertson to join Colonel Brind, on which duty the former remained until the few guns opposed to the latter were silenced; after this Colonel Brind communicated with Lieutenant Robertson through his Brigade Major, Captain Tucker, Deputy Assistant Quartermaster-General, having been in the Lieutenant's battery when he was engaged in enfilading the enemy's guns above mentioned. Thus was my division robbed of its light guns, which reinforced the main body of that arm under Colonel Brind, without any information being given to me at the time or afterwards, or any report from the Commanding Officer of your artillery or your Staff Officer being made to you, to whose division they went. Both you and I have to complain of these gentlemen on this account. I will not permit myself to enlarge on the conduct or motives of those, whosoever they may be, who have concealed the part they took in giving this order, and on the consequent injury they did to my division and to me personally. It is principally on that account that I have been misrepresented, not only in the public prints, but in the highest and most influential quarters.

"You dwell on the fact that in your opinion, and in that of your staff, the estimate of distance between your division and mine made by Lieutenant Robertson is excessive. Allow me to say, my dear General, that this has nothing to do with the matter. There was a *large interval* between us in a *thick* jungle—and I was as much separated from you therefore, *de facto*, as if you had been in Calcutta. Your own appreciation of this, you having been in the field with your own division, shows how little is known accurately of the movements of mine on the 13th January. Owing to this ignorance on the part of the public, I have had much to endure. With others whom I do not know, and with anonymous writers, I have no concern. But I cannot let this opportunity pass without setting you right on the absolute state of isolation in which I was left, till, after a long action, and having traversed a wide space, I finally rolled up the enemy's line from the point of attack of my left brigade to that of my junction with Mountain's force belonging to the right wing of the army, my own right brigade (or Pennycuick's) having met with a disaster of which I was not cognisant till near the end of the day, as you will see by what follows."

There is then given a long account of the movements of Hoggan's brigade led by Campbell, prefaced by the statement that when ordered to advance he at the same time received two messages from Lord Gough, to the effect that his left flank would be supported by Brind's guns; that his Brigade Major, Keiller, was sent to find the horse artillery, but failed to do so; and that Campbell therefore had to advance at once, not knowing where these supporting guns were. At the same time, apparently, Robertson left him in obedience to the orders of the mysterious staff officer, and did not rejoin the Third Division until Campbell had effected his junction with Gilbert's left brigade. Campbell concludes his letter to Sir Joseph by saying:

"Such, my dear General, is the true relation of the manner in which the left brigade of my division was left to fight its way, without any portion of the artillery support promised by Lord Gough, being at the same time divested of its own guns, as shown above. I think you will now agree with me that the estimate of distance between your right and my left, to a hundred yards or so, is not of any consequence—the separation of our two divisions was absolute; combination and its results were precluded."

By this time General Thackwell had received communications from Brind, Tucker, and Lieutenant Thackwell, to all of whom he had written.

Colonel Brind, who had evidently also heard from Campbell—for he speaks of "letters from Peshawar"—has a good deal to say on the subject.

"The infantry," he writes: "moved from Chillianwala a little before us. We were detained by a staff officer to wait orders, which came, and we moved on a few hundred yards, perhaps two hundred, after the infantry. We did not move very far when the enemy saw us, being mounted, and fired. I continued advancing, and inclined a little to the left to get a good position, and then opened fire. Our position was thus."

Sikh Guns

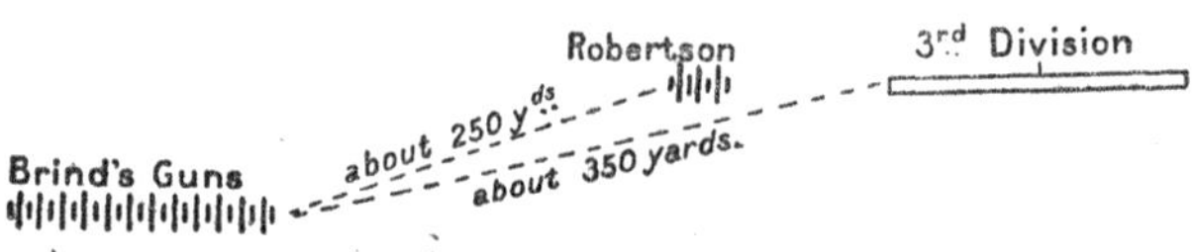

"At the time I came into action on the 13th January, I do not think the left of the Third Division was more than 300 or 400 yards from me, and Robertson's three guns were, so it appeared to me, in line with their infantry, and came into action about the same time that we did, being about 250 yards to our right and in advance of us. I soon after lost sight of the infantry, who appeared, while we were engaging the enemy's batteries, to take ground towards their right flank, leaving a considerable gap—how great I cannot specify. A few hundred yards in a jungle seems a great distance, and I should doubt that the Third Division was at any time 1,200 yards from their guns, judging from the extent of the field and our subsequent position at the centre of the line. Robertson must have received orders, or he would not, I am certain, have left the line he was attached to. In my belief, he merely halted during the advance, but did not take ground to the left: had he done so he would have got in front of my guns. I am not aware who gave him the order said to have been given him by some staff officer. He was not very long in action before he was called away to rejoin his own infantry. His guns did good service where they were, as they brought a cross fire on the enemy's guns—their strongest battery, I believe—while we had a direct fire on these guns. Robertson never joined me or was under my orders. He appeared to be firing at the same guns we were engaged with, and I remember sending Bruce to point out that if he fired more obliquely he would do more good, as we could only bring a direct fire on these guns. After Bruce came back I observed the disappearance of the infantry."

"It struck me," says Brind elsewhere, "that Robertson, not getting any orders from Campbell, naturally fired at the guns which fired at him. I fancy Campbell ran on without giving any orders to his guns."

In another communication to Sir Joseph Thackwell, Colonel Brind wrote:

"On the 13th January he" (Campbell) "did not exercise the command over more than his left brigade; he gave no instructions to the artillery attached to his division. If my guns were of no use to him, they were to the field, and, I think, saved his left brigade from being cut to pieces. I had just sent Bruce to Robertson to point out he had better fire across our front, if he had no other orders, for his direct fire was lost, when the M.B. (Keiller), Bruce says, came up and took the guns away. Who ordered him to fire at all, I don't know—the mysterious staff officer! I think from 16 to 18 guns opened on us. I was not disregarding Campbell's infantry, though not under his orders, or in any way able to act otherwise than I did."

There is also extant a letter from Tudor Tucker, the Deputy Assistant Quartermaster-General, wherein he says:

"Quite at the commencement of the day, before Brind's horse artillery was in position, you ordered me to see that we did not get too far from the line of infantry, because being thrown back at an angle, every forward movement, either of theirs or ours, increased considerably the interval from flank to flank. I galloped by the rear across the angle to their left, and returned by Robertson's battery and Brind's three troops of horse artillery, to the front of the line of cavalry, where I found you. Robertson's battery was getting into action as I came up to it. He had two guns and a howitzer. He had moved from the infantry towards Brind's guns, and was nearer to the latter than to the former when he opened fire, but not fronting the same way. Brind fronted with the cavalry, Robertson with the infantry. I spoke to Robertson and stayed with him while he fired several rounds. His howitzer did great execution on the Sikh horsemen—they were in masses when he first opened, and he soon dispersed them. The [our] infantry advanced while I was with Robertson. I did not ask Robertson why he was where I found him, nor under whose orders he was; I knew he was attached to an infantry division, and concluded he had received his orders from his General. I beg you to believe that *I gave him no orders of any sort.* The distance between the infantry and your division was widening very much, owing to the advance of the infantry."

Lieutenant Thackwell, A.D.C., also denied having given any orders whatever to Lieutenant Robertson.

Having received the above letters, Sir Joseph replied to Colin Campbell's letter of the 25th November, thanking him for his—

"long account of the battle of Chillianwala, though I must differ with you in some of its conclusions. I was fully aware of the gallantry with which Hoggan's brigade, under your able direction, rolled up the enemy after breaking up their line, and that it was left quite to its own exertions for some time, and that its noble bearing repulsed all the attacks of the enemy upon it; but I cannot agree with you that Brind's guns and the first brigade of cavalry were of no use to you, and might as well have been at Calcutta, for this force had the whole of Atar Singh's troops, which had been entrenched at Fatehshah-ke-Chuk, opposed to it, composed of 16 or 18 guns and a large body of cavalry and infantry. Lord Gough's information, given on the 12th January, stated their artillery to be 17 guns—consequently, had all my troops been away you would have had some" (? several) "thousand men with the above-mentioned guns on your flank and rear, and you might then have wished Brind's guns and White's cavalry to have been near you. Lieutenant-Colonel Brind and all my staff, as I have before stated, positively deny having given any orders to Lieutenant Robertson. For my part, I consider him not free from blame in having left your division unsanctioned by you. Lieutenant-Colonel Brind's guns were attached to the first brigade of cavalry and under my orders, and were intended to act on your left flank, and they did so most effectually, according to my opinion. But when you took ground to the right in rolling up the enemy's line, of course the gap between us became greater every instant, until, when I joined you, it was very considerable. When your movement was discovered, Brind's guns and White's brigade took ground to the right, but two troops of artillery and all the cavalry, save one squadron, went to the other side of Chillianwala, without my knowledge, by an order brought by Brigadier Brooke, as was said, from the Commander-in-Chief."

"I do not believe," says Sir Joseph in another communication, "they were the Bunnu troops in this position, but those of Atar Singh, who was entrenched in front of Fatehshah-ke-Chuk, and his works had 17 or 18 gun embrasures in them, which I counted on the 15th January[1] and traced

[1] I think the 19th is meant, from an entry in the diary.—H. C. W.

the wheels of the guns from the entrenchment to where they formed the right Sikh battery opposed to Brind."

Robertson, in his much-referred-to letter, wrote to Campbell:

"I received the following order: 'Take your guns more to the left and assist the horse artillery in silencing those guns,' from a staff officer who rode up to me, when I had advanced with the line about 200 yards, if so much. The order was given to me in a very distinct tone, and although Heath" (his brother subaltern) "and myself have frequently since endeavoured to recall to our recollection the appearance of the officer, we have failed. It was conveyed with such an air of authority, and its nature at such a time was such, that it never entered my judgment to question its authenticity, the more especially as, having called in question the authenticity of an order received that same morning, which deprived me of the use of three of my guns, I found myself mistaken when appearances justified my suspicions. I accordingly trotted out to the left—probably 500 yards. Tucker of the Quartermaster-General's department, must, I think, be accurately acquainted with my position there as relates to your left, for he was in the battery when I opened fire at this point; certain I am that we could not have seen each other. I dispersed a considerable body of horsemen here, and after having fired about 20 rounds, I limbered up and proceeded in the direction indicated by the mysterious staff officer as the whereabouts of the horse artillery. A shot or two, evidently fired either at the infantry of your left or your right brigade, informed me of the position of the enemy at last, and as these shots hopped along my front, I brought up my right shoulders and, unlimbering, found myself opposed to a string of guns which were busily engaged directly with the horse artillery, but occasionally wheeled round a gun in acknowledgment of my attacks on their left flank, which, being unprotected, enabled me to enfilade the whole string of them. I think we were engaged here half an hour, and then the enemy's fire slackened and ceased. Bruce, the adjutant of the artillery, rode up from the horse artillery (which I never saw, though I knew its position) and told me that my fire had been of great service to them, that Brind was about to advance and wished to know what I intended to do. Having done as I was ordered, I said I would rejoin my own division. When we limbered up we heard the rattle of the musketry on the left, as we guessed, at 1,200 yards' distance."

At the time of the opening of this campaign Campbell had had but little or no experience of the command of a brigade in the field, and it was only since Ramnuggar that he had held charge of a division. At Chillianwala nothing could have been more admirable than his leading of Hoggan's brigade; but it is at least open to question whether, by thus confining himself to the duties of a brigadier, he can be absolved from some responsibility for the disaster which overtook one half of his division. His action prevented his devoting his whole attention to the proper combination of the component parts of his division; the too rapid advance of his right brigade precluded its due support by the field battery specially told off to accompany it; while the withdrawal of the three field guns intended to accompany Hoggan, could hardly have escaped the notice of a divisional leader duly exercising the ordinary functions of his command. Campbell was certainly deprived during a considerable part of the action of the immediate support of the three guns told off to accompany Hoggan, but he makes but slight mention of the very great obligations due by this brigade to the six guns of Mowatt, when the hurried advance of the right brigade prevented Mowatt from giving Pennycuick the intended support, and obliged that artillery officer to turn his guns upon the enemy in Campbell's front.

That Hoggan's brigade, led by Campbell, experienced great opposition there is no doubt; but that is no reason for minimising the assistance indirectly given to the troops under Campbell's immediate command, through the pressure exercised upon the remainder of the enemy's line by the units on the right and left of the brigade, or for belittling the strength of the hostile forces with which other commanders had to contend.

The British attacks upon the Sikh line were to all intents and purposes simultaneous, and although contact was not established at precisely the same moment at every point, it is evident that the Sikhs could nowhere and at no time concentrate overpowering forces upon any one portion of the British attack, even when certain units of the advancing line had suffered repulse or even disaster. The defeated units were still there; they had not been driven from the field; they were still to reckon with; and consequently the force of the enemy immediately opposed to Campbell's left brigade was never overpoweringly greater than he could, from the outset, have anticipated. The continued presence in the field of the defeated British units must have exercised a certain deterrent effect upon the Sikhs, in so far as concerns the permanent reinforcement of any other part of their extended line, already rather over-strained for their numbers.

In the map accompanying Sir Colin Campbell's account of the battle, to be found in Shadwell's "Life of Lord Clyde," as, indeed, in almost every known plan of the battle of Chillianwala, it appears that while the British right was practically opposite the left of the Sikh position at Russool, the southernmost flank of the enemy very greatly overlapped Lord Gough's left, and consequently the British cavalry and horse artillery on this flank were not only opposed to the Sikhs immediately in their front, but to those who were able to menace their flank and even their rear. Malleson says that the cavalry and horse artillery on the left "had performed the great service of keeping in check the centre right division of the enemy." Sir Charles Gough has described how Brind's horse artillery came into action against "a powerful battery of Sikh guns." The evidence of Brind himself, and of Sir Joseph, is in agreement as to the number of guns composing this battery; and from Brind's sketch, and from Robertson's statement that, even while heavily engaged with our artillery, the Sikhs were able occasionally to throw a round shot at Campbell's infantry, it is plain how greatly Hoggan would have suffered during his advance, had the horse artillery of the left not taken some of the pressure off him on to their own shoulders.

Sir Colin himself does not seem quite clear as to what extent, if any, Brind's horse artillery was at his disposal, for he speaks indiscriminately of having been informed that the horse artillery "was to accompany and assist" his advance, of the "artillery support promised by Lord Gough," and of having been told that his "left flank would be supported by Brind's guns"; but from the general tone of his complaint it appears that he expected the horse artillery was intended to advance with him, engaging any part of the enemy's force to which the Third Division—or that portion of it led by Campbell himself—was directly opposed. It seems clear that this view of the proposed action of the artillery cannot be justified; and that if the two young staff officers quoted gave Sir Colin that impression, they were not correctly communicating the intentions of the Commander-in-Chief, who, in his dispatch of the 16th January, when describing his order of battle, merely says that "Brigadier-General Campbell's division formed the left, *flanked* by Brigadier White's brigade of cavalry and three troops of horse artillery under Lieutenant-Colonel Brind." No orders reached either Sir Joseph or Colonel Brind from Lord Gough respecting any fresh disposition of the horse artillery, nor did Campbell himself send any orders or message to Brind which could lead that officer to suppose that he was to *accompany* the advance of any other arm than that with which horse artillery usually act.

Brind is quite correct when he says that he was "not disregarding Campbell's infantry," and there can be no doubt that he afforded very material support to the Third Division, to which, moreover, Sir Joseph at once dispatched a troop of horse artillery when he found how wide the interval had become, and how greatly Hoggan's flank was exposed by the continued movement of the infantry to the right.

Splendidly as Sir Colin Campbell led the left brigade at Chillianwala, and greatly as his reputation as a leader in action was thereby deservedly enhanced, it must be agreed that for the "absolute state of isolation" in which his brigade found itself, he himself must accept the sole responsibility, which cannot in fairness be transferred to the shoulders of any other individual.

CHAPTER XIX

THE Sikhs had withdrawn after the battle of Chillianwala to Tupai, on the Jhelum, and it has been generally admitted that had the British been able to advance on the morning of the 14th the enemy might well have been driven into the river. That night, however, and the next day, and the day after, the rain descended in torrents, and movement was impossible, so that while the British remained inactive in the camp which they entrenched at Chillianwala, the Sikhs were able to draw off to their left and occupy the very strong position on the hills about Russool. The First Brigade of cavalry, with one troop of horse artillery, was employed during the whole of the 14th in scouring the battlefield in the hope of discovering abandoned guns, but none were found; all not removed by us had, as already stated, been carried off by the Sikhs.

During the days of perpetual rain, reinforcements continued to join the enemy, and consequently Lord Gough made up his mind to wait until the fall of Multan—now only a matter of days—should permit of his force being swelled by the division under Whish, and the Bombay brigade under

Dundas; meanwhile "he would strengthen his bridge-head opposite Ramnuggar, maintain his present position, and, by starving out Sher Singh and all his hosts, compel him to move towards food supplies" and the open country, where the Sikh leader and his following might more easily be crushed.

A few days after the battle of Chillianwala, Chutter Singh, accompanied by the Afghan levies, arrived in the Sikh camp, and was received with a royal salute. He had brought with him Major Lawrence and others, who had been captured on the fall of Attock and Peshawar, and on the 27th Sir Joseph writes in his diary:

"Mr. Bowie, who was the Company's artillery officer who disciplined the Sikh artillery under Major Lawrence at Peshawar, came into camp by Sikh permission yesterday, and he is to return at 4 p.m. to-day. He says the Sikh army is 60,000 men, with 70 guns."

Lord Gough was now anxious about a convoy coming from the direction of Gujerat.

"30*th January*.—A wing of irregular cavalry went to Dinghi; two squadrons of the 14th Light Dragoons, two of light cavalry, and three guns went to Noorjemal at half-past 8 for the protection of the Gujerat convoy."

"1*st February*.—Four squadrons of dragoons, the 1st Light Cavalry and four guns went from camp to Dinghi to protect the convoy from Gujerat."

"2*nd*.—The Commander-in-Chief very anxious respecting the safety of the convoy coming from Gujerat, having four squadrons of European cavalry, five squadrons of native cavalry, and four guns to protect it, apprehending that Sher Singh had gone by Moong to turn our left flank, whilst Chutter Singh was in great force at Pooran to turn our right. My patrols found out that no enemy had passed through Moong, and that no force was threatening our right, and at about 4 p.m. the convoy came into camp safely."

"*3rd.*—An absurd report of the advance of the enemy by the Dinghi road, and that they occupied the mound at Magnawala. No truth in this, for we had occupied the mound before sunrise, though the enemy have 300 regular cavalry and a few hundred irregulars at the foot of the hills. It is said, but whether truly or not, God knows, that Chutter Singh and his troops, including the Afghans, are at Pooran . . . also Sher Singh and his troops, Ram Singh, Lal Singh, Khan Singh, Dhokul Singh's regiment, and Bahadur Singh's regiment. At Russool are Atar Singh, and Soorut Singh, with their troops."

"*4th.*—I went to Noorjemal, and up on the mound at Magnawala. Saw a good many Sikh Horse and footmen, but they did not attempt to molest our people on the mound."

"*5th.*—The enemy emerged from the pass in large force and got possession of the mound and of Noorjemal. All the divisional generals and the Commander-in-Chief's staff and engineers waited on the Commander-in-Chief to consult what is best to be done. I rode out towards Noorjemal, and reconnoitred the Sikh position at Khorre, Moong, and Meanee Chuk, and I should think eight or ten thousand men had encamped or bivouacked near these villages, and the villagers of Noorjemal said they had ten guns through the pass."

"*6th.*—The Sikhs this morning had occupied Noorjemal, and had advanced large bodies towards

Dinghi. I therefore marched at 12 with three squadrons of Dragoons and Lancers, a regiment of native cavalry, two *risalas* of irregulars and six guns, towards Dinghi, the whole under command of Brigadier Bradford, and on my left was Lieutenant-Colonel King, with two squadrons of Dragoons and Lancers, three of light cavalry, and six guns. When arrived near Dinghi, I found about 500 regulars and 1,000 others had plundered the town. A few stragglers were taken. The column made a detour to the left to within a mile and a half of the Sikh position, and returned to camp."

"*7th.*—Major Yerbury (3rd Light Dragoons) and six squadrons of cavalry were at Bazarwala to-day with six guns. The Sikh cavalry, about, as said, 4,000, drew out to oppose them, but did not approach nearer than about a mile. They fired a few shots at vedettes, and when they found out that we had guns they retired to their camp."

"*11th.*—The enemy in force on the left, and drove in our post at Bazarwala. I had out 12 guns and nearly two brigades of cavalry, and about a mile from the camp the advance had a good deal of skirmishing, but little execution was done. The enemy began to retire about half-past 2, and the supports came into camp."

"*12th.*—The enemy seems to have commenced a retreat from Russool during last night. No tents are to be seen, but a good many men were still visible, but they soon retired, and the Commander-in-Chief went to Russool. . . . Our sick and wounded went under escort to-day at 3 p.m. to Ramnuggar."

(The Sikhs were moving towards Gujerat, where they took up a position between the town and a ford on the Chenab above Wazirabad.)

"*13th.*—Visited the Sikh position at Russool.

It stands close to the Jhelum, from which it is separated by a narrow plain. Russool is upon a high cliff, surrounded by deep ravines. The passage to the village is by a narrow dam about eight feet broad. The Sikhs had three lines of batteries to the east of it, with a gentle slope towards the south; at the south-west side it rested on the Jhelum, with a salient battery of five guns about 500 yards from the river. . . . The Jhelum only one stream about 80 yards broad."

"15*th*.—Marched to Lussoorie in consequence of the threatened march of the Sikh army upon Lahore; the road lay through jungle, leaving Sura to the left and Gunderwala to the right, and then to Hussnawala."

"16*th*.—Marched at half-past 6 in one column, same as yesterday, by Panewala, to an encampment half a mile in front of Sadulapore—encamped in one line. Scarcely a vestige of the late action remaining."

"17*th*.—Marched at half-past 8 in brigade columns at deploying distance about three and a half miles towards Kunjah, which I went to with three squadrons of cavalry and two guns. The enemy had retired. The country flat, with many babul trees near Kunjah. In most places fine for cavalry, but in parts intersected with shallow nullahs with water."

"18*th*.—At a quarter-past 10 the army moved in the same order as yesterday, in column of brigades, about four miles. The right advanced towards Wazirabad, and within a short distance of the river, the left resting on Kunjah. The Sikhs moved from Gujerat to a nullah two miles in advance, towards Kunjah. All quiet."

"19*th*.—A report in the morning that the enemy were advancing . . . The Chief ordered out two strong reconnoitring parties of four guns each, to which

I added four squadrons of cavalry to each. They advanced—that is, Lieutenant-Colonels Fullerton and Bradford—towards the Sikh position, but only discovered their picquets, which seemed strong and disposed to maintain their ground. In the afternoon I visited the picquets on the right, and advanced that of the 3rd Light Dragoons. Went to the mound in front of Shadiwal, and ordered it to be occupied to-morrow morning. . . . Saw Lord Gough, and met Lieutenant-Colonel Franks, Major Miller, and another officer of the gallant Queen's Tenth. The 10th Queen's, and 8th and 52nd Native Infantry arrived yesterday, and Brigadier-General Dundas with the European Bombay Fusiliers, 60th Rifles, and two Bombay regiments arrived in camp about 6 o'clock."

"*20th.*—The army marched in column of brigades towards the enemy's position—about two miles—and encamped in front of the village of Shadiwal. I took a reconnoitring party of four guns and six squadrons to Narawala, within less than three miles of Gujerat. This caused their whole line to turn out, and they displayed a good line of men. However, they did not dare to advance, and the whole returned to camp unmolested."

"Throughout the 20th the two armies lay opposite to each other, the Sikhs round the town of Gujerat, the British about three miles to the south, their left resting on the small town of Kunjah, their centre on the large village of Shadiwal, and their right extending to the low alluvial lands on the Gujerat side of the ford at Wazirabad. The whole country was perfectly flat, open, and cultivated, dotted by populous villages, and covered with young spring crops, chiefly wheat and barley, at that time of the year standing a few inches high. Had Lord Gough searched all India for a battlefield better adapted for the overthrow of the Sikhs, he could have found none more suitable."[1]

The town of Gujerat and the position to the south of it taken up by the Sikh army, were flanked by two nullahs running down towards the Chenab; the nullah to the east was wet, that to the west was dry. The latter made a slightly easterly bend below the town, and then ran almost due south towards Shadiwal. The two wings of the Sikh infantry found shelter in the two nullahs, while the centre of the line occupied three entrenched villages, known as Burra and Chota Kalra. On each flank, outside either nullah, were strong bodies of the Sikh and Afghan cavalry. The total Sikh force was estimated at close upon 60,000 men, with 59 guns, but of these latter very few were of heavier metal than 9-pounders.

The reinforcements which had reached Lord Gough had raised his army to 24,000, of whom probably 20,000 could be placed in line of battle. Although his numbers were thus less than half those of the enemy, he was now very strong in artillery, having, besides heavy guns, thirteen other batteries, of which all but one—which remained with the baggage escort in rear of the army—were brought into action. The Commander-in-Chief could now dispose of 18 heavy guns, viz. ten 18-pounders and eight 8-inch howitzers, while he had a large number of 9-pounders. In all, and allowing for the guns lost at Chillianwala, Lord Gough was able to bring 94 guns of varying calibre into action, *including* the Bombay Light Field Battery with the baggage guard.

[1] Thorburn.

There has been considerable difference of opinion among various writers as to the number of British guns actually present at the battle of Gujerat, the total being given as 106, 96, and 88, while Sir Joseph gives the number as 90, viz. 18 heavy guns, 42 9-pounders, and 30 6-pounders. Taking the reports of the officers commanding respectively the Bengal and Bombay Artillery, there would appear to have been in all 18 heavy guns, 9 troops of horse artillery, and 4 light field batteries—making a total of 96 guns which should have been present. Stubbs, however, in his "History of the Bengal Artillery," writes of "the fire of the 88 guns," from which it is apparent that, as the Bengal Artillery could only number at most 84 guns, he must be *including* the Bombay horse battery which was engaged, and *excluding* the Bombay light field battery, which, being with the baggage guard, did not fire. It is evident also from Stubbs's figures that two at least of the guns lost at Chillianwala had not been replaced; but as all accounts seem in agreement that Duncan and Huish on the left had 12 guns between them, it would appear that while Huish's troop had been again made up to six guns, that of Christie—now Kinleside—had not. If these figures and deductions are correct, there must have been altogether 94 guns present in the field.

The 18 heavy guns seem to have been divided into four batteries, or among four artillery companies, viz. six in one, and four in each of the other three.

Some changes had naturally taken place in the higher commands; Godby had been transferred to the command at Lahore, and Penny, from Campbell's division, now commanded Godby's brigade under Gilbert. Carnegie had taken Penny's place, and McLeod had been appointed to the command of Pennycuick's brigade. The two brigades in the First Division were commanded respectively by Markham and Hervey, while the Bombay column, containing four regiments, remained as an independant unit under Dundas. The cavalry division under Sir Joseph Thackwell now comprised some fourteen regiments of cavalry, of a total strength of close upon 7,000 sabres, and was organised as follows:

1st Cavalry Brigade.

Brigadier M. White, C.B.
3rd Light Dragoons.
5th Light Cavalry.
8th Light Cavalry.
Attached on the 21st, Scinde Horse.

3rd Cavalry Brigade.

Brigadier Bradford.
11th Irregular Cavalry.
14th Irregular Cavalry.

2nd Cavalry Brigade.

Brigadier G. H. Lockwood, C.B.[1]
9th Lancers.
14th Light Dragoons.
1st Light Cavalry.
6th Light Cavalry.

4th Cavalry Brigade.

Brigadier J. B. Hearsey, C.B.
11th Light Cavalry.
3rd Irregular Cavalry.
9th Irregular Cavalry.
13th Irregular Cavalry.

There was also the 12th Irregular Cavalry, but this regiment had been detached with other details

[1] This officer—of the 3rd Light Dragoons—only rejoined from England on the 7th February.

to Wazirabad, while the army still lay at Chillianwala, to resist any attempt of the Sikhs to cross the Chenab.

For the battle on the 21st some modifications were made in the above-mentioned composition of brigades: the Third Cavalry Brigade was broken up, and of the two regiments which it contained, the 11th Irregular Cavalry was sent to Hearsey's brigade, while the 14th was at first posted to the First Brigade, but remained during the action on the right at Hearsey's disposal. The 9th Lancers and Scinde Horse—the latter about 250 strong—were temporarily placed under Brigadier White, and the 5th Light Cavalry of the First and the 6th Light Cavalry of the Second Brigade formed part of the baggage guard for the day.

From Sir Joseph Thackwell's despatch it will be seen that he was not only in command of the cavalry division on this day, but that the whole of the troops on the left or west of the nullah which bisected the British force were also placed under his immediate supervision as second in command under Lord Gough.

It was 7 o'clock on a cloudless morning when the veteran Commander-in-Chief drew out his army for his "last and his best battle"—one, moreover, which was finally to crush the power of the Khalsa, which was, in the words of the great Sikh maharaja, to paint the map of India "all red," and which saved the old soldier who won it from being sacrificed to the popular outcry raised by the previous indecisive and bloody battles of the campaign.

"There was no dust," says Durand, who accompanied the Headquarters, "to cloud the purity of the air and sky. The snowy ranges of the Himalaya, forming a truly magnificent background to Gujerat and the village-dotted plain, seemed on that beautiful morning to have drawn nearer, as if, like a calm spectator, to gaze on the military spectacle."

And as Lord Gough rode down his line from right to left, he was received with a tremendous outburst of enthusiasm by the officers and men of his army, filled with the confidence of victory.

Soon after 7 a.m., Sir Joseph's diary tells us, that the troops moved in the order of battle as set forth in the accompanying plan.

The two batteries of Kinleside and Lane were sent to the right later on in the morning. The 11th and 14th Irregular Cavalry had only two *risalas* each present, and the 13th Irregulars but one squadron.

The Commander-in-Chief had determined to attack the Sikh centre and left, and drive them back on their right, when the British left wing, having advanced, would complete the work of destruction. About 8.30 a.m. the Sikhs opened fire, thereby disclosing the position of every one of their guns, and the Commander-in-Chief sending forward the whole of his powerful artillery, covered by infantry skirmishers, the guns took up a position in one long line, and for nearly three hours poured a most destructive fire upon the enemy.

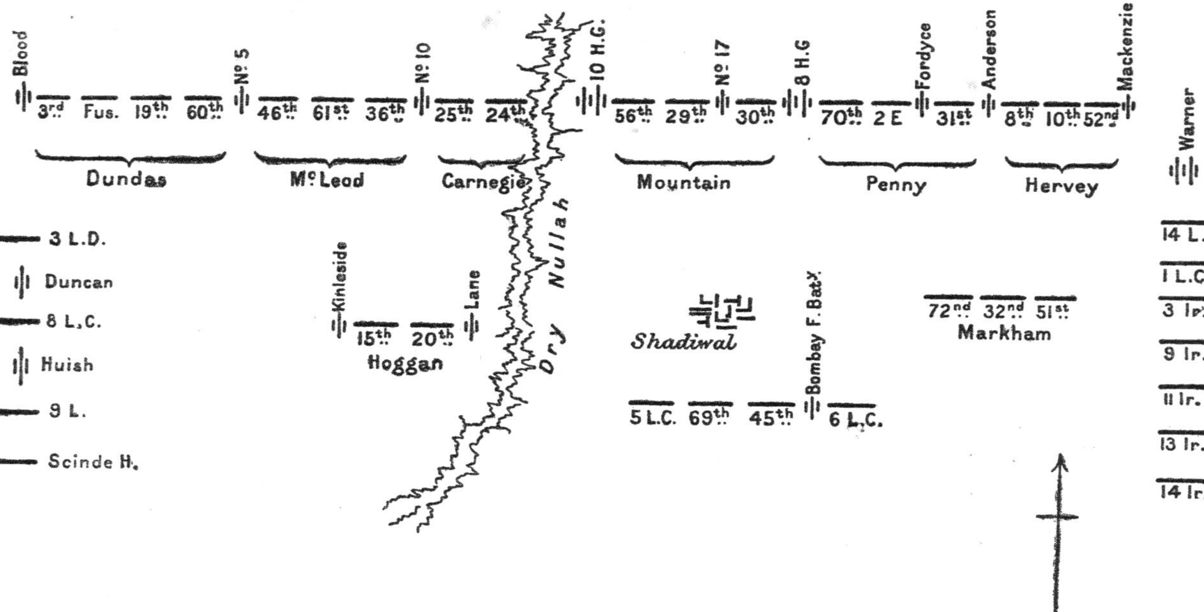

Blood
3rd
Fus.
19th
60th
No 5
46th
61st
36th
No 10
25th
24th
10 H.G.
56th
29th
No 17
30th
8 H.G
70th
2 E
Fordyce
31st
Anderson
8th
10th
52nd
Mackenzie
Warner
Dundas
McLeod
Carnegie
Mountain
Penny
Hervey
3 L.D.
Duncan
8 L.C.
Huish
9 L.
Scinde H.
Kinleside
15th
20th
Lane
Hoggan
Dry Nullah
Shadiwal
72nd
32nd
51st
Markham
5 L.C.
69th
45th
Bombay F. Batx.
6 L.C.
14 L.D.
1 L.C.
3 Ir.
9 Ir.
11 Ir.
13 Ir.
14 Ir.

About midday Lord Gough ordered a general advance of the whole line, and General Gilbert, thinking the village of Burra Kalra had been evacuated, sent Penny's brigade to occupy it. The Sikhs, however, were still in possession, and a tremendous musketry fire was opened from the loopholed walls of the village, which was eventually carried by the 2nd Europeans, supported by the 31st and 70th Native Infantry, all these regiments suffering severely. Chota Kalra was also found to be held by the Sikhs, and the brigade under Hervey—the only one of Whish's division then in the front line—diverging slightly to the right in order to attack it, a gap was thus left between the First and Second Divisions, of which the enemy showed a disposition to take advantage, until Markham, advancing, filled the vacant space in the line. The enemy's horsemen were in considerable force opposite the cavalry brigades of Lockwood and Hearsey, but they were kept at a distance by the fire of Warner's battery, by the wet nullah on the right, and by the swampy nature of the ground. A small body of the enemy, however, made a wide sweep to the east, and, bringing up their left shoulders, approached the position occupied by Lord Gough and the Head-quarter staff. Here they were at once attacked and cut to pieces by the escort of the 5th Light Cavalry under Lieutenant Stannus.

As soon, however, as the infantry carried the Sikh position, the whole of the cavalry and horse artillery of the right were launched in pursuit and pressed the retreat of the enemy until dark, compelling them to disperse, capturing many guns, and inflicting immense loss.

On the left of the dry nullah dividing the British advance the infantry of the Third Division were not called upon to fire a shot, but Campbell's artillery was of material assistance to the troops under Dundas, when threatened by a determined rush of the Sikh cavalry and infantry from the shelter of the nullah immediately in front of the enemy's position. Of great importance was the work of the cavalry on the extreme left, the operations of which are thus noted in the diary of their Commander:

"Large bodies of Sikh cavalry showed themselves in front and on the rising ground on our left flank, and this occasioned the cavalry to form line, as the infantry had already deployed . . . this induced the enemy to open fire from four guns, which obliged White to retire his left. I, however, soon remedied the mischief by directing Duncan's troop to advance within four or five hundred yards of the enemy's cavalry, and he opened a well-directed fire upon them. Huish's troop also advanced and opened fire, and as the enemy's cavalry were advancing to outflank my left, the Lancers, who had thrown back the left, were formed to the front, and I ordered the Scinde Horse and a squadron of Lancers, supported by another squadron in échelon to the right, to charge the enemy's right, which they did in fine style, and drove this Afghan body and Goorchurras opposed to them back with considerable loss." (In this magnificent charge the opposing squadrons were shattered and driven in headlong flight, while two standards were captured from the enemy.)

"During this time our line and guns were advanced, and cannonaded the enemy with good effect, and the Sikh cavalry were driven with loss beyond the Bara Darri, losing a gun which had belonged to Captain Huish's troop. The infantry of the left being well up, I continued my movement to the left of the above place, where the tents of Sher Singh and others, with their entire camp, were left standing, and opened my twelve guns upon the retreating enemy, and continued thus, inclining well to the left, by which movement an immense body was cut off from the road to Jhelum, and eventually from that to Bimber. They were several times charged by the 9th Lancers and 8th Light Cavalry, and a good deal cut up by Duncan's, Huish's, and latterly Blood's battery, until the horses had no longer a trot in them. A great number of men were killed in the pursuit, which lasted for nearly twelve miles from the Bara Darri, and ended at twenty minutes after 4 p.m., when no enemy was in sight, and the villagers reported that they had taken away only three horse artillery guns."

General Thackwell intended to bivouac for the night and resume the pursuit next morning, having now been joined by Lockwood's and Hearsey's brigades from the right, but the cavalry was recalled by Lord Gough and rejoined the army in front of Gujerat between 10 and 11 o'clock at night. The enemy never halted until they reached the Jhelum, thirty miles distant, while Akram Khan, brother of the Ameer Dost Muhamed, who had commanded the Afghan Horse at Gujerat, and who was wounded in the action, never drew rein until he arrived near Rawul

Pindi. The author of the "Narrative of the Second Seikh War" writes of the pursuit, in which, as aide-de-camp to Sir Joseph, he took part:

"The slaughter, perpetrated by the British Dragoons in the pursuit, was fearful. No quarter was given, and every Sikh, armed or unarmed, fell under the sword or the pistol."

Hope Grant, who was with the 9th Lancers, says:

"It was horrible work slaughtering these wretched fugitives, who had taken refuge in trees and in the standing corn hoping to evade discovery. . . . Our men were enraged with the Sikhs, owing to the brutal manner in which they had slaughtered our wounded at Chillianwala."

Full justice was done in Lord Gough's despatch to the work of the Cavalry Division:

"I feel indebted," he wrote, "to Major-General Sir Joseph Thackwell for the able and judicious manner he manœuvred the cavalry, with horse artillery attached, on the left, keeping in check the immensely superior force of the enemy, whose main object was to turn my flanks."

Sir Charles Gough, in his account of the "Battles of Chillianwala and Gujerat," says:

"It was a glorious sight to see the Scinde Horse and the 9th Lancers sweeping forward over the open plain, and in a few minutes the whole force of Afghan Cavalry turned and fled, and Thackwell . . . found that his advance had completely turned the right of the Sikh line."

In another place he writes:

"Thackwell's handling of the cavalry on the left was perfect, and precisely at the right time and in the right way he delivered his attack. Supported by the fire of two troops of horse artillery and the advance of the remainder of the brigade, the two flank regiments—the Scinde Horse in first line and the 9th Lancers in second—charged the Sikh and Afghan Horse, who were completely defeated and fled from the field. This success placed him on the right flank of the Sikh infantry and in an awkward position for those attempting to resist the advance of our centre and precipitated their retreat." Had the ground not been so broken and intersected by nullahs, "which favoured the retirement of their infantry and hindered the cavalry in pursuit, they would scarcely have been able to get away any body of men at all."

The triumph of the British was complete and their loss had been comparatively small. The total casualties of the army under Lord Gough's command were but 96 killed and less than 700 wounded. Fifty-three guns and several stand of small arms were taken, while the actual losses of the enemy in men were most severe—the line of retreat for twelve miles being strewn with killed and wounded, with guns, small arms, and baggage.

"*22nd.*—Rode over the principal part of the Sikh position: a good many of the Sikhs lay dead, but I think as many have been killed in the pursuit. Sir Walter Gilbert, his division, the Bombay Division, the 14th Light Dragoons, the 3rd and 9th Irregular Cavalry, and the 11th and the Scinde Horse marched to-day towards Jhelum, and Brigadier-General Campbell, with his division, the 9th Lancers and 8th Light Cavalry, towards Bimber, a little in front of where I halted last night or rather evening. I rode over the field of battle of our right and centre. . . . This morning the Commander-in-Chief read me some letters to the Governor-General and his replies ; foreseeing that a strong force would have to cross the Jhelum, he had, in one of these letters, named me and Gilbert to command this force, and no one else. He gave us both strong recommendations, but left it for the Governor-General to decide. He read his reply, to the effect that 'as but few of the cavalry would be employed, he thought it but right that Gilbert should have the command'—which, perhaps, I have no right to be dissatisfied with."

(Elsewhere Sir Joseph gives the composition of Gilbert's force as follows: three troops of horse artillery, two field batteries, six 18-pounders, four companies of sappers and miners, 14th Light Dragoons, 1st Light Cavalry, 13th and 14th Irregular Cavalry, 280 of the Scinde Horse (the above cavalry under Brigadier Hearsey), First, Second and Third Infantry Brigades, and the Bombay Brigade.)

"24*th.*—Nine guns taken from the Sikhs were brought in to-day by Lieutenant-Colonel Bradford and the 1st Light Cavalry."

"25*th.*—Brigadier-General Campbell's division, the 9th Lancers and 8th Light Cavalry returned to camp, bringing with them two more captured guns."

Sir Walter Gilbert, moving by Dinghi and the Khoree Pass, followed up the enemy with great rapidity; on the 6th the Sikhs gave up their prisoners; a few days later, at Rawul Pindi, the Sirdars surrendered, and the disarmament of the soldiers of the Khalsa was completed; 49 guns with many thousand stand of arms had been taken; and by the 21st of March—exactly a month after the battle of Gujerat—Peshawar had been reoccupied and Dost Muhamed had been chased back to the Khyber; the Afghans "had ridden down through the hills like lions and ran back into them like dogs."

On the 22nd the Sirdars, Sher Singh and Chutter Singh, came into camp, and on the following day part of the army which was to occupy Wazirabad crossed the Chenab and went into camp near the town—Sir Joseph, with his staff, taking up his quarters in a house which had formerly belonged to General Avitabili—and on the 28th the force broke up, and all, except those units which were to garrison Wazirabad, marched to their destinations.

Sir Joseph had hoped that on the army breaking up he might be transferred from Meerut and permitted to remain in command of the Punjab division, and he had indeed applied in this sense to Lord Gough; but the "imperious little man," who then in India ruled all men and all things, had decided that any arrangements made at the present time must be altogether of a temporary character and that "the commands should remain as they are, until a more deliberate and careful consideration can be given to the subject of divisional commands in Upper India generally."

At the end of March, then, Sir Joseph left

Wazirabad for a short tour in the Punjab, revisiting some of the battlefields, and then travelled quietly down to Meerut, which was reached on the 23rd of April, and where Lady Thackwell and his daughter were anxiously awaiting him.

During the next few months rewards and congratulations came in fast, and among the latter few can have given Sir Joseph more real pleasure than the letters he received from those comrades and commanders with whom and under whom he had seen his early service. There are letters from old General Sir John Slade, who had commanded the brigade in which the 15th Hussars had served in the Corunna campaign, and who writes:

"Trusting that my name may not be quite forgotten by you, whose early career I am so well acquainted with personally, and whose recent successes it has given me so much pleasure to trace—marking how the early promise has been surpassed by the later stirring events of Indian warfare."

Lord Anglesey, too, wrote:

"I have had my eye upon you all, and have often been rejoiced by the brilliant success that has attended you and others of my good and gallant old friends."

His lordship asks Sir Joseph to help a grandson, William Paget, in the 54th Native Infantry, to see something of active service.

"He," writes Lord Anglesey, "is anxious for a staff situation, but this (as I have told him) is by no means my wish, for, to say the truth, I have little opinion of staff-hunters to form good soldiers."

Sir Joseph seems to have shared these sentiments, for in reply to the Marquis of Anglesey's letter he writes:

"The routine of duty in a regiment forms a young soldier for any higher employment, and, for myself, I have reason to rejoice that no staff avocations took me away from the fine school of instruction, both at home and abroad, the 15th Hussars, with other regiments, were so fortunate as to have under your lordship's command and guidance."

General Thackwell's old Peninsular and Indian comrade, Sir Harry Smith, wrote from the Cape:

"I had been watching your operations with every possible interest, and I congratulate you first upon your whole *skin*, next upon the *honours* your sword had previously won and richly merited at Sobraon. Your campaign at one time looked blue . . . you began with too much. Multan ought to have been reduced before you moved on the numerous Singhs. In all campaigns, however, there are errors, but none ever ended more gloriously than yours. When you were detached after poor Cureton's death, you were placed in a most peculiar position, and you were the best and only judge of what you ought and could do. Every ass in a difficulty shuts up, having no opinion, but so soon as a result attaches, then they all know. . . . 'If' and 'if.' You feel you did right, and the end of the campaign shows it, for anything you might with your force have effected would have been at the best only a partial success and without any effect on the positive dissolution of the enemy in the finale. . . . God bless you, my gallant dragoon."

On the 17th May the Duke of Wellington wrote to Sir Joseph Thackwell, acquainting him that—

"the Secretary of State has, upon my recommendation, submitted to the Queen your appointment to be a Knight Grand Cross of the Most Honourable Military Order of the Bath, of which Her Majesty has been graciously pleased to approve."

In reply Sir Joseph said—

"To be promoted to the highest military honour is most flattering to the pride of an old soldier, and to have been recommended to Her Most Gracious Majesty's favour by your Grace, under whose guiding wand I had served some campaigns, is a matter of congratulation of which I may well be proud."

"Allow me to congratulate you," wrote Lord Gough, "upon, I hope, but the first instalment of our Gracious Sovereign's estimate of your long and brilliant services. Lady Gough warmly unites in the hope that you and Lady Thackwell may long live to enjoy the present and, I trust, many future marks of your Sovereign's high estimation."

The thanks of both Houses of Parliament were unanimously accorded to all the Commanders and subordinates of the army of the Punjab for the indefatigable zeal and exertions exhibited by them throughout the recent campaign, and the thanks of the Court of Directors were in like manner given to all those who had taken part in the military operations on the Chenab. Finally, towards the end of this year Sir Joseph was further gratified by being appointed Colonel of the 16th Lancers—a regiment which had served under his command in the army of the Indus, at Maharajpore, and on the Sutlej. On hearing of this appointment Lord Charles Manners, the Colonel of the 3rd Light Dragoons, wrote to Sir Joseph begging that—

"however distinguished is your regiment, I trust you will not allow the 3rd Dragoons to be entirely obliterated from your recollection, and that you will not cast away all regard for your old love now that you have formed connection with a new one."

Sir Joseph was now entitled to supporters to his coat of arms, and, with the permission of the Colonel of the 15th Hussars, he selected the figure of a hussar for his dexter supporter, and having obtained the approval of Lord Charles Manners, he took a dragoon for the sinister supporter.

CHAPTER XX

THE command at Meerut was no sinecure; the district was a very large one, embracing the stations of Agra, Delhi, Muttra, Alighur, Bareilly, Shahjehanpore, and the posts about Almora; and in the days when the second half of the last century was only commencing, the travelling necessitated by "inspections" was continuous and protracted, and Sir Joseph Thackwell was frequently absent from Meerut inspecting regiments and batteries in the outlying garrisons of his widespread command. For the greater part of the four years which General Thackwell still remained in India, his aide-de-camp was Lieutenant Gibney, of the 50th Native Infantry, whose father was an old 15th Hussars comrade of Sir Joseph; but in the year preceding the expiration of the General's period of command, Lieutenant Gibney returned to regimental duty, and his place was taken by Sir Joseph's nephew, Captain Thackwell of the 22nd Regiment.[1]

The summer of 1849 Lady Thackwell and her daughter spent at Mussoorie; the following hot weather, and that of 1851, they passed at Simla. Sir Joseph was only able to visit these hill

[1] Afterwards General Thackwell, C.B.

stations for very brief periods, and while in these highlands nearly the whole of his short leave was passed in travelling among the hills . . . among mountain villages and along paths which, however well-known now to the majority of Anglo-Indians, might then almost be described as "the roads that only Rumour knows." In the summer of 1850 Sir Charles Napier was in Simla as Commander-in-Chief, but even then he was probably meditating, if he had not indeed already resolved on, the resignation of his high post, for in December of the same year we read in Sir Joseph's diary that on the 17th "the troops paraded to hear the General Order read, inducting Sir William Gomm into the office of Commander-in-Chief."

Thus Sir Joseph Thackwell had already served in India under no fewer than five commanders-in-chief—Sir Henry Fane, Sir Jasper Nicholls, Lord Gough, Sir Charles Napier, and now Sir William Gomm.

It is only on the 8th February 1850 that Sir Joseph acknowledges the receipt of the medal and clasps for the Peninsula campaign!

Major-General Thackwell was now anxious to return home, and he was looking forward to leaving India in 1852, when he was greatly disappointed to hear from Lord Fitzroy Somerset, that the period of service of a major-general on the staff in India was not five years as he had imagined, but six, and that he would consequently not have completed his service in that country until May 1853.

He was kept busy, however, up to the last with reports and committees—reports called for, wherein his opinion was invited as to the best type of horse for cavalry and light batteries, and reports on the arming of native cavalry; while he was appointed president of a committee sanctioned by the Government of India to go exhaustively into, and report fully upon, the whole system of military prisons in India—their management and maintenance, their plan of construction; and when the work was at last finished and his report was sent in, Colonel Mountain, Adjutant-General of Queen's troops in India, wrote to him:

"I congratulate you on the conclusion of your labours, which will, I have no doubt, be ultimately attended with the most beneficial results to Her Majesty's service in India. When the committee was sanctioned by Government, with you for president, my mind was relieved of a care that had dwelt upon it for many years."

Early in March 1853 Lady Thackwell and her daughter proceeded to England, but it was not until May that Sir Joseph Thackwell's long, eventful, and honourable period of Indian service was to come to an end—when, leaving Meerut on the 3rd, he arrived at Calcutta, accompanied by his nephew, on the 13th May, embarking for England four days later. He was not, however, allowed to leave India without receiving the following appreciative letter from the Commander-in-Chief:

"SIMLA, *27th April* 1853.

"MY DEAR SIR JOSEPH,—

"I cannot allow the time of your departure from among us to draw near without writing you a few lines expressive of my own regret at losing your assistance in whatever may be in store for us during the remainder of my service in this country. And I feel very confident that those among us who have partaken of your long course of brilliant and valuable service in India, or any portion of it, will regret equally with myself this privation.

"Your known desire to return home after so long an absence, and private affairs requiring your personal supervision, are the best arguments with which we can furnish ourselves to reconcile us in any way to the change.

"We have been comrades in fields of old, and that is an additional reason with me for grudging the losing sight of you for so long as I am in this command. Moreover, you have been doing right good service in another and not less valuable way—and that under my own eye—as president of the Military Prison Committee—and we all owe you much for the active and judicious assistance you have invariably afforded for bringing its labours to a satisfactory conclusion.

"On all these accounts, I beg, my dear Sir Joseph, you will accept my very best thanks and a brother soldier's heartiest wishes for your prosperous journey and that of Lady Thackwell and your daughter homeward—in all which Lady Gomm most cordially joins with me, and also that we may find all in good health on our return.

"Believe me always, my dear Sir Joseph,

"Yours most faithfully,

"WM. GOMM."

It was not, however, only from those of his own profession that expressions of sorrow were to be heard on Sir Joseph Thackwell's severance of his connection with "the Land of Regrets," for an Indian newspaper of the day records:

"We have never heard Sir Joseph Thackwell's name mentioned except with reverence and respect, no matter what the rank or grade of the person who had occasion to speak of him. Independent of his intimate knowledge of every branch of his profession and his wonderful activity, General Thackwell has an amount of good sense, good feeling, good taste, and good humour, which peculiarly fit him for command."

England was reached on the 5th July, and Sir Joseph proceeded direct to Bristol, where he found all his family assembled.

From all sides he received a cordial welcome home; from Lord Hardinge, who now filled the place of the Duke of Wellington; from Lord Ellenborough, who, as Governor-General, had been present with Sir Joseph at the battle of Maharajpore; from his old comrade Sir Harry Smith, who wrote on the 20th July, saying:

"As I think by this time the fiery edge of your congratulations on reaching England are taken off, I write to say none of your old friends are more happy than I am to know you are once more in old England after your unprecedented career in India; and long may you live to enjoy your repose."

When he went over to Ireland his diary records that—

"bonfires along the coast announced my arrival, and Admiral Purvis sent his barge, with Lieutenant Twyford and twelve men, who took me on board his yacht and carried me to Whitegate, and the barge took me ashore. I was cordially cheered by those on board the steamer on my departure, and by a brig full of people at Whitegate, and by all the inhabitants of the village, who lighted bonfires, decorated their houses with green boughs, garlands, and arches, and illuminated them at night . . . indicative of the good-will of the people."

Early in September he visited Gloucester and attended a public dinner given in his honour by two hundred gentlemen of the county, with Lord Fitzhardinge in the chair.

It might have been expected that one who had now for years been so continually in harness would have welcomed some respite from military occupations, and would have been glad, at least for a time, of perfect rest and change; but Sir Joseph Thackwell had only been in England a very few weeks when he visited the camp at Chobham, where a considerable force was undergoing its first experience of camp life in England, preparatory to the more realistic work which the British Army was to see in the following year.

"Went by train to Chertsey in one hour," writes General Thackwell, "and walked thence to the camp at Chobham, about four miles. Walked for upwards of a mile about the camp."

He then walked back to London, and by the end of the day had got over "between seventeen and eighteen miles, and without fatigue."

In February 1854 the strained relations between the Western Powers and Russia, and the prospect of an expeditionary force proceeding to Turkey, brought General Thackwell up to London to inquire what chance there might be for him of employment with the cavalry.

"*9th February.*—Called at St. James's on the Duke of Cambridge, and sat with him for about twenty minutes. He was very kind, and asked many questions, spoke highly of the 16th Lancers and 3rd Light Dragoons, and seemed to entertain a good opinion of Colonel White. Called on the Duke of Newcastle and on Lord Palmerston. Saw Lord Hardinge at the Horse Guards, and had a most kind interview, but with but little chance of employment in Turkey. A lieutenant-general is to command, and an inferior rank as brigadier would not suit me. He thinks it most likely the troops will not go beyond Malta."

Early in April, however, Sir Joseph was appointed—

"to serve as Inspecting General of Cavalry in Great Britain during the temporary absence of Major-General His Royal Highness the Duke of Cambridge, K.G.," and he took over the duties of the appointment on the 4th May.

The Horse Guards had apparently mislaid Sir Joseph's address, for the letter was sent, not to his house, but to his club, where it seems to have lain for some time, and the Duke of Cambridge had to embark without handing over personally to Sir Joseph, to whom, however, he wrote the following characteristic letter:

"St. James's Palace,
"*April* 10*th* 1854.

"My dear Sir Joseph,—

"I was in hopes to have seen you before my departure, but I have not been able to ascertain your whereabouts, and so I must write. I was anxious to have said a few words to you on the subject of the command you are about to assume, and to wish you joy upon having been selected for it. You will find Lieutenant-Colonel Cotton, the Cavalry Brigade Major, a most attentive and zealous officer. He is well aware of all my views and feelings as regards the cavalry, and if you would like him to name these points to you, he would do so with pleasure. My great anxiety has been to keep the regiments as efficient as possible for their reduced strength. The riding is a point to be especially attended to. I have introduced a system of jumping at Maidstone which answers to perfection, and I hope you will carry it out fully throughout the service. Pray try to get the depôt squadrons as efficient, both in men and horses, as possible.

"Griffiths at Maidstone is an excellent fellow, and so is my little friend Major Meyer, the army riding-master. I will now not trouble you further, and remain, my dear Sir Joseph,

"Yours very sincerely,
"George.

"I expect the officers to attend to *everything* themselves, and not to leave things to sergeants."

Sir Joseph and Lady Thackwell took a house in London, in Montague Square, and on the 20th June—after just over fifty-four years' service—he was at last promoted to the rank of Lieutenant-General.

For the few months which his new appointment

lasted, General Thackwell was kept very busy, constantly travelling to inspect depôts of cavalry regiments in England, Scotland, and Ireland, and sending men and remounts to the regiments in the Crimea. On the 1st February 1855, however, Major-General Lord Cardigan was appointed to be Inspector-General of Cavalry, and Sir Joseph was then offered by Lord Hardinge the Lieutenant-Governorship of Chelsea Hospital, which, however, he declined.

Previous to leaving India he had purchased a property called Conneragh in county Waterford, and in 1853 he had bought another place, called Aghada Hall, in county Cork, and here he established his headquarters and passed the greater part of the last quiet years of his busy and eventful life—shooting, carrying out improvements in his property, and running over from time to time to England to visit his nephew, John Cam Thackwell, J.P., D.L., the son of his eldest brother, who occupied the family house at Wilton Place, Dymock, in Gloucestershire. All Sir Joseph's sons had gone into the Army, and of the eldest we have seen something during the campaign in the Punjab ; he died in 1903. William de Wilton Roche had served in the Crimea with the 39th Regiment and, exchanging to the 38th, rose to command, and retired, with the rank of Major-General. Osbert D'Abitôt had entered the Indian Army and had greatly distinguished himself against the rebels in Rajputana and Neemuch, after his regiment, the 15th Native Infantry, had mutinied ; he was present at the siege of Lucknow under Lord Clyde, and was killed in the streets after its capture on the 21st March 1858. Francis John Roche entered the 5th Royal Irish Lancers and rose to the rank of Captain, but died in 1869 from the effects of wounds received from a tiger. Of the daughters, all three—Elizabeth Cranbourne, Annie Esther, and Maria Roche—survived their father. The second was the wife of the Rev. T. P. Little, M.A., Vicar of The Edge, Gloucestershire, and died in 1902; the youngest married Lieutenant-Colonel James Bennett. Lady Thackwell survived her husband until the 21st June 1874.

Early in 1856 there seemed some possibility that the great services of Sir Joseph Thackwell might be further recognised by the grant of a baronetcy, but in a letter, dated the 15th March, Sir Charles Yorke wrote that the Field Marshal Commanding-in-Chief had submitted Sir Joseph's name for the dignity of a baronet to Lord Palmerston,—

"stating at the same time the high opinion which he entertained for your services—particularly in the wars in India, when you held a large command of cavalry. . . that Lord Hardinge has received in reply an answer from Lord Palmerston, in which he expresses how fully he appreciates the merit and value of your services, but regrets that he could not recommend you to the Queen for the honour in question without creating an inconvenient precedent."[1]

[1] Lord Hardinge had asked for the baronetcy on the grounds that Sir Joseph had been second in command to Lord Gough: there was of course ample precedent for conferring even a higher title upon the second in command of a successful general.

Towards the end of 1858 and in the beginning of the following year there were signs that Sir Joseph's health was beginning to fail, and that even his iron constitution was at last affected by nearly sixty years of war and strife. The journal which he had commenced with his entrance into the army is kept up to the end, and it is only on the 6th April that the last entry is recorded, and that the pen drops from the fingers. Two days later—on the 8th April—heart failure had supervened, and the brave life had come to an end.

The services of Sir Joseph Thackwell speak for themselves; it cannot be denied that they were great, and it must be admitted that even for those days—when reward was more sparingly bestowed and praise less loudly declaimed than in modern times—these services were by no means lavishly recompensed. It is, however, to be hoped that those who have followed to the end this memoir of the military life of one whose experience of the employment of the mounted arm was greater and more varied than perhaps any other British cavalry man, will have recognised that not only was Sir Joseph Thackwell one of the most loyal of men—alike towards his superiors, his contemporaries, and his subordinates—but that he had a soul far above self-advertisement. If one may venture to suggest another characteristic for consideration, it seems to be that upon which the greatest of living British cavalry leaders has recently laid much stress. Sir John French has lately said that "comradeship and unselfishness" form the keynote of true military endeavour; that—

"the motto of the soldier should be 'to love his comrade as himself,' and that he believed that this forgetfulness of self and unalloyed striving for the good of the whole to be more difficult to practise in the military than in any other profession."

Surely these were the principles which Sir Joseph Thackwell had striven throughout his sixty years of arduous service to live up to; to sink all idea of individual gain, of personal distinction, in the general good of the noble service to which he was so proud to belong; and if the striving after these ideals should result in a certain curtailment of all that has perhaps in other ways been hoped for and desired, the knowledge and the memory of their attainment must "bring a man peace at the last."

APPENDIX A [1]

DESPATCHES. AFGHANISTAN

SECRET DEPARTMENT, SIMLAH, *August 18th*, 1839.

The Right Honourable the Governor-General of India has great gratification in publishing for general information a copy of a report this day received from His Excellency Lieutenant-General Sir J. Keane, K.C.B., etc., Commander-in-Chief of the Army of the Indus, announcing the capture, by storm, on the 23rd ultimo, of the important fortress of Ghuznee.

A salute of 21 guns will be fired on the receipt of this intelligence at all the principal stations of the army in the three Presidencies.

By order, etc.

(*Signed*) T. H. MADDOCK,
Officiating Secretary to Government of India,
With the Governor-General.

" HEADQUARTERS CAMP, GHUZNEE, *July 24th*, 1839.

" TO THE RIGHT HONOURABLE LORD AUCKLAND, G.C.B., ETC., ETC., ETC.

" MY LORD,

1. " I have the satisfaction to acquaint your Lordship that the army under my command has succeeded in performing one of the most brilliant acts it has ever been my lot to witness during my service of forty-five years, in the four quarters of the globe, in the capture, by storm, of the strong and important fortress and citadel of Ghuznee, yesterday.

2. " It is not only that the Afghan nation, and, I understand, Asia generally, have looked upon it as impregnable, but it is in reality a place of great strength, both by nature and art, far more so than I had reason to suppose from any description that I have received of it; although some are from officers from our own service who had seen it in their travels.

3. " I was surprised to find a high rampart in good repair, built on a scarped mound, about 35 feet high, flanked by numerous towers, and surrounded by a 'faussebraye' and a wet ditch, whilst the height of the 'citadel' covered the interior from the commanding

[1] In these appendices only those despatches have been included wherein General Thackwell's name is mentioned, and all detailed lists of casualties have been omitted.

fire of the hills from the north, rendering it nugatory. In addition to this, screen walls had been built before the gates; the ditch was filled with water, and unfordable, and an outwork built on the right bank of the river, so as to command the bed of it.

4. "It is therefore the more honourable to the troops, and must appear to the enemy out of all calculation extraordinary, that a fortress and citadel, to the strength of which, for the last 30 years, they had been adding something each year, and which had a garrison of 3,500 Afghan soldiers, commanded by Prince Mahomed Hyder, the son of Dost Mahomed Khan, the ruler of the country, with a commanding number of guns, and abundance of ammunition and other stores, provisions, etc., for a regular siege, should be taken by British science and British valour, in less than two hours from the time the attack was made, and the whole, including the Governor and garrison, should fall into our hands.

5. "My despatch of the 20th inst. from Nannee will have made known to your Lordship that the camp of His Majesty Shah Shoojah-ool-Moolk, and of Major-General Willshire, with the Bombay troops, had there joined me, in accordance with my desire; and the following morning we made our march of 12 miles to Ghuznee. The line of march being over a fine plain, the troops were disposed in a manner that would have enabled me at any moment, had we been attacked, as was probable from the large bodies of troops moving on each side of us, to have placed them in position to receive the enemy. They did not, however, appear, but on our coming within range of the guns of the citadel and fortress of Ghuznee, a smart cannonade was opened on our leading columns, together with a heavy fire of musketry from behind garden walls and temporary fieldworks thrown up, as well as the strong outwork I have already alluded to, which commanded the bed of the river. From all but the outworks the enemy were driven in, under the walls of the fort, in a spirited manner by parties thrown forward by Major-General Sir W. Cotton, of the 16th and 48th Bengal Native Infantry, and H.M.'s 13th Light Infantry under Brigadier Sale. I ordered forward three troops of horse artillery, the camel-battery, and one foot-battery, to open upon the citadel and fortress by throwing shrapnel shells, which was done in a masterly style, under the direction of Brigadier Stevenson. My object in this was to make the enemy show their strength in guns, and in other respects, which completely succeeded, and our shells must have done great execution and occasioned great consternation. Being perfectly satisfied on the point of their strength, in the course of half an hour I ordered the fire to cease, and placed the troops *en bivouac*. A close reconnaissance of the place all round was then undertaken by Captain Thomson, the chief engineer, and Captain Peat of the Bombay Engineers, accompanied by Major Garden, the Deputy Quartermaster-General of the Bengal Army, supported by a strong party of H.M.'s 16th Lancers,[1] and one of H.M.'s 13th Light Infantry. On this party a steady fire was kept up, and some casualties occurred. Captain Thomson's report was very clear (he found the fortifications equally strong all round), and as my own opinions coincided with his, I did not hesitate a moment as to the manner in which our approach and attack upon the place should be made; notwithstanding the march the troops had performed in the morning, and their having been a considerable time engaged with the enemy, I ordered the whole to move across the river (which runs close under the fort walls), in columns to the right and left of the town, and they were placed in position on the north side,

[1] A mistake for the 2nd Bengal Light Cavalry.

on more commanding ground, and securing the Cabool road. I had information that a night attack upon the camp was intended from without. Mahomed Ufzul Khan, the eldest son of Dost Mahomed Khan, had been sent by his father with a strong body of troops from Cabool to his brother's assistance at Ghuznee, and was encamped outside the walls, but abandoned his position on our approach, keeping, however, at the distance of a few miles from us. The two rebel chiefs of the Gilzie tribe, men of great influence, viz., Abdool Rhuman and Gool Mahomed Khan, had joined him with 1,500 Horse, and also a body of 3,000 Ghazees from Zeinat under a mixture of chiefs and Moolahs, carrying banners, and who had been assembled on the cry of a religious war. In short, we were, in all directions, surrounded by enemies. These last actually came down the hills on the 22nd, and attacked the part of the camp occupied by His Majesty Shah Shoojah and his troops; but were driven back with considerable loss, and banners taken.

6. "At daylight on the 22nd I reconnoitred Ghuznee, in company with the chief engineer and the Brigadier commanding the Artillery, with the Adjutant and Quartermaster-General of the Bengal Army, for the purpose of making all arrangements for carrying the place by storm, and these were completed in the course of the day. Instead of the tedious process of breaching (for which we were ill prepared), Captain Thomson undertook, with the assistance of Captain Peat of the Bombay Engineers, Lieutenants Durand and MacLeod, of the Bengal Engineers, and other officers under him (Captain Thomson), to blow in the Cabool gate (the weakest point) with gunpowder; and so much faith did I place on the success of the operation, that my plans for the assault were immediately laid down, and the orders given.

7. "The different troops of horse artillery, the camel and foot batteries, moved off to their ground at 12 o'clock that night, without the slightest noise, as had been directed, and in the most correct manner took up the position assigned to them, about 250 yards from the walls; in like manner, and with the same silence, the infantry soon after moved from their ground, and all were at their post at the proper time. A few minutes before 3 o'clock in the morning the 'explosion' took place, and proved completely successful. Captain Peat of the Bombay Engineers was thrown down and stunned by it, but shortly after recovered his senses and feeling. On hearing the advance sounded by the bugles (being the signal for the gate having been blown in) the artillery, under the able directions of Brigadier Stevenson, consisting of Captain Grant's troop of Bengal Horse Artillery, the camel-battery under Captain Abbott, both superintended by Major Pew, Captains Martin's and Cotgrave's troops of Bombay Horse Artillery, and Captain Lloyd's battery of Bombay Foot Artillery, all opened a terrific fire upon the citadel and ramparts of the fort, and in a certain degree paralysed the enemy.

8. "Under the guidance of Captain Thomson of the Bengal Engineers, the chief of the Department, Colonel Dennie, of H.M.'s 13th Light Infantry commanding the advance, consisting of the light companies of H.M.'s 2nd and 17th Foot, and of the Bengal European regiment, with one company of H.M.'s 13th Light Infantry, proceeded to the gate, and with great difficulty, from the rubbish thrown down, and the determined opposition offered by the enemy, effected an entrance and established themselves within the gateway, closely followed by the main column, led in a spirit of great gallantry by Brigadier Sale, to whom I had entrusted the important post of commanding the 'storming party,' consisting (with the advance above mentioned)

of H.M.'s 2nd Foot under Major Carruthers, the Bengal European Regiment under Lieutenant-Colonel Orchard, followed by H.M.'s 13th Light Infantry under Major Tronson, and H.M.'s 17th Regiment under Lieutenant-Colonel Croker. The struggle within the fort was desperate for a considerable time; in addition to the heavy fire kept up, our troops were assailed by the enemy, sword in hand, and with daggers, pistols, etc., but British courage, perseverance, and fortitude overcame all opposition, and the fire of the enemy in the lower area of the fort being nearly silenced, Brigadier Sale turned towards the citadel, from which could now be seen men abandoning their guns, running in all directions, throwing themselves down from immense heights, endeavouring to make their escape, and on reaching the gate, with H.M.'s 17th, under Lieutenant-Colonel Croker, followed by the 13th, forced it open; at 5 o'clock in the morning the colours of H.M.'s 13th and 17th were planted on the citadel of Ghuznee, amid the cheers of all ranks. Instant protection was granted to the women found in the citadel (amongst whom were those of Mahomed Hyder, the Governor), and sentries placed over the magazine for its security. Brigadier Sale reports having received much assistance from Captain Kershaw, of H.M.'s 13th Light Infantry, throughout the whole of the service of the storming.

9. "Major-General Sir W. Cotton executed, in a manner much to my satisfaction, the orders he had received. The Major-General followed closely the assaulting party into the fort, with the 'Reserve,' namely, Brigadier Roberts, with the only available regiment in his brigade, the 35th Native Infantry, under Lieutenant-Colonel Monteath; part of Brigadier Sale's brigade, the 16th Native Infantry under Major McLaren, and 48th Native Infantry under Lieutenant-Colonel Wheeler; and they immediately occupied the ramparts, putting down opposition wherever they met any, and making prisoners until the place was completely in our possession. A desultory fire was kept up in the town long after the citadel was in our hands, from those who had taken refuge in houses, and in desperation kept firing on all that approached them. In this way several of our men were wounded and some killed, but the aggressors paid dearly for their bad conduct in not surrendering when the place was completely ours. I must not omit to mention that the three companies of the 35th Native Infantry under Captain Hay, ordered to the south side of the fort, to begin with a false attack, to attract attention to that side, performed that service at the proper time, and greatly to my satisfaction.

10. "As we were threatened with an attack for the relief of the garrison, I ordered the 19th Bombay Native Infantry, under the command of Lieutenant-Colonel Stalker, to guard the Cabool road, and to be in support of the cavalry division. This might have proved an important position to occupy; but as it was, no enemy appeared.

11. "The cavalry division under Major-General Thackwell, in addition to watching the approach of the enemy, had directions to surround Ghuznee and to sweep the plain, preventing the escape of runaways from the garrison. Brigadier Arnold's brigade (the brigadier himself, I deeply regret to say, was labouring under very severe illness, having shortly before burst a blood-vessel internally, which rendered it wholly impossible for him to mount a horse that day), consisting of H.M.'s 16th Lancers under Lieutenant-Colonel Persse, momentarily commanding the brigade, and Major McDowel, the junior major, the regiment; the senior major of the 16th Lancers, Major Cureton, an officer of great merit, being actively engaged in the execution of his duties as Assistant Adjutant-General of the cavalry division; the 2nd Cavalry under Major Salter, and the 3rd under Lieutenant-Colonel Smyth, were

ordered to watch the south and west sides. Brigadier Scott's brigade was placed on the Cabool road, consisting of H.M.'s 4th Light Dragoons under Major Daly, and the 1st Bombay Cavalry under Lieutenant-Colonel Sandwith, to watch the north and east sides. This duty was performed in a manner greatly to my satisfaction.

12. " After the storming, and that quiet was in some degree restored within, I conducted His Majesty Shah Shoojah-ool-Moolk, and the British Envoy and Minister, Mr. Macnaghten, round the citadel and a great part of the fortress. The king was perfectly astonished at our having made ourselves masters of a place conceived to be impregnable, when defended, in the short space of two hours, and in less than 48 hours after we came before it. His Majesty was of course greatly delighted at the result. When I afterwards, in the course of the day, took Mahomed Hyder Khan, the Governor, first to the British Minister, and then to the king, to make his submission, I informed His Majesty that I had made a promise that his life should not be touched, and the king in very handsome terms assented, and informed Mahomed Hyder in my presence, that although he and his family had been rebels, yet he was willing to forgive and forget all.

13. " Prince Mahomed Hyder, the Governor of Ghuznee, is a prisoner of war in my camp, and under the surveillance of Sir A. Burnes; an arrangement very agreeable to the former.

14. " From Major-General Sir W. Cotton, commanding the First Infantry Division (of the Bengal Army), I have invariably received the strongest support, and on this occasion his exertions were manifest in support of the honour of the profession and of our country.

15. " I have likewise at all times received able assistance from Major-General Willshire, commanding the Second Infantry Division (of the Bombay Army), which it was found expedient on that day to break up, some for the storming party, and some for other duties; the Major-General, as directed, was in attendance upon myself.

16. " To Brigadier Sale I feel deeply indebted for the gallant and soldier-like manner in which he conducted the responsible and arduous duty entrusted to him, in command of the storming party, and for the arrangements he made in the citadel immediately after taking possession of it. The sabre wound which he received in the face did not prevent his continuing to direct his column until everything was secure, and I am happy in the opportunity of bringing to your Lordship's notice the excellent conduct of Brigadier Sale on this occasion.

17. " Brigadier Stevenson, in command of the artillery, was all I could wish; and he reports that Brigade Majors Backhouse and Coghlan ably assisted him; his arrangements were good, and the execution done by the arm he commands was such as cannot be forgotten by those of the enemy who have witnessed and survived it.

18. " To Brigadier Roberts, to Colonel Dennie (who commanded the advance), and to the different officers commanding regiments already mentioned, as well as to the officers and gallant soldiers under them, who so nobly maintained the honour and reputation of our country, my best acknowledgments are due.

19. " To Captain Thomson, of the Bengal Engineers, the chief of the department with me, much of the credit of the success of this brilliant 'coup-de-main' is due; a place of the same strength, and by such simple means as this highly talented and scientific officer recommended to be tried, has perhaps never before been taken, and I feel I cannot do sufficient justice to Captain Thomson's merits for his conduct throughout. In the execution he was ably supported by the officers already mentioned; and so eager were the other officers of the Engineers, of both Presidencies, for the honour of carrying the powder

bags, that the point could only be decided by seniority, which shows the fine feeling by which they are animated.

20. "I must now inform your Lordship that since I joined the Bengal column in the valley of Shawl, I have continued my march with it in the advance, and it has been my good fortune to have had the assistance of two most efficient staff officers, in Major Craigie, Deputy Adjutant-General and Major Garden, Deputy Quartermaster-General. It is but justice to these officers that I should state to your Lordship the high satisfaction I have derived from the manner in which all their duties have been performed up to this day; and I look upon them as promising officers to fill the higher ranks. To the other officers of both departments I am also much indebted for the correct performance of all duties appertaining to their situations.

21. "To Major Keith, the Deputy Adjutant-General, and Major Campbell, the Deputy Quartermaster-General of the Bombay Army, and to all the other officers of both departments under them, my acknowledgments are also due, for the manner in which their duties have been performed during this campaign.

22. "Captain Alexander, commanding the 4th Local Horse, and Major Cunningham, commanding the Poonah Auxiliary Horse, with the men under their orders, have been of essential service to the army in this campaign.

"The arrangements made by Superintending Surgeons Kennedy and Atkinson previous to the storming, for affording assistance and comfort to the wounded, met with my approval.

23. "Major Parsons, the Deputy Commissary-General in charge of the department in the field, has been unremitting in his attention to keep the troops supplied, although much difficulty is experienced, and he is occasionally thwarted by the nature of the country and its inhabitants.

24. "I have, throughout this service, received the utmost assistance I could desire from Lieutenant-Colonel Macdonald, my officiating military secretary, and Deputy Adjutant-General H.M.'s Forces, Bombay; from Captain Powell, my Persian interpreter, and the other officers of my personal staff. The nature of the country in which we are serving prevents the possibility of my sending a single staff officer to deliver this to your Lordship, otherwise I should have asked my aide-de-camp, Lieutenant Keane, to proceed to Simla, to deliver this despatch in your hands, and to have afforded any further information that your Lordship could have desired.

25. "The brilliant triumph we have obtained, the cool courage displayed, and the gallant bearing of the troops I have the honour to command, will have taught such a lesson to our enemies in the Afghan nation, as will make them hereafter respect the name of a British soldier.

26. "Our loss is wonderfully small, considering the occasion; the casualties in killed and wounded amount to about 200.

27. "The loss of the enemy is immense; we have already buried of their dead nearly 500, together with an immense number of horses.

28. "I enclose a list of the killed, wounded, and missing. I am happy to say that although the wounds of some of the officers are severe, they are all doing well.

29. "It is my intention, after selecting a garrison for this place, and establishing a general hospital, to continue my march to Cabool forthwith.

"I have, etc.,

(*Signed*) "J. Keane, *Lieutenant-General.*"

General Orders by the Commander of the Forces: Headquarters, Meerut, November 22nd, 1839. *By the Right Honourable the Governor-General, Camp Somalka, November 19th,* 1839.

The following General Orders, issued by the Right Honourable the Governor-General in the Secret Department under date the 18th instant, are published for general information to the army.

"General Orders by the Right Honourable the Governor-General of India.

"Secret Department, Camp, Paniput, *November 18th*, 1839.

1. "Intelligence was this day received of the arrival, within the Peshawar territory, of His Excellency Lieutenant-General Sir John Keane, K.C.B. and G.C.H., Commander-in-Chief of the Army of the Indus, with a portion of that force on its return to the British provinces. The military operations under the direction of His Excellency having now been brought to a close, the Right Honourable the Governor-General has, on the part of the Government of India, to acquit himself of the gratifying duty of offering publicly his warmest thanks to His Excellency, and to the officers and men who have served under his command, for the soldierlike spirit and conduct of all ranks throughout the late campaign, and he again cordially congratulates them on the attainment of the great objects of national security and honour for which the expedition was undertaken.

2. "The plans of aggression, by which the British empire in India was dangerously threatened, have, under Providence, been arrested. The Chiefs of Cabool and Candahar, who had joined in hostile designs against us, have been deprived of power, and the territories which they ruled have been restored to the Government of a friendly monarch. The Ameers of Scinde have acknowledged the supremacy of the British Government, and ranged themselves under its protection; their country will now be an outwork of defence, and the navigation of the Indus within their dominions, exempt from all duties, has been opened to commercial enterprise. With the allied Government of the Seikhs, the closest harmony has been maintained; and on the side of Herat, the British alliance has been courted, and a good understanding, with a view to common safety, has been established with that power.

3. "For these important results the Governor-General is proud to express the acknowledgments of the Government to the Army of the Indus, which, alike by its valour, its discipline, and cheerfulness under hardships and privations, and its conciliatory conduct to the inhabitants of the countries through which it passed, has earned respect for the British name, and has confirmed in central Asia a just impression of British energy and resources.

4. "The native and European soldiers have vied with each other in effort and endurance. A march of extraordinary length,[1] through difficult and untried countries, has been within a few months successfully accomplished; and in the capture of the one stronghold where resistance was attempted, a trophy of victory has been won, which will add a fresh lustre to the reputation of the armies of India.

5. "To Lieutenant-General Sir John Keane, the Commander-in-Chief of the Army, the Governor-General would particularly declare his thanks for his direction of these honourable achievements. He would especially acknowledge the marked forbearance and just appreciation of the views of the Government which guided His

[1] More than 1,700 miles; on arriving at Ferozepoor, 2,070 miles.

Excellency in his intercourse with the Ameers of Scinde. He feels the Government to be under the deepest obligations to His Excellency, for the unshaken firmness of purpose with which, throughout the whole course of the operations, obstacles and discouragements were disregarded, and the prescribed objects of policy were pursued, and, above all, he would warmly applaud the decisive judgment with which the attack upon the fortress of Ghuznee was planned, and its capture effected; nor would he omit to remark upon that spirit of perfect co-operation with which His Excellency gave all support to the political authorities with whom he was associated. Mr. Macnaghten, the Envoy and Minister at the Court of Shah Shoojah-ool-Moolk, and Colonel Pottinger, the Resident in Scinde, have been chiefly enabled by the cordial good understanding which has throughout subsisted between them and His Excellency, to render the important services by which they have entitled themselves to the high approbation of the Government; and his Lordship has much pleasure in noticing the feelings of satisfaction with which His Excellency regarded the valuable services of Lieutenant-Colonel Sir A. Burnes, who was politically attached to him in the advance upon Ghuznee.

6. "The Governor-General would follow His Excellency the Commander-in-Chief in acknowledging the manner in which Major-General Sir Willoughby Cotton, K.C.B. and K.C.H., exercised his command of the Bengal division throughout the campaign, and supported the honour of his country on July 23rd; and His Lordship would also offer the thanks of the Government to Major-General Willshire, C.B., commanding the Second Infantry Division; to Major-General Thackwell, C.B. and K.H., commanding the cavalry division; to Brigadier Roberts, commanding the 4th Infantry Brigade; to Brigadier Stevenson, commanding the artillery of the army; to Brigadier Scott, commanding the Bombay Cavalry brigade ; and to Brigadier Persse, upon whom, on the lamented death of the late Brigadier Arnold, devolved the command of the Bengal Cavalry brigade; as well as to the commandants of corps and detachments, with the officers and men under their respective commands; and to the officers at the head of the several departments, with all of whom His Excellency the Commander-in-Chief has expressed his high satisfaction.

7. "To Brigadier Sale, C.B., already honunrably distinguished in the annals of Indian warfare, who commanded the storming party at Ghuznee; to Lieutenant-Colonel Dennie, C.B., who led the advance on the same occasion; and to Captain George Thomson, of the Bengal Engineers, whose services in the capture of that fortress have been noticed in marked terms of commendation by His Excellency the Commander-in-Chief; and to Captain Peat, of the Bombay Engineers, and Lieutenants Durand and MacLeod, of the Bengal Engineers, and the other officers and men of the Bengal and Bombay Engineers under their command, the Governor-General would especially tender the expression of his admiration of the gallantry and science which they respectively displayed in the execution of the important duties confided to them in that memorable operation.

8. "In testimony of the services of the Army of the Indus, the Governor-General is pleased to resolve that all the corps, European and native, in the services of the East India Company, which proceeded beyond the Bolan Pass, shall have on their regimental colours the word 'Afghanistan,' and such of them as were employed in the reduction of the fortress of that name, the word 'Ghuznee' in addition.

"In behalf of the Queen's regiments, the Governor-General will recommend to Her Majesty, through the proper channel, that the same distinction may be granted to them.

9. "The Governor-General would here notice with approbation the praiseworthy conduct, during this expedition, of the officers and men attached to the disciplined force of His Majesty Shah Shoojah-ool-Moolk. This force was newly raised, and opportunities had not been afforded for its perfect organisation and instruction; but it shared honourably in the labours and difficulties of the campaign, and it had the good fortune, in repelling an attack made by the enemy in force, on the day prior to the storming of Ghuznee, to be enabled to give promise of the excellent services which may hereafter be expected from it.

10. "His Lordship has also much satisfaction in adding that the best acknowledgments of the Government are due to Lieutenant-Colonel Wade, who was employed upon the Peshawar frontier, and who, gallantly supported by the officers and men of all ranks under him, and seconded by the cordial aid of the Seikh Government—an aid the more honourable because rendered at a painful crisis of its affairs—opened the Khyber Pass, and overthrew the authority of the enemy in that quarter, at the moment when the advance of the forces of the Shah Zadah Tymoor could most conduce to the success of the general operations.

"By command, etc.

(*Sgd.*) "T. H. Maddock, *Officiating Secretary to Government of India, with the Governor-General.*

(*Sgd.*) "J. Stewart, *Lieutenant-Colonel, Secretary to the Government of India, Military Department, with the Right Honourable the Governor-General.*

"By order of the Commander of the Forces.

(*Sgd.*) "J. R. Lumley, *Major-General, Adjutant-General of the Army.*"

APPENDIX B

DESPATCHES. GWALIOR

"GENERAL ORDERS BY THE RIGHT HONOURABLE THE GOVERNOR-GENERAL OF INDIA.

"CAMP, GWALIOR RESIDENCY, *January 4th*, 1844.

"The Governor-General directs the publication of the annexed despatch, from His Excellency the Commander-in-Chief, reporting the operations of the corps under His Excellency's immediate command, and of that under the command of Major-General Grey, against the mutinous troops which overawed and controlled the Government of His Highness the Maharajah Jyajee Rao Scindiah, and attacked the British forces, on their advance to Gwalior to His Highness's support.

"The Governor-General deeply laments the severe loss in killed and wounded which has been sustained in these operations; but it has been sustained in the execution of a great and necessary service, and the victories of Maharajpore and Punniar, while they have shed new glory upon the British army, have restored the authority of the Maharajah, and have given new security to the British empire in India.

"The Governor-General cordially congratulates His Excellency the Commander-in-Chief upon the success of his able combinations, by which two victories were obtained on the same day, and the two wings of the army, proceeding from distant points, have been now united under the walls of Gwalior.

"To His Excellency, and to Major-General Grey, and to all the generals and other officers, and to all the soldiers of the army, the Governor-General, in the name of the Government, and of all the people of India, offers his most grateful acknowledgments of the distinguished service they have performed; nor can he withhold the tribute of his admiration justly due to the devoted courage manifested by all ranks in action with brave enemies, who yielded their numerous and well-served artillery only with their lives.

"The Governor-General's special thanks are due to Her Majesty's 39th and 40th Regiments, to the 2nd and 16th Regiments of Native Grenadiers, to the 56th Regiment of Native Infantry, which took with the bayonet the batteries in front of Maharajpore.

"Her Majesty's 39th Regiment had the peculiar fortune of adding to the honour of having won at Plassy, the first great battle which laid the foundation of the British empire in India, the further honour of thus contributing to this, as it may be hoped, the last and crowning victory by which that empire has been secured.

"Her Majesty's 40th Regiment, and the 2nd and 16th Regiments of Native Grenadiers, again serving together, again displayed their pre-eminent qualities as soldiers, and well supported the character of the ever-victorious army of Candahar.

"The corps of Major-General Grey, suddenly attacked at Punniar, after a long march, carried the several strong positions of the enemy with a resolution no advantage of ground could enable him to withstand; and Her Majesty's 3rd Buffs and 50th Regiment added new lustre to the reputation they gained in the Peninsular War.

"Everywhere, at Maharajpore and at Punniar, the British and the native troops, emulating each other, and animated by the same spirit of military devotion, proved that an army so composed, and united by the bonds of mutual esteem and confidence, must ever remain invincible in Asia.

"The Government of India will, as a mark of its grateful sense of their distinguished merit, present to every general and other officer, and to every soldier engaged in the battles of Maharajpore and Punniar, an Indian star of bronze, made out of the guns taken at those battles, and all officers and soldiers in the service of the Government of India will be permitted to wear the star with their uniforms.

"His Excellency the Commander-in-Chief is requested to furnish the Governor-General with nominal rolls of all the officers and soldiers engaged in the two battles respectively, in order that the star presented to each may be inscribed with the name of the battle in which he was engaged.

"A triumphal monument, commemorative of the campaign of Gwalior, will be erected at Calcutta, and inscribed with the names of all who fell in the two battles.

"The Governor-General directs that the words 'Maharajpore' and 'Punniar' shall be borne upon the colours or standards and appointments of the several regiments, troops, and companies named below, as respectively engaged in those battles:

"MAHARAJPORE.

"2nd troop Second Brigade Horse Artillery; 3rd ditto, ditto, ditto; Second ditto Third Brigade ditto; 1st company 1st Battalion of Foot Artillery; 1st ditto, 4th ditto, ditto; Her Majesty's 16th Lancers; Governor-General's Bodyguard; 1st Regiment Light Cavalry; 4th ditto, ditto; 10th ditto, ditto; 4th Regiment Irregular Cavalry; Detachment of 5th Light Cavalry; ditto, 8th ditto; 3rd, 4th, and 5th companies of Sappers and Miners; Her Majesty's 39th Regiment of Foot; ditto, 40th ditto; 2nd Regiment of Grenadiers; 14th Regiment of Native Infantry; 16th ditto of Grenadiers; 31st ditto of Native Infantry; 43rd ditto of Light Infantry; 56th ditto of Native Infantry; 6th Company 39th Native Infantry; flank companies of the Khelat-i-Ghilzie Regiment. The few men of No. 1 company of the Khelat-i-Ghilzie Regiment, and of the cavalry and infantry, of the Bundelcund legion, who were present, will receive stars.

"PUNNIAR.

"1st troop Third Brigade Horse Artillery; 3rd ditto, ditto, ditto; 6th company 6th Battalion Foot Artillery; 2 squadrons Her Majesty's 9th Lancers; 2 ditto 5th Regiment Light Cavalry; 2 ditto 11th ditto, ditto; 8th Regiment Light Cavalry; 8th Regiment Irregular Cavalry; 1st Company of Sappers and Miners; Her Majesty's 3rd Buffs; Her Majesty's 50th Foot; 39th Regiment Native Infantry; 50th ditto, ditto; 51st ditto, ditto; 58th ditto, ditto. The cavalry, infantry, and artillery of the Seepree contingent.

"A royal salute and a *feu-de-joie* will be fired at all the stations of the army, on the receipt of this Order.

"By order of the Right Honourable the Governor-General of India.

(*Signed*) "F. CURRIE,
"*Secretary to the Government of India,*
"*with the Governor-General.*"

"FROM HIS EXCELLENCY GENERAL SIR H. GOUGH, BART., G.C.B., *Commander-in-Chief, East Indies.*

"TO THE RIGHT HONOURABLE THE GOVERNOR-GENERAL, ETC., ETC., ETC.

DATED HEADQUARTERS CAMP, BEFORE GWALIOR, *January 4th*, 1844.

"MY LORD,

"Your Lordship having witnessed the operations of the 29th, and being in possession, from my frequent communications, of my military arrangements for the attack on the Mahratta army in its strong position of Chonda, I do not feel it necessary to enter much into detail, either as to the enemy's position or the dispositions I made for attacking it; I shall here merely observe that it was peculiarly well-chosen, and most obstinately defended; indeed, I may safely assert that I never witnessed guns better served, nor a body of infantry apparently more devoted to the protection of their regimental guns, held by the Mahratta corps as objects of worship.

"I previously communicated to your Lordship that my intention was to have turned the enemy's left flank by Brigadier Cureton's brigade of cavalry, consisting of Her Majesty's 16th Lancers under Lieutenant-Colonel McDowell, your Lordship's Bodyguard under Captain Dawkins, 1st Regiment of Light Cavalry under Major Crommelin, C.B., 4th Irregular Cavalry under Major Oldfield, C.B., with Major Lane's and Major Alexander's troops of horse artillery under Brigadier Gowan; the whole under the orders of Major-General Sir J. Thackwell, K.C.B. and K.H., commanding the cavalry division.

"With this force the Third Brigade of Infantry under Major-General Valiant, K.H., was directed to co-operate, consisting of Her Majesty's 40th Regiment of Foot under Major Stopford, 2nd Regiment of Grenadiers under Lieutenant-Colonel Hamilton, and the 16th Regiment of Grenadiers under Lieutenant-Colonel McLaren, C.B.

"It is equally known to your Lordship that I proposed to have attacked the centre, with Brigadier Stacy's brigade of the Second Division of Infantry, which consisted of the 14th Regiment Native Infantry under Lieutenant-Colonel Gairdner, the 31st Native Infantry under Lieutenant-Colonel Weston, and the 43rd Light Infantry under Major Nash, to which I attached No. 17 Light Field Battery under Captain Browne; the whole under Major-General Dennis, commanding the Second Division of Infantry.

"The force was to have been supported by Brigadier Wright's brigade, composed of Her Majesty's 39th Regiment commanded by Major Bray, and the 56th Native Infantry under Major Dick, with No. 10 Light Field Battery under Brevet-Major Sanders; Major-General Little, commanding the Third Division of Infantry, superintended the movement of this column.

"On the left of this force I placed the Fourth Brigade of Cavalry under Brigadier Scott, C.B., of Her Majesty's 9th Lancers, consisting of the 4th Light Cavalry (Lancers) under Major Mactier, and the 10th Light Cavalry under Lieutenant-Colonel Pope, with Captain Grant's troop of horse artillery; with this force I proposed threatening the right flank of the enemy's position.

"Your Lordship is aware of the extreme difficulty of the country, intersected by deep and almost impassable ravines, which were only made practicable by the unremitting labour of Major Smith, with the Sappers; and that I had to pass the Koharee river in three columns at considerable distances, on the morning of the 29th; but, by the judicious movements of their respective leaders, the whole were in the appointed positions by 8 o'clock a.m., about a mile in front of Maharajpore.

"I found the Mahrattas had occupied this very strong position during the previous night, by seven regiments of infantry, with their guns, which they entrenched; each corps having four guns, which opened on our advances. This obliged me to alter, in some measure, my disposition.

"Major-General Littler's column being exactly in front of Maharajpore, I ordered it to advance direct, while Major-General Valiant's brigade took it in reverse, both supported by Major-General Dennis's column, and the two light field batteries.

"Your Lordship must have witnessed, with the same pride and pleasure that I did, the brilliant advance of these columns under their respective leaders; the European and native soldiers appearing emulous to prove their loyalty and devotion. And here I must do justice to the gallantry of their opponents, who received the shock without flinching; their guns doing severe execution as we advanced, but nothing could withstand the rush of British soldiers.

"Her Majesty's 39th Foot, with their accustomed dash, ably supported by the 56th Regiment Native Infantry, drove the enemy from their guns into the village, bayoneting the gunners at their posts. Here a most sanguinary conflict ensued: the Mahratta troops, after discharging their matchlocks, fought sword in hand with the most determined courage.

"General Valiant's brigade, with equal enthusiasm, took Maharajpore in reverse, and 28 guns were captured by this combined movement; so desperate was the resistance, that very few of the defenders of this very strong position escaped. During these operations Brigadier Scott was opposed by a body of the enemy's cavalry on the extreme left, and made some well-executed charges, most ably supported by Captain Grant's troop of horse artillery and 4th Lancers, capturing some guns and taking two standards, thus threatening the right flank of the enemy.

"In conformity with previous instructions, Major-General Valiant, supported by the Third Cavalry Brigade, moved on the right of the enemy's position at Chonda. During the advance Major-General Valiant had to take in succession three strong entrenched positions, where the enemy defended their guns with frantic desperation; Her Majesty's 40th Regiment losing two successive commanding officers, Major Stopford and Captain Coddington, who fell wounded at the very muzzles of the guns, and capturing four regimental standards. This corps was ably and nobly supported by the 2nd Grenadiers, who captured two regimental standards, and by the 16th Grenadiers, under Lieutenant-Colonels Hamilton and McLaren. Too much praise cannot be given to these three regiments.

"Major-General Littler, with Brigadier Wright's brigade, after dispersing the right of the enemy's position at Maharajpore, steadily advanced to fulfil his instructions of attacking the main position at Chonda in front, supported most ably by Captain Grant's troop of horse artillery and the 1st Regiment of Light Cavalry. This column had to advance under a very severe fire over very difficult ground, but when within a short distance, again the rush of the 39th

Regiment, as before, under Major Bray, gallantly supported by the 56th Regiment under Major Dick, carried everything before them, and thus gained the entrenched main position of Chonda. In this charge the 39th Regiment lost the services of its brave commanding officer, Major Bray, who was desperately wounded by the blowing up of one of the enemy's tumbrils in the midst of the corps, and were ably brought out of action by Major Straubenzee. This gallant corps on this occasion captured two regimental standards.

"A small work of four guns on the left of this position was long and obstinately defended, but subsequently carried, and the guns captured by the Grenadiers of the 39th under Captain Campbell, admirably supported by a wing of the 56th Native Infantry under Major Phillips.

"Brigadier Cureton's brigade of cavalry taking advantage of every opportunity, manœuvred most judiciously on the right, and would have got in rear of the position, and cut off the retreat of the whole, had they not been prevented by an impassable ravine. I witnessed with much pride the rapidity of movement of the three troops of horse artillery, which bore a conspicuous part in this well-contested action; their leaders promptly brought them forward in every available position, and the precision of their fire was admirable. With the two light field batteries I have every reason to be pleased: they well supported the high character of the Bengal artillery.

"I was greatly gratified with a spirited charge made by Major Oldfield, C.B., of the 4th Irregular Cavalry, who had been left to cover Major Alexander's troop of horse artillery, and who charged, by my orders, a considerable body of the enemy's infantry, who were moving off from the right position of Maharajpore. Two guns and two standards rewarded this charge.

"I was likewise much pleased with a charge made by my personal escort under Cornet Stannus, which did great credit to himself and the small body of the 5th Light Cavalry which formed it.

"Several acts of individual heroism occurred on this day; none exceeded those of Major-General Churchill, C.B., Quartermaster-General of Her Majesty's forces in India, and Captain Somerset, of the Grenadier Guards, your Lordship's military secretary, whom you kindly allowed to act on my staff, and whom I sent with Brigadier Cureton's brigade, to communicate to me the movements of that corps. These two gallant officers nobly fell, having received several wounds in personal rencontre. In Major-General Churchill, Her Majesty's service will experience a great loss; he died during the night, after amputation of the leg, but I am glad to add there is every hope that Captain Somerset will do well, though severely wounded.

"I regret to say that our loss has been very severe, infinitely beyond what I calculated on; indeed I did not do justice to the gallantry of my opponents. Their force, however, so greatly exceeded ours, particularly in artillery; the position of their guns was so commanding; they were so well served, and determinedly defended, both by their gunners and their infantry, and the peculiar difficulties of the country giving the defending force so great advantages, that it could not be otherwise.

"In the return of killed I greatly deplore the loss of Lieutenant-Colonel E. Sanders, C.B., of the Engineers, than whom this army, with its numerous list of devoted soldiers, could not boast a more promising, nor a more enthusiastic officer.

"It is also my painful duty to record my deep regret at the loss of a most valuable officer, Major Crommelin, C.B., of the 1st Light Cavalry, who died of wounds received when nobly leading his corps in support of the Fifth Brigade of Infantry.

"Your Lordship is aware that I had collected a strong force in Bundlecund under Major-General J. Grey, C.B., to co-operate with me, and that both corps crossed the Scindian frontier, from the north-east and south-west, at the same time. It may, therefore, be necessary here to observe that, on finding that all your Lordship's strenuous attempts to maintain those friendly relations which had hitherto existed between the two Governments were fruitless, and that the Mahratta army, the ruling power, appeared determined to rest the fate of the country on the hazard of a general action, I instructed Major-General Grey, with the troops under his command, to push on as rapidly as practicable to Punniar, twelve miles south-west of Gwalior, thus placing the Mahratta army between two corps capable of supporting each other, should it remain in the vicinity of its capital, or of sub-dividing that army to repel or attack these two columns. They adopted the latter alternative, and, under Divine guidance, the consequence has been most decisive and honourable to the British arms.

"I beg now to enclose Major-General Grey's report, detailing his movements and operations, which were as creditable to this meritorious officer as the result has proved highly honourable to the brave troops he had under his command. I beg earnestly to draw your Lordship's favourable attention to their conduct, and to the recommendations of Major-General Grey.

"I beg also to bring to your Lordship's notice the several officers named by those in command of the separate columns. In this list I shall not re-name these, whom it is my intention to bring to your Lordship's favourable consideration as commanding divisions and brigades, and on the staff, at the conclusion of this report.

"Major-General Sir Joseph Thackwell, commanding the cavalry division, mentions having received every assistance from Captain Pratt, 16th Lancers, Assistant Adjutant-General; Captain Clayton, 4th Light Cavalry, Assistant Quartermaster-General, and Lieutenant Pattinson, 16th Lancers, Brigade-Major of Cavalry; Lieutenant Cowell, 3rd Light Dragoons, aide-de-camp, and Captain Herries, 3rd Light Dragoons, your Lordship's aide-de-camp, who acted as aide-de-camp to Sir J. Thackwell on this occasion, and Lieutenant Renny, of the Engineers, attached to the cavalry division.

"Major-General Dennis mentions having derived every assistance from Captain McKie, H.M.'s 3rd Buffs, Assistant Adjutant-General; Lieutenant Sneyd, 57th Native Infantry, Acting Deputy Assistant Quartermaster-General of the Second Division of Infantry, and Lieutenant Dowman, Her Majesty's 40th Foot, acting aide-de-camp.

"Major-General Littler strongly brings to notice Major Bray, commanding 39th Regiment (severely wounded), and Major Straubenzee, of H.M.'s 39th Regiment; Majors Dick and Phillip, 56th Native Infantry; Major Ryan, Her Majesty's 50th Regiment; Captain Nixon, H.M.'s 39th Regiment, Brigade-Major; Captain Campbell, H.M.'s 39th Grenadier company; Captain Graves, 16th Grenadiers, Assistant Adjutant-General, and Lieutenant Croker, H.M.'s 39th Regiment, Assistant Quartermaster-General, Fifth Brigade of Infantry; Captain Harris, 70th Native Infantry, his aide-de-camp, and Captain Alcock, 46th; Captain Johnstone, 46th, and Lieutenant Vaughan, 21st Native Infantry, acting aide-de-camp.

"Major-General Valiant has also strongly brought to my notice the conduct of Lieutenant-Colonel McLaren, C.B., 16th Grenadiers; Lieutenant-Colonel Hamilton, 2nd Grenadiers; Major Stopford and Captain Coddington (both severely wounded), and Captain Oliver, successively commanding Her Majesty's 40th Regiment; Captain Manning, 16th Grenadiers; Captain Young, 2nd Grenadiers; Brigade Major, Captain

Abercrombie, Engineers, and Lieutenant Nelson, H.M.'s 40th Foot, his aide-de-camp.

"Brigadier Gowan strongly recommends Captain McDonald, Deputy Assistant Adjutant-General of Artillery.

"Although it was not their good fortune to come into action, I feel it is due to Brigadier Tennant, and the officers and men of the foot artillery, to express my thanks for the great exertions they made to bring up the heavy ordnance, which alone was prevented from opening on Maharajpore by the rapidity of the movement of the attacking columns, and from the action having commenced earlier than I expected. My thanks are likewise due to Brigadier Riley, and that portion of the Sixth Brigade of Infantry which formed the rearguard, for the manner in which he protected and brought forward the immense train of stores, provisions, and baggage which accompanied the army, over so great an extent of the enemy's country.

"I have thus, my Lord, attempted to detail a series of movements, many of which came under your Lordship's observation. It now becomes my duty, as Commander-in-Chief of this army, to do justice to those gallant men who, I feel, I can conscientiously assure your Lordship, merited my warmest approbation; and although it fell to the lot of some, as in all actions it must, to be more prominently forward than others, yet I am proud to say I found in all arms the most animated enthusiasm and the most able support.

"To Major-General Sir Joseph Thackwell, K.C.B., to Major-Generals Dennis and Littler, and to Brigadier Gowan, commanding divisions, my best thanks are due, for the manner in which they conducted and led their respective divisions.

"To Major-General Valiant, K.H., and Brigadiers Scott, Stacy, Cureton, and Wright, I am equally indebted, for their gallantry and exertions in command of brigades; and to the several officers commanding corps and troops of horse artillery and light field batteries, named in a foregoing part of this despatch, I feel called on to express my warmest approval. They nobly led and were gallantly supported by the officers and men of their respective corps.

"I beg to bring to your Lordship's notice the high professional talents of Major Smith, from whom I derived every assistance, as well as from several officers of the Engineer department.

"To the general staff I am greatly indebted: from Major-General Smith, C.B., Adjutant-General of Her Majesty's forces in India, and Major-General Lumley, Adjutant-General of the army, I experienced the most cordial support.

"To Lieutenant-Colonel Garden, Quartermaster-General of the army, whose exertions were as creditable as they were unceasing, I am very much indebted; to Major Grant, Deputy Adjutant-General of the army; to Major Barr, Assistant Adjutant-General, Her Majesty's forces, and to Major Drummond, Deputy Quartermaster-General of the army, I feel under much obligation, for their zeal and assistance.

"The arrangements of Lieutenant-Colonel Burlton, Commissary-General, who accompanied me in the field, and Captain Ramsay, Assistant Commissary-General, and principal executive officer, have been very judicious.

"Lieutenant-Colonel Birch, Judge-Advocate-General, accompanied me in the field, and rendered me his assistance. The exertions of the junior officers in the several departments well justify my most favourable report of them, especially Lieutenant W. Fraser Tytler, Deputy Assistant Quartermaster-General, and Captain Ekins, Assistant Adjutant-General of the army.

"I have every reason to be satisfied with the conduct of Superintending Surgeon Wood, and Field-Surgeon Chalmers. This latter officer peculiarly and most creditably exerted himself in the field hospital at Maharajpore.

"To Lieutenant-Colonel Gough, C.B., H.M.'s 3rd Light Dragoons, my military secretary, and the other officers composing my personal staff; Major Havelock, C.B., H.M.'s 13th Light Infantry, Persian interpreter; my aides-de-camp, Lieutenant Frend, of H.M.'s 31st Foot (who was severely wounded, and had his right arm amputated in the field); Captain R. Smith, 28th Regiment of Native Infantry; Captain Evans, 26th Light Infantry, and Lieutenant Bagot, of the 15th Native Infantry, I am greatly indebted, for their alacrity and zeal in carrying my orders; also my best thanks are due to Captain Sir R. Shakespear, Bt., of the artillery; to Captain Curtis, 37th Native Infantry; Sub-Assistant Commissary-General; to Lieutenant Macdonald, of the 2nd Madras Light Cavalry, and to Lieutenant Hayes, of the 62nd Native Infantry, all of whom acted as my aides-de-camp during the day.

"I must not omit to mention Assistant-Surgeon Stephens, M.D., H.M.'s 62nd Regiment, surgeon on my personal staff, who accompanied me in the field, and was attentive and useful to the wounded in the field.

"I have the honour to enclose a list of our killed and wounded; that of the enemy must have been exceedingly great.

"By the accompanying returns your Lordship will perceive we have captured in the action of Maharajpore, 56 guns, and the whole of the enemy's ammunition waggons.

"I feel I have been led into a much longer detail than I had intended, and have brought to your Lordship's notice a very lengthened list of officers who led, and troops who achieved a victory that, in one day, has brought a once powerful nation, and undoubtedly brave army, to feel the indomitable power of the British arms, thus securing the internal peace of central India.

"I have the honour to be, my Lord,

"Your Lordship's obedient and humble servant,

(*Signed*) "H. Gough, *General*,

"*Commander-in-Chief, East Indies.*"

APPENDIX C

DESPATCHES. SUTLEJ CAMPAIGN

SOBRAON

"General Order by the Right Honourable the Governor-General of India.

"Camp, Kussoor, *February 24th*, 1846.

"The Governor-General, having received from His Excellency the Commander-in-Chief, the despatch annexed to this paper, announces to the army and the people of India, for the fourth time during this campaign, a most important and memorable victory obtained by the army of the Sutlej over the Sikh forces at Sobraon, on the 10th instant.

"On that day the enemy's strongly entrenched camp, defended by 35,000 men, and 67 pieces of artillery, exclusive of heavy guns on the opposite bank of the river, was stormed by the British army, under the immediate command of His Excellency Sir Hugh Gough, Bart., G.C.B., and in two hours the Sikh forces were driven into the river with immense loss, 67 guns being captured by the victors.

"The Governor-General most cordially congratulates the Commander-in-Chief and the British army on this exploit, one of the most daring ever achieved, by which, in open day, a triple line of breastworks, flanked by formidable redoubts, bristling with artillery, manned by 32 regular regiments of infantry, was assaulted and carried by forces under His Excellency's command.

"This important operation was most judiciously preceded by a cannonade from the heavy howitzers and mortars, which had arrived from Delhi on the 8th instant, the same day on which the forces under Major-General Sir Harry Smith, which had been detached to Loodiana, and had gained the victory of Aliwal, rejoined the Commander-in-Chief's camp.

"The vertical fire of the heavy ordnance had the effect intended by His Excellency; it shook the enemy's confidence in works so well and so laboriously constructed, and compelled them to seek shelter in the broken ground within their camp.

"The British infantry, formed on the extreme left of the line, then advanced to the assault, and, in spite of every impediment, cleared the entrenchments, and entered the enemy's camp. H.M.'s 10th, 53rd, and 80th Regiments, with the 33rd, 43rd, 59th, and 63rd Native Infantry, moving at a firm and steady pace, never fired a shot till they had passed the barriers opposed to them, a forbearance much to be commended and most worthy of constant imitation, to which may be attributed the success of their first effort, and the small loss they sustained. This attack was crowned with the success it deserved, and (led by its gallant commander, Major-General Sir Robert Dick) obtained the admiration of the army, which witnessed its disciplined valour. When checked by the formidable obstacles and superior numbers to which the attacking division was exposed, the Second Division, under Major-General Gilbert, afforded the most opportune assistance by rapidly advancing to the attack of the enemy's batteries, entering their fortified position after a severe struggle, and sweeping through the interior of the camp. This division inflicted a very severe loss on the retreating enemy.

"The same gallant efforts, attended by the same success, distinguished the attack of the enemy's left, made by the First Division, under the command of Major-General Sir Harry Smith, K.C.B., in which the troops nobly sustained their former reputation.

"These three divisions of infantry, concentrated within the enemy's camp, drove his shattered forces into the river, with a loss which far exceeded that which the most experienced officers had ever witnessed.

"Thus terminated, in the brief space of two hours, this most remarkable conflict, in which the military combinations of the Commander-in-Chief were fully and ably carried into effect with His Excellency's characteristic energy. The enemy's select regiments of regular infantry have been dispersed, and a large proportion destroyed, with the loss, since the campaign began, of 220 pieces of artillery taken in action.

"The same evening, six regiments of native infantry crossed the Sutlej; on the following day the bridge of boats was nearly completed by that able and indefatigable officer, Major Abbott, of the Engineers, and the army is this day encamped at Kussoor, 32 miles from Lahore.

"The Governor-General again most cordially congratulates the Commander-in-Chief on the important results obtained by this memorable achievement. The Governor-General, in the name of the Government and of the people of India, offers to His Excellency the Commander-in-Chief, to the general officers, and all the other officers and troops under their command, his grateful and heartfelt acknowledgments for the services they have performed.

"To commemorate this great victory, the Governor-General will cause a medal to be struck, with 'SOBRAON' engraved upon it, to be presented to the victorious army in the service of the East India Company, and requests His Excellency the Commander-in-Chief to forward the lists usually furnished of those engaged.

"The Governor-General deeply regrets the loss of the brave officers and men who have fallen on this occasion. Major-General Sir Robert Dick, K.C.B., who led the attack, received a mortal wound after he had entered the enemy's entrenchments. Thus fell, most gloriously, at the moment of victory, this veteran officer, displaying the same energy and intrepidity as when, 35 years ago, in Spain, he was the distinguished leader of the 42nd Highlanders.

"The army has also sustained a heavy loss by the death of Brigadier Taylor, commanding the Third Brigade of the Second Division, a most able officer, and very worthy to have been at the head of so distinguished a corps as H.M.'s 29th Regiment, by which he was beloved and respected.

"The Company's service has lost an excellent officer in Captain Fisher, who fell at the head of the brave Sirmoor Regiment, which greatly distinguished itself.

"The Governor-General has much satisfaction in again offering to

Major-General Sir Harry Smith, K.C.B., commanding the First Division of infantry, his best thanks for his gallant services on this occasion, by which he has added to his well-established reputation.

"The Governor-General acknowledges the meritorious conduct of Brigadier Penny and Brigadier Hicks, commanding brigades in the First Division.

"Her Majesty's 31st and 50th Regiments greatly distinguished themselves, as well as the 42nd and 47th Native Infantry and the Nusseree battalion.

"The Governor-General's thanks are also due to Lieutenant-Colonel Ryan, commanding H.M.'s 50th, who, he regrets to hear, has been severely wounded.

"To Major-General Gilbert, commanding the Second Division, the Governor-General is most happy to express his acknowledgments for the judgment, coolness, and intrepidity displayed by him on every occasion since the campaign opened; and on the present, the promptitude and energy of his attack essentially contributed to ensure the success of the day.

"The Governor-General trusts that the wound received by Brigadier McLaren will not long deprive the service of one of its best officers.

"H.M.'s 29th, and the 1st European Regiments, and the 16th, 48th, 61st Native Infantry, and the Sirmoor battalion, have entitled themselves, by their gallant conduct, to the thanks of the Government.

"To Brigadier Stacy, on whom the command of the Second Division devolved, the Governor-General's thanks are especially due, for the able manner in which the attack within the enemy's camp was directed.

"The Governor-General is also glad to have this opportunity of acknowledging the services of Brigadier Wilkinson, commanding the Sixth Brigade of the attacking division.

"The brigade composed of H.M.'s 9th and 62nd Regiments, and the 26th Native Infantry, under the command of Brigadier the Honourable T. Ashburnham, placed in support of the attacking division, by its firm and judicious advance contributed to the success of the assault.

"The cavalry, under the command of Major-General Sir J. Thackwell, K.C.B., Brigadiers Cureton, Scott, and Campbell, were well in hand, and ready for any emergency. H.M.'s 3rd Light Dragoons, as usual, were in the foremost ranks, and distinguished themselves under their commanding officer, Lieutenant-Colonel White.

"Brigadier Smith, the commanding Engineer, fully accomplished the Commander-in-Chief's instructions; and to Captain Baker and Lieutenant Becher of the Engineers, the Governor-General's acknowledgments are due, for leading the division of attack into the enemy's camp. These officers will maintain the reputation of their corps whenever gallantry or science may be required from its members.

"Major Abbott, of the Engineers, exclusive of his exertions in constructing the bridge of boats, displayed much intelligence in the field. The merits of Major Reilly, commanding that most useful corps the Sappers and Miners, are acknowledged. The ability and zeal of Brigadier Irvine, the senior officer of the Engineer corps, are well known to the Governor-General; and his forbearance in not assuming the command, having reached the camp on the preceding evening, is duly appreciated.

"Brigadier Gowan, commanding the artillery, ably directed the practice of the heavy artillery on the left, assisted by Lieutenant-Colonel Biddulph, Lieutenant-Colonel Brooke, Lieutenant-Colonel Wood, and Captain Pillans.

"On the right, the howitzer practice was well sustained by Major Grant.

"The troops of horse artillery of Lieutenant-Colonel Lane and Captain Fordyce greatly assisted the attack of our infantry on the left; and whilst the enemy were crossing the river, the fire of Lieutenant-Colonel Alexander's troop was most effective.

"The troops of Captain Horsford and Captain Swinley also did good service.

"The Governor-General's acknowledgments are due to Major Grant, Deputy Adjutant-General, and to his department generally, for their ability and intelligence. To the Quartermaster-General the service is much indebted for the judgment and zeal which mark all the proceedings of that officer, and the Governor-General offers acknowledgments to him, to the Deputy Quartermaster-General, Lieutenant-Colonel Drummond, and the officers of that department.

"To Lieutenant-Colonel Barr, Acting Adjutant-General, and to Lieutenant-Colonel Gough, Acting Quartermaster-General, Queen's service, the Governor-General's thanks are due. He regrets the temporary privation of the services of these officers by the wounds they have received.

"To Lieutenant-Colonel Birch, Judge-Advocate-General, the Governor-General again has to repeat his thanks for his intelligence and gallantry.

"To Lieutenant-Colonel Havelock, Persian interpreter, the Governor-General offers his best thanks.

"The Governor-General desires to record his obligations to Count Ravensburg, and to the officers of His Royal Highness's staff, Count Oriola and Count Greuben. This gallant and amiable Prince, with his brave associates of the Prussian army, has shared all the dangers and secured for himself the respect and admiration of the British army; and the Governor-General begs to convey to His Royal Highness and to his staff, his cordial thanks for the ready offers of their services on the field of battle.

"The Governor-General has now to acknowledge the services rendered by the officers attached to his own staff.

"He renews to Lieutenant-Colonel Benson, of the Military Board, his strong sense of the important services rendered by that officer during the whole of this campaign, whose general information in military details, and cool judgment in action, deserve this acknowledgment.

"Lieutenant-Colonel Wood, the Governor-General's Military Secretary, displayed on the 10th instant the same intelligence and gallantry as on former occasions.

"Major Lawrence, the Governor-General's Political Agent, has, throughout these operations, afforded most useful assistance by his ability, zeal, and activity in the field, as well as on every other occasion.

"Captain Mills, Assistant Political Agent, and Honorary Aide-de-camp to the Governor-General, has shown the most unwearied devotion to the service as well in the field as in the exercise of his personal influence in the protected Sikh States.

"The Governor-General's thanks are also due to Captain Cunningham, Engineers, Assistant Political Agent.

"The Governor-General's Aides-de-camp, Captain Grant, Lord Arthur Hay, Captain Peel, and Captain Hardinge, by their gallantry and intelligence, rendered themselves most useful.

"In the operations of this campaign, in which officers of the civil service have accompanied the camp, and participated in the risks incidental to active warfare, the Governor-General's thanks are due for their readiness in encountering these risks, and their endurance of privations.

"The Governor General acknowledges the able assistance he has at all times received from the Political Secretary, F. Currie, Esq. His acknowledgments are also due to his Private Secretary, C. Hardinge, Esq., and to the Assistant Political Agent, R. Cust, Esq.

"Lieutenant-Colonel Parsons, Deputy Commissary-General, has succeeded in keeping the Army well supplied; and the Governor-General is much satisfied with his exertions, and those of the officers under his command. The army took the field under circumstances of great difficulty; and, by strenuous exertions, and good arrangements on the part of the Lieutenant-Colonel, the army has now a large supply in reserve, a result very favourable to the Chief of the Commissariat Department. The manner in which Captain Johnson has conducted the commissariat duties entrusted to him has also met with the Governor-General's approbation.

"To Dr. Macleod, superintending surgeon, and to Dr. Graham, as well as to the officers of the medical department generally, the Governor-General offers his acknowledgments.

"His thanks are due to Dr. Walker, surgeon to the Governor-General, whose ability is only to be equalled by his zeal and humanity.

"A salute of 21 guns will be fired in celebration of the victory of Sobraon at all the usual stations of the army.

"By order of the Right Honourable the Governor-General of India.

(*Signed*) "F. Currie,

"*Secretary to the Government of India,*

"*with the Governor-General.*"

"General Sir Hugh Gough, Bart., G.C.B., Commander-in-Chief of the Forces in India, to the Governor-General of India.

"Headquarters, Army of the Sutlej, Camp, Kussoor, *February* 13*th*, 1846.

"Right Honourable Sir,

"This is the fourth despatch which I have had the honour of addressing to you since the opening of the campaign. Thanks to Almighty God, whose hand I desire to acknowledge in all our successes, the occasion of my writing now is to announce a fourth and most glorious and decisive victory.

"My last communication detailed the movements of the Sikhs and our counter-manœuvres since the great day of Ferozeshah. Defeated on the Upper Sutlej, the enemy continued to occupy his position on the right bank, and his formidable *tête-du-pont* and entrenchments on the left bank of the river, in front of the main body of our army. But on the 10th instant, all that he held of British territory, which was comprised in the ground on which one of his camps stood, was stormed from his grasp, and his audacity was again signally punished by a blow, sudden, heavy, and overwhelming. It is my gratifying duty to detail the measures which have led to this glorious result.

"The enemy's works had been repeatedly reconnoitred during the time of my headquarters being fixed at Nihalkee, by myself, my departmental staff, and my engineer and artillery officers. Our observations, coupled with the reports of spies, convinced us that there had devolved on us the arduous task of attacking a position covered with formidable entrenchments, not fewer than 30,000 men, the best of the Khalsa troops, with 70 pieces of cannon, united by a good bridge to a reserve on the opposite bank, on which the enemy had a considerable camp and some artillery, commanding and flanking his fieldworks on our side. Major-General Sir Harry Smith's division having rejoined me on the evening of the 8th, and part of my siege train having come up with me, I resolved, on the morning of the 10th, to dispose our mortars and battering guns on the alluvial land, within good range of the enemy's picquets at the post of observation in front of Kodeewalla, and at the Little Sobraon. It was directed that this should be done during the night of the 9th, but the execution of this part of the plan was deferred, owing to misconceptions and casual circumstances, until near daybreak. The delay was of little importance, as the event showed that the Sikhs had followed our example, in occupying the two posts in force by day only. Of both, therefore, possession was taken without opposition. The battering and disposable field artillery was then put in position on an extended semicircle, embracing within its fire the works of the Sikhs. It had been intended that the cannonade should have commenced at daybreak; but so heavy a mist hung over the plain and river that it became necessary to wait until the rays of the sun had penetrated it and cleared the atmosphere. Meanwhile on the margin of the Sutlej, on our left, two brigades of Major-General Sir Robert Dick's division, under his personal command, stood ready to commence the assault against the enemy's extreme right. His Seventh Brigade, in which was the 10th Foot, reinforced by the 53rd Foot, and led by Brigadier Stacy, was to head the attack, supported, at 200 yards' distance, by the Sixth Brigade, under Brigadier Wilkinson. In reserve was the Fifth Brigade, under Brigadier the Honourable T. Ashburnham, which was to move forward from the entrenched village of Kodeewalla, leaving, if necessary, a regiment for its defence. In the centre, Major-General Gilbert's division was deployed for support or attack, its right resting on and in the village of the Little Sobraon. Major-General Sir Harry Smith's division was formed near the village of Guttah, with its right thrown up towards the Sutlej. Brigadier Cureton's cavalry threatened, by feigned attacks, the ford at Hurrekee and the enemy's horse, under Rajah Lal Singh Misr, on the opposite bank. Brigadier Campbell, taking an intermediate position in the rear between Major-General Gilbert's right and Major-General Sir Harry Smith's left, protected both. Major-General Sir Joseph Thackwell, under whom was Brigadier Scott, held in reserve on our left, ready to act as circumstances might demand, the rest of the cavalry.

"Our battery of nine-pounders, enlarged into twelves, opened near the Little Sobraon with a brigade of howitzers formed from the light field batteries and troops of horse artillery, shortly after daybreak. But it was half-past 6 before the whole of our artillery fire was developed. It was most spirited and well-directed. I cannot speak in terms too high of the judicious disposition of the guns, their admirable practice, or the activity with which the cannonade was sustained. But, notwithstanding the formidable calibre of our iron guns, mortars, and howitzers, and the admirable way in which they were served, and aided by a rocket battery, it would have been visionary to expect that they could, within any limited time, silence the fire of 70 pieces behind well-constructed batteries of earth, plank, and fascines, or dislodge troops, covered either by redoubts or epaulements, or within a treble line of trenches. The effect of the cannonade was, as has been since proved by an inspection of the camp, most severely felt by the enemy, but it soon became evident that the issue of this struggle must be brought to the arbitrament of musketry and the bayonet.

"At 9 o'clock, Brigadier Stacy's brigade, supported on either flank by Captains Horsford's and Fordyce's batteries, and Lieutenant-Colonel Lane's troop of horse artillery, moved to the attack in admirable order. The infantry and guns aided each other correlatively. The former marched steadily on in line, which they halted only to correct when necessary. The latter took up successive positions at the gallop, until at length they were within three hundred yards of the heavy batteries of the Sikhs; but, notwithstanding the regularity and coolness and scientific character of this assault, which Brigadier Wilkinson well supported, so hot was the fire of cannon, musketry, and zumboorucks, kept up by the Khalsa troops, that it seemed for some moments impossible that the entrenchments could be won under it; but soon, persevering gallantry triumphed, and the whole army had the satisfaction to see the gallant Brigadier Stacy's soldiers driving the Sikhs in confusion before them within the area of their encampment. The 10th Foot, under Lieutenant-Colonel Franks, now for the first time brought into serious contact with the enemy, greatly distinguished themselves. The regiment never fired a shot until it had got within the works of the enemy. The onset of Her Majesty's 53rd Foot was as gallant and effective. The 43rd and 59th Native Infantry, brigaded with them, emulated both in cool determination.

"At the moment of the first success I directed Brigadier the Honourable T. Ashburnham's brigade to move on in support, and Major-General Gilbert's and Sir Harry Smith's divisions to throw out their light troops to threaten the works, aided by artillery. As these attacks of the centre and right commenced, the fire of our heavy guns had first to be directed on the right, and then gradually to cease; but at one time the thunder of full 120 pieces of ordnance reverberated in this mighty combat through the valley of the Sutlej; and as it was soon seen that the weight of the whole force within the Sikh camp was likely to be thrown upon the two brigades that had passed its trenches, it became necessary to convert into close and serious attacks the demonstrations with skirmishers and artillery of the centre and right; and the battle raged with inconceivable fury from right to left. The Sikhs, even when at particular points their entrenchments were mastered with the bayonet, strove to regain them by the fiercest conflict, sword in hand. Nor was it until the cavalry of the left, under Major-General Sir Joseph Thackwell, had moved forward and ridden through the openings in the entrenchments made by our Sappers in single file, and re-formed as they passed them, and the 3rd Dragoons, whom no obstacle usually held formidable by horse appears to check, had on this day, as at Ferozeshah, galloped over and cut down the obstinate defenders of batteries and fieldworks, and until the full weight of three divisions of infantry, with every field artillery gun which could be sent to their aid, had been cast into the scale, that victory finally declared for the British. The fire of the Sikhs first slackened, and then nearly ceased; and the victors then pressing them on every side, precipitated them in masses over their bridge, and into the Sutlej, which a sudden rise of seven inches had rendered hardly fordable. In their efforts to reach the right bank through the deepened water, they suffered from our horse artillery a terrible carnage. Hundreds fell under this cannonade; hundreds upon hundreds were drowned in attempting the perilous passage. Their awful slaughter, confusion, and dismay were such as would have excited compassion in the hearts of their generous conquerors, if the Khalsa troops had not, in the earlier part of the action, sullied their gallantry by slaughtering and barbarously mangling every wounded soldier whom, in the vicissitudes of attack, the fortune of war left at their mercy. I must pause in this narrative especially to notice the determined hardihood and bravery with which our two battalions of Ghoorkhas, the Sirmoor and Nusseree, met the Sikhs, wherever they were opposed to them. Soldiers of small stature but indomitable spirit, they vied in ardent courage in the charge with the Grenadiers of our own nation, and, armed with the short weapon of their mountains, were a terror to the Sikhs throughout this great combat.

"Sixty-seven pieces of cannon, upwards of 200 camel-swivels (zumboorucks), numerous standards, and vast munitions of war, captured by our troops, are the pledges and trophies of our victory. The battle was over by 11 in the morning; and in the forenoon I caused our Engineers to burn a part and to sink a part of the vaunted bridge of the Khalsa army, across which they had boastfully come once more to defy us, and to threaten India with ruin and devastation.

"We have to deplore a loss severe in itself, but certainly not heavy when weighed in the balance against the obstacles overcome and the advantages obtained. I have especially to lament the fall of Major-General Sir Robert Dick, K.C.B., a gallant veteran of the Peninsula and Waterloo campaigns. He survived only until the evening the dangerous grape-shot wound which he received close to the 80th Regiment, in their career of noble daring. Major-General Gilbert, to whose gallantry and unceasing exertions I have been so deeply indebted, and whose services have been so eminent throughout this eventful campaign, and Brigadier Stacy, the leader of the brigade most hotly and successfully engaged, both received contusions. They were such as would have caused many men to retire from the field, but they did not interrupt for a moment the efforts of these heroic officers. Brigadier McLaren, so distinguished in the campaigns in Afghanistan, at Maharajpore, and now again in our conflicts with the Sikhs, has been badly wounded by a ball in the knee. Brigadier Taylor, C.B., one of the most gallant and intelligent officers in the army, to whom I have felt deeply indebted on many occasions, fell in this fight, at the head of his brigade, in close encounter with the enemy, and covered with honourable wounds. Brigadier Penny, of the Nusseree battalion, commanding the Second Brigade, has been wounded, but not, I trust, severely. I am deprived for the present of the valuable services of Lieutenant-Colonel J. B. Gough, C.B., Acting Quartermaster-General of Her Majesty's troops, whose aid I have so highly prized in all my campaigns in China and India. He received a wound from a grape-shot, which is severe, but I hope not dangerous. Lieutenant-Colonel Barr, Acting Adjutant-General of Her Majesty's forces, whose superior merit as a staff-officer I have before recorded, has suffered a compound fracture in the left arm by a ball. It is feared that amputation may be necessary. Lieutenant-Colonels Ryan and Petit, of the 50th Foot, were both badly wounded with that gallant regiment. Captain John Fisher, commandant of the Sirmoor battalion, fell at the head of his valiant little corps, respected and lamented by the whole army.

"I have now to make the attempt, difficult, nay, impracticable I deem it, of expressing in adequate terms my sense of obligation to those who especially aided me by their talents and self-devotion in the hard-fought field of Sobraon.

"First, Right Honourable Sir, you must permit me to speak of yourself. Before the action, I had the satisfaction of submitting to you my plan of attack, and I cannot describe the support which I derived from the circumstances of its having in all its details met your approbation. When a soldier of such sound judgment and experience as your Excellency assured me that my projected operation deserved success, I could not permit myself to doubt that, by the blessing of Divine Providence, the victory would be ours. Nor did your assistance stop here; though suffering severely from the effects of a fall, and unable to mount on horseback without assistance, your uncontrollable desire to see this army once more triumphant carried you into the hottest of the fire, filling all who witnessed your exposure to such peril at once with admiration of the intrepidity that prompted it, and anxiety for your personal safety, involving so deeply in itself the interests and happiness of British India. I must acknowledge my obligation to you for having, whilst I was busied with another portion of our operations, superintended all the arrangements that related to laying our bridge across the Sutlej, near Ferozepore. Our prompt appearance on this side of the river, after victory, and advance to this place, which has enabled us to surprise its fort, and encamp without opposition in one of the strongest positions in the country, is the result of this valuable assistance.

"The Major-Generals of the divisions engaged deserve far more commendation that I am able, within the limits of a despatch, to bestow. Major-General Sir Robert Dick, as I have already related, has fallen on a field of renown worthy of his military career and services, and the affectionate regret of his country will follow him to a soldier's grave.

"In his attack on the enemy's left, Major-General Sir Harry Smith displayed the same valour and judgment which gave him the victory of Aliwal. A more arduous task has seldom, if ever, been assigned to a division. Never has an attempt been more gloriously carried through.

"I want words to express my gratitude to Major-General Gilbert. Not only have I to record that in this great fight all was achieved by him which, as Commander-in-Chief, I could desire to have executed. Not only on this day was his division enabled, by his skill and courageous example, to triumph over obstacles from which a less ardent spirit would have recoiled as insurmountable; but, since the hour in which our leading columns moved out to Umballa, I have found in the Major-General an officer who has not merely carried out all my orders to the letter, but whose zeal and tact have enabled him in a hundred instances to perform valuable services in exact anticipation of my wishes. I beg explicitly to recommend him to your Excellency's especial notice as a divisional commander of the highest merit.

"Major-General Sir Joseph Thackwell has established a claim on this day to the rare commendation of having achieved much with a cavalry force, where the duty to be done consisted of an attack on fieldworks, usually supposed to be the particular province of infantry and artillery. His vigilance and activity throughout our operations, and the superior manner in which our outpost duties have been carried on under his superintendence, demand my warmest acknowledgments.

"Brigadier Stacy, C.B., I must commend to your special protection and favour. On him devolved the arduous duty of leading the first column to the attack, turning the enemy's right, encountering his fire, before his numbers had been thinned, or his spirit broken, and, to use a phrase which a soldier like your Excellency will comprehend, taking off the rough edge of the Sikhs in the fight. How ably, how gallantly, how successfully this was done, I have before endeavoured to relate. I feel certain that Brigadier Stacy and his noble troops will hold their due place in your Excellency's estimation, and that his merits will meet with fit reward.

"Brigadier Orchard, C.B., in consequence of the only regiment under his command that was engaged in the action being with Brigadier Stacy's brigade, attached himself to it, and shared all its dangers, glories, and success.

"I beg as warmly and sincerely to praise the manner in which Brigadier Wilkinson supported Brigadier Stacy, and followed his lead into the enemy's works.

"Brigadier the Honourable T. Ashburnham manœuvred with great coolness and success as a reserve to the two last-mentioned brigades.

"Brigadier Taylor, of H.M.'s 29th, fell nobly, as has already been told, in the discharge of his duty. He is himself beyond the reach of earthly praise; but it is now my earnest desire that his memory may be honoured in his fall, and that his regiment, the army with which he served, and his country, may know that no officer held a higher place in my poor estimation, for gallantry or skill, than Brigadier C. C. Taylor.

"Brigadier McLaren, C.B., in whom I have ever confided, as one of the ablest of the senior officers of this force, sustained on this day, as I have before intimated, his already enviable reputation. I trust he may not long be kept by his wound out of the sphere of active exertion, which is his natural element.

"Brigadiers Penny and Hicks commanded the two brigades of Major-General Sir Harry Smith's division, and overcame at their head the most formidable opposition. I beg to bring both, in the most earnest manner, to your notice, trusting that Brigadier Penny's active service will soon become once more available.

"The manœuvres of Brigadier Cureton's cavalry, in attracting and fixing the attention of Rajah Lal Singh Misr's horse, fulfilled every expectation which I had formed, and were worthy of the skill of the officer employed, whose prominent exploits at the battle of Aliwal I have recently had the honour to bring to your notice.

"Brigadier Scott, C.B., in command of the First Brigade of cavalry, had the rare fortune of meeting and overcoming a powerful body of in antry in the rear of a line of formidable fieldworks. I have to congratulate him on the success of the noble troops under him, and to thank him for his own meritorious exertions. I am quite certain that your Excellency will bear them in mind.

"Brigadier Campbell's brigade was less actively employed; but all that was required of it was most creditably performed. The demonstration on the enemy's left by the 9th Lancers, towards the conclusion of the battle, was made in the best order under a sharp cannonade.

"Brigadier Gowan, C.B., deserves my best thanks for his able arrangements, the value of which was so fully evinced in the first hour and a half of this conflict, when it was almost exclusively an artillery fight. Brigadiers Biddulph, Brooke, and Dennis supported him in the ablest way throughout the day, and have given me the most effectual assistance under every circumstance of the campaign.

"The effective practice of our rockets, under Brigadier Brooke, elicited my particular admiration.

"Brigadier Smith, C.B., had made all the dispositions in the Engineer department, which were in the highest degree judicious, and in every respect excellent. On the evening of the 9th instant Brigadier Irvine, whose name is associated with one of the most brilliant events in our military history, the capture of Bhurtpore, arrived in camp. The command would, of course, have devolved on him, but with that generosity of spirit which ever accompanies true valour and ability, he declined to assume it, in order that all the credit of that work which he had begun might attach to Brigadier Smith. For himself, Brigadier Irvine sought only the opportunity of sharing our perils in the field, and he personally accompanied me throughout the day. Brigadier Smith has earned a title to the highest praise I can bestow.

"To the general staff I am in every way indebted. Nothing could surpass the activity and intelligence of Lieutenant-Colonel Garden and Major Grant, who are the heads of it, in the discharge of the duties of their departments, ever very laborious, and during this campaign almost overwhelming. Both yet suffer under the effects of wounds previously received. Lieutenant-Colonel Drummond, C.B., Deputy Quartermaster-General, and Lieutenant Arthur Becher, Deputy

Assistant Quartermaster-General, ably supported the former; and the exertions of Captains Anson and Tucker, Assistant Adjutants-General, have been most satisfactory to the latter and to myself.

"Lieutenant-Colonel Parsons, Deputy Commissary-General, has evinced the most successful perseverance in his important endeavours to supply the army. He has been ably aided at headquarters by Major W. J. Thompson, C.B., and Major Curtis, Sub-Assistant Commissary-General; all three of these officers were most active in conveying my orders in the battle of Sobraon in the face of every danger. I have, in the most explicit way, to record the same intelligence and ability, and the same activity and bravery, in the case of Lieutenant-Colonel Birch, Judge-Advocate-General, both as respects departmental duties and active attendance on me in the field. I have already spoken of the loss which I have sustained by Lieutenant-Colonels Gough and Barr being wounded. The exertions of both in animating our troops in moments of emergency were laudable beyond my power to praise. Lieutenant Sandys, 55th Regiment Native Infantry, postmaster of the force, assisted in conveying my orders.

"Superintending Surgeon B. Macleod, M.D., has been indefatigable in the fulfilment of every requirement of his important and responsible situation. I am entirely satisfied with his exertions and their results. I must bring to notice also the merits of Field-Surgeon J. Steel, M.D., and Surgeon Graham, M.D., in charge of the depôt of sick.

"I was accompanied during the action by the following officers of my personal staff: Captain the Honourable C. R. Sackville West, H.M.'s 21st Foot, officiating military secretary (Captain Haines, for whom he acts, still being disabled by his severe wound); Lieutenant-Colonel H. Havelock, C.B., H.M.'s 39th Foot, Persian interpreter; Lieutenant Bagot, 15th Native Infantry; Lieutenant Edwards, 1st European Light Infantry; and Cornet Lord James Browne, 9th Lancers, my aides-de-camp; and Assistant Surgeon J. E. Stephens, M.D., my medical officer, assisted in conveying my orders to various points, in the thickest of the fight and the hottest of the fire, and to all of them I feel greatly indebted.

"I have to acknowledge the services in the command of regiments, troops, and batteries, or on select and particular duties in the Engineer department, of the following officers, and to recommend them to your Excellency's special favour, viz.: Major F. Abbott, who laid the bridge by which the army crossed into the Punjaub, and who was present at Sobraon, and did excellent service; Captain Baker and Lieutenant John Beecher, Engineers, who conducted Brigadier Stacy's column (the last of these was wounded); Lieutenant Colonel Wood, artillery, commanding the mortar battery; Major Lawrenson, commanding the 18-pounder battery; Lieutenant-Colonel Huthwaite, commanding the 8-inch howitzer battery; and Lieutenant-Colonel Geddes, commanding the rockets; Captain R. Waller, horse artillery; Captain G. H. Swinley, Captain E. F. Day, Captain J. Turton, Brevet-Major C. Grant, Brevet-Lieutenant-Colonel J. Alexander, Brevet-Major J. Brind, Brevet-Lieutenant-Colonel J. T. Lane, Brevet-Major J. Campbell, Captain J. Fordyce, Captain R. Horsford, and Lieutenant G. Holland, commanding troops and batteries; Major B. Y. Reilly, commanding Sappers and Miners; Lieutenant-Colonel White, C.B., commanding the 3rd Light Dragoons; Captain Nash, 4th Light Cavalry; Major Alexander, 5th Light Cavalry; Captain Christie, 9th Irregular Cavalry; Lieutenant-Colonel Fullerton, 9th Lancers; Captain Leeson, 2nd Irregular Cavalry; Brevet-Captain Beecher, 8th Irregular Cavalry; Captain Pearson, 16th Lancers; Brevet-Captain Quin, Governor-General's Bodyguard; Brevet-Major Angelo, 3rd Light Cavalry;

Lieutenant-Colonel Spence, 31st Foot; Captain Corfield, 47th Native Infantry; Brevet-Lieutenant-Colonel Ryan, and Brevet-Lieutenant Colonel Petit, and Captain Long, 50th Foot; Major Polwhele, 42nd Regiment Native Infantry; Captain O'Brien, and Lieutenant Travers, Nusseree battalion; Captain Stepney, 29th Foot; Major Sibbald, 41st Regiment Native Infantry; Major Birrell and Brevet-Captain Seaton, 1st European Light Infantry; Brevet-Major Graves, 16th Grenadiers; Lieutenant Reid, Sirmoor battalion; Lieutenant-Colonel Davis, 9th Foot; Major Hanscomb, 26th Regiment Native Infantry; Lieutenant-Colonel Bunbury, 80th Foot; Captain Hoggan, 63rd Regiment Native Infantry; Captain Sandeman, 33rd Regiment Native Infantry; Lieutenant-Colonel Franks, 10th Foot; Brevet-Lieutenant-Colonel Nash, 43rd Regiment Native Infantry; Brevet-Lieutenant-Colonel Thompson, 59th Regiment Native Infantry; Lieutenant-Colonel Phillips, 53rd Foot; Major Shortt, 62nd Foot; Brevet-Major Marshall, 68th Regiment Native Infantry; and Captain Short, 45th Regiment Infantry.

"The following staff and engineer officers I have also to bring to your especial notice, and to pray that their services may be favourably remembered, and the survivors duly rewarded, viz.: Captain E. Christie, Deputy Assistant Adjutant-General, and Lieutenant Maxwell, Deputy Assistant Quartermaster-General of Artillery, and Captain Pillans and Brevet-Captain W. K. Warner, Commissaries of Ordnance; Brevet-Captain M. Mackenzie and Brevet-Captain E. G. Austen, and First Lieutenant E. Kaye, Artillery, Majors of Brigade; Napier, Major of Brigade of Engineers; Captain Tritton, 3rd Light Dragoons, Deputy Assistant Adjutant-General; Lieutenant E. Roche, 3rd Dragoons, aide-de-camp to Major-General Sir J. Thackwell, and Officiating Deputy Assistant Quartermaster-General of Cavalry, in the place of Captain Havelock, 9th Foot, who was present in the field, but unable, from the effects of a wound, to discharge the duties of his office; Captain E. Lugard, 31st Foot, Deputy Assistant Adjutant-General; Lieutenant A. S. Galloway, 3rd Light Cavalry, Deputy Assistant Quartermaster-General; Lieutenant E. A. Holdich, 80th Foot, aide-de-camp to Major-General Sir Harry Smith; Lieutenant F. M'D. Gilbert, 2nd Grenadiers, acting aide-de-camp to Major General Gilbert; Captain R. Houghton, 63rd Regiment Native Infantry, Officiating Assistant Adjutant-General; Lieutenant Rawson, Deputy Assistant Quartermaster-General, killed; Lieutenant R. Bates, 82nd Foot, aide-de-camp to the late Major-General Sir R. Dick; Captain J. R. Pond, 1st European Light Infantry, Deputy Assistant Adjutant-General; Lieutenant J. S. Paton, 14th Regiment Native Infantry, Officiating Deputy Assistant Quartermaster-General; Brevet-Captain Harrington, 5th Light Cavalry; Captain A. Spottiswoode, 9th Lancers; Lieutenant R. Pattinson, 16th Lancers; Captain J. Garvock, 31st Foot; Lieutenant G. H. M. Jones, 29th Foot; Captain J. L. Taylor, 26th Light Infantry; Lieutenant H. F. Dunsford, 59th Regiment Native Infantry, Majors of Brigade; Captain Combe, 1st European Light Infantry, Major of Brigade Second Brigade; Captain Gordon, 11th Native Infantry, Major of Brigade, Sixth Brigade; Captain A. G. Ward, 68th Native Infantry, Major of Brigade, and Lieutenant R. Hay, Major of Brigade, killed.

"Having ventured to speak of your Excellency's own part in this action, it would be most gratifying to me to go on to mention the brilliant share taken in it by Lieutenant-Colonel Wood and the officers of your personal staff, as well as by the civil, political, and other military officers attached to you. But as these were all under your own eyes, I cannot doubt that you will yourself do justice to their exertions.

"We were in this battle again honoured with the presence of Prince Waldemar of Prussia, and the two noblemen in his suite, Count Oriola and Count Greuben. Here, as at Moodkee and Ferozeshah, these distinguished visitors did not content themselves with a distant view of the action, but, throughout it, were to be seen in front wherever danger most urgently pressed.

"The loss of the enemy has been immense; an estimation must be formed with a due allowance for the spirit of exaggeration which pervades all statements of Asiatics where their interest leads them to magnify numbers; but our own observation on the river banks and in the enemy's camp, combined with the reports brought to our Intelligence Department, convince me that the Khalsa casualties were between 8,000 and 10,000[1] men killed and wounded in action and drowned in the passage of the river. Amongst the slain are Sirdar Sham Singh Attareewalla, Generals Gholab Singh Koopta and Heera Singh Topee, Sirdar Kishen Singh, son of the late Jemadar Kooshall Singh; Generals Mobaruck Ally and Illahee Buksh, and Shah Newaz Khan, son of Futtehood-deen, Khan of Kussoor. The body of Sham Singh was sought for in the captured camp by his followers; and respecting the gallantry with which he is reported to have devoted himself to death rather than accompany the army in its flight, I forbade his people being molested in their search, which was finally successful.

"The consequences of this great action have yet to be fully developed. It has at least, in God's providence, once more expelled the Sikhs from our territory, and planted our standards on the soil of the Punjaub. After occupying their entrenched position for nearly a month, the Khalsa army had perhaps mistaken the caution which had induced us to wait for the necessary material, for timidity. But they must now deeply feel that the blow which has fallen on them from the British arm has only been the heavier for being so long delayed.

"I have, etc.

(*Signed*) "H. GOUGH, *General*,
"*Commander-in-Chief, East Indies.*"

'MAJOR-GENERAL SIR JOSEPH THACKWELL, K.C.B., K.H., COMMANDING CAVALRY DIVISION, TO THE ADJUTANT-GENERAL OF THE ARMY OF THE SUTLEJ.

"SIR, "CAMP, KUSSOOR, *February 11th*, 1846.

"I have the honour to report, for the information of His Excellency the Commander-in-Chief, the operations of that part of the cavalry division under my command which had the happiness in participating in the glorious victory obtained over the Sikhs on February 10th, 1846.

"The Third Brigade of Cavalry, under Brigadier Cureton, consisting of the 16th Lancers, commanded by Captain Pearson, the 3rd Light Cavalry, under Major Angelo, and the Bodyguard, under the command of Captain Quin, with a troop of horse artillery, had been directed to make a demonstration in the direction of Hurreekee Ghat. The Second Brigade, under Brigadier Campbell, consisting of H.M.'s 9th Lancers, under the command of Lieutenant-Colonel Fullerton, and the 2nd Irregular Cavalry, under the command of Captain Leeson, and Major Campbell's troop of horse artillery, had been ordered to support the Second Division of Infantry near Sobraon. The First Brigade of Cavalry, under the command of Brigadier Scott, C.B., composed of H.M.'s 3rd Light Dragoons, under the command of Lieutenant-Colonel White, C.B.,

[1] We have since ascertained, from undoubted authority, that the Sikhs acknowledged they had 37,000 men engaged in this battle, exclusive of the large force, particularly of cavalry, at this side of the river, and that their loss on this occasion was from 13,000 to 14,000 men.

the 4th Native Cavalry, under that of Captain Nash, the 5th Light Cavalry, under Major Alexander, and the 9th Irregular Cavalry, under Captain Christie, with the 8th Irregular Cavalry, under Captain Beecher, had been ordered to take post in front of the village of Asyah. The troops arrived on their ground at the time ordered, and it becomes my duty to state the operations of the troops immediately under my command, the reports of the brigadiers of the Second and Third Brigades being best calculated to show theirs. I will therefore advert shortly to those operations of the First Brigade. Early in the morning, on a party of the enemy's cavalry showing themselves, the 8th Irregular Cavalry and a squadron of the 3rd Light Dragoons were detached three-quarters of a mile in front of Alawalee, and the enemy retired and made no further demonstration on that point. On the advance of Sir Robert Dick's division the cavalry under Brigadier Scott supported it at a short distance, and on his entering the enemy's entrenchments they made a flank movement to their right, at first intended to support the Second Division ; but ere I advert to the effects of this movement I must bear testimony to the gallant conduct of Lieutenant-Colonel Lane's troop of horse artillery, which came into action within a short distance of the enemy's entrenchments. Part of Brigadier Scott's brigade was ordered to support, and I took two squadrons of the 3rd Light Dragoons, followed by the 4th and 5th Light Cavalry, along the ravine or nullah in front of the enemy's line till the head of the column came upon a part which he still defended, flanked by a battery of three guns. But the firm appearance of the cavalry made the Sikhs gradually give way on the right, and on my discovering a passage into the entrenchment the troops were passed through in single file.

"On the first squadron being formed I ordered it to charge, and led it over broken ground against the enemy's retiring infantry (many of whom were sabred) nearly to the ford. The difficulty of the ground prevented support arriving in due time, and the squadron, on being pressed upon by large masses of infantry, was obliged to repass the difficult ground it had before gone over. In a short time the second squadron of the 3rd came up, together with the 4th and 5th Native Cavalry, and the enemy were charged, pressed upon by the leading squadrons, into the river.

"On entering the entrenchments I had sent for a troop of artillery to play upon the ford, and Major Grant's troop and, near the end, Lieutenant-Colonel Lane's, were employed on this service.

"I beg to assure His Excellency that I am well satisfied with the conduct of all the officers and men engaged under my own observation, and the gallant manner in which the 3rd Light Dragoons and the other troops above mentioned entered the entrenchments deserves my warmest commendations, and I beg to state that I derived every assistance from Brigadier Scott, who conducted his brigade much to my satisfaction during the conflict, as did also commanding officers their respective regiments. I beg to bring to His Excellency's notice the highly satisfactory conduct of Captain Hale at the head of the right squadron of the 3rd Light Dragoons in the attack on the enemy, and I am greatly indebted to the able assistance of Captain Tritton, the Deputy Assistant Adjutant-General of the division; to Lieutenant Roche, 3rd Light Dragoons, my aide-de-camp, and Officiating Deputy Assistant Quartermaster-General, and to Lieutenant Francis of the 9th Lancers, who acted as my extra aide-de-camp.

"I have, etc.,

(*Signed*) "JOS. THACKWELL, *Major-General.*

"*To the Adjutant-General of the Army,*
"*Commanding Cavalry Division.*"

APPENDIX D

DESPATCHES. PUNJAB CAMPAIGN

NOTIFICATION

FOREIGN DEPARTMENT, CAMP, UMBALLAH, *December 8th*, 1848.

The Right Honourable the Governor-General has much pleasure in publishing, for general information, the following despatch from His Excellency the Commander-in-Chief.

"HEADQUARTERS, FLYING CAMP, HILLAH, *December 5th*, 1848.

"MY LORD,

"It has pleased Almighty God to vouchsafe to the British arms the most successful issue to the extensive combinations rendered necessary for the purpose of effecting the passage of the Chenab, the defeat and dispersion of the Sikh force under the insurgent Rajah Shere Sing, and the numerous Sikh sirdars, who had the temerity to set at defiance the British power. This force, from all my information, amounted to from 30,000 to 40,000 men, with 28 guns, and was strongly entrenched on the right bank of the Chenab, at the principal ford, about two miles from the town of Ramnuggar.

"My despatch of November 23rd will have made your Lordship acquainted with the motives which induced me to penetrate thus far into the Punjaub, and the occurrences of the previous day, when the enemy were ejected from the left bank of the Chenab. My daily private communications will have placed your Lordship in possession of the difficulties I had to encounter in a country so little known, and in the passage of a river, the fords of which were most strictly watched by a numerous and vigilant enemy, and presenting more difficulties than most rivers, whilst I was surrounded by a hostile peasantry.

"Finding that to force the passage at the ford in my front must have been attended with considerable loss, from the very strong entrenchments and well-selected batteries which protected the passage, I instructed the field-engineer, Major Tremenhere, in co-operation with the Quartermaster-General's department, to ascertain (under the difficulties before noticed) the practicability of the several fords reported to exist on both my flanks, while I had batteries erected and made demonstrations so as to draw the attention of the enemy to the main ford in my front, and with the view, if my batteries could silence their guns, to act simultaneously with the force I proposed to detach under an officer of much experience in India, Major-General Sir Joseph Thackwell.

"On the night of November 30th, this officer, in command of the following force, and more particularly detailed in the accompanying memorandum:

	European.	Native.	Total.
3 Troops Horse Artillery ..	3	0	3
2 Light Field Batteries ..	2	0	2
1 Brigade of Cavalry	1	4	5
3 Brigades of Infantry ..	2	6	8

". . . two eighteen-pounders with elephant draft and detail artillery, pontoon train, with two companies, moved up the river in light marching order, without tents and with three days' provisions, upon a ford which I had every reason to consider very practicable (and which I have since ascertained was so), but which the Major-General deemed so dangerous and difficult that he proceeded (as he was instructed should such turn out to be the case) to Wuzeerabad, a town 22 miles up the river, where Lieutenant Nicholson, a most energetic assistant to the Resident at Lahore, had secured 16 boats, with the aid of which this force effected the passage on the evening of the 1st and morning of the 2nd instant.

"Upon learning from an aide-de-camp sent for the purpose that the Major-General's force had crossed and was in movement, I directed a heavy cannonade to commence upon the enemy's batteries and encampment at Ramnuggar, which was returned by only a few guns, which guarded effectually the ford, but so buried that, although the practice of our artillery was admirable under Major Mowatt and Captain Sir Richmond Shakespear, we could not, from the width of the river, silence them. This cannonade, however, inflicted very severe loss to the enemy in their camp and batteries, and forced him to fall back with his camp about two miles, which enabled me, without the loss of a man, to push my batteries and breastworks, on the night of the 2nd, to the bank of the river, the principal ford of which I then commanded. By this I was enabled to detach another brigade of infantry, under Brigadier Godby, at daylight on the 3rd, which effected the passage, with the aid of pontoon train, six miles up the river, and got into communication with Major-General Sir Joseph Thackwell.

"The cannonade and demonstration to cross at Ramnuggar were kept up on the 2nd and 3rd, so as to fix a large portion of the enemy there to defend that point. Having communicated to Sir Joseph my views and intentions, and although giving discretionary power to attack any portion of the Sikh force sent to oppose him, I expressed a wish that when he covered the crossing of Brigadier Godby's brigade he should await their junction, except the enemy attempt to retreat; this induced him to halt within about two or three miles of the left of their position. About 2 o'clock on the 3rd, the principal part of the enemy's force, encouraged by the halt, moved to attack the detached column, when a smart cannonade on the part of the enemy took place, and an attempt to turn both Major-General Sir Joseph Thackwell's flanks by numerous bodies of cavalry was made. After about one hour's distant cannonade on the part of the Sikhs, the British artillery never returning a shot, the enemy took courage and advanced, when our artillery, commanded by that excellent officer Lieutenant-Colonel C. Grant, poured in upon them a most destructive fire, which soon silenced all their guns and frustrated all their operations, with very severe loss upon their side; but the exhausted state of both man and horse induced the Major-General to postpone the attack upon their flank and rear, as he was directed, until the following morning, the day having nearly closed when the cannonade ceased.

"I regret to say that during the night of the 3rd the whole of the Sikh force precipitately fled, concealing or carrying with them their artillery, and exploding their magazines. I immediately pushed across the river the 9th Lancers and 14th Light Dragoons in pursuit, under that most energetic officer, Major-General Sir Walter Gilbert. The Sikhs, it appears, retreated in the greatest disorder, leaving in the villages numerous wounded men. They have subdivided into three divisions, which have become more a flight than a retreat; and I understand a great portion of those not belonging to the revolted Khalsa army have dispersed and returned to their homes, thus, I trust, effectually frustrating the views of the rebel Shere Sing and his rebel associates.

"I have not received Major-General Sir Joseph Thackwell's report, nor the returns of his loss, but I am most thankful to say that our whole loss, subsequent to November 22nd, does not much exceed 40 men; no officers have been killed, and but three wounded. Captain Austin, of the artillery, only appears severely so.

"I have to congratulate your Lordship upon events so fraught with importance, and which will, I have no doubt, with God's blessing, tend to most momentous results. It is, as I anticipate, most gratifying to me to assure your Lordship that the noble army under my command has, in these operations, upheld the well-established fame of the arms of India, both European and native, each vying who should best perform his duty. Every officer, from the general of division to the youngest subaltern, well supported their Commander-in-Chief, and cheerfully carried out his views, which at a future period, and when we shall have effected the views of the Government, I shall feel proud in bringing to your Lordship's notice.

"I have, etc.

(*Signed*) "GOUGH."

FOREIGN DEPARTMENT, FORT WILLIAM, *February 2nd*, 1849.

The President of the Council of India in Council is pleased to direct the publication of the following General Order by the Right Honourable the Governor-General, with the Commander-in-Chief's despatch, dated January 16th, detailing the operations of the army under His Excellency's command at Chillianwallah.

By order of the President of the Council of India in Council.

(*Signed*) FRED. JAS. HALLIDAY,

Officiating Secretary to the Government of India.

"GENERAL ORDER BY THE RIGHT HONOURABLE THE GOVERNOR-GENERAL OF INDIA.

"FOREIGN DEPARTMENT, CAMP, MUKKO, *January 24th*, 1849.

"The Governor-General, having received from the Commander-in-Chief in India a despatch, dated the 16th instant, directs that it shall be published for the information of the army and of the people of India.

"In this despatch His Excellency reports the successful operations of the troops under his immediate command, on the afternoon of the 13th instant, when they attacked and defeated the Sikh army under the command of Rajah Shere Sing.

"Notwithstanding great superiority in numbers, and the formidable position which he occupied, the enemy, after a severe and obstinate resistance, was driven back, and retreated from every part of his position in great disorder, with much slaughter, and with the loss of 12 pieces of artillery.

"The Governor-General congratulates the Commander-in-Chief on the victory so obtained by the army under his command; and, on behalf of the Government of India, he desires cordially to acknowledge the gallant services which have been rendered on this occasion, by His Excellency the Commander-in-Chief, the generals, the officers, non-commissioned officers, and soldiers of the army in the field.

"The Governor-General offers his thanks to Major-General Sir Joseph Thackwell, K.C.B. and K.H., for his services, and to Brigadier White, for his conduct of the brigade of cavalry on the left.

"Major-General Sir W. R. Gilbert, K.C.B., and Brigadier-General Campbell, C.B., are entitled to the especial thanks of the Governor-General, for the admirable manner in which they directed the divisions under their orders.

"To Brigadier Mountain, C.B., and to Brigadier Hoggan, the Governor-General tenders his acknowledgments for the gallant example they offered in the lead of their men; and to them, to Brigadier Godby, C.B., and Brigadier Penny, C.B., for their able conduct of their respective brigades.

"The warm thanks of the Governor-General are due to Brigadier General Tennant, commanding the artillery division; to Brigadier Brooke, C.B., and Brigadier Huthwaite, C.B., for their direction of the operations of that distinguished arm, and for the effective service which it rendered.

"To the heads of the various departments, and to the officers of the general and personal staff, whose services are acknowledged by the Commander-in-Chief, the Governor-General offers his thanks.

"The Governor-General deeply regrets the loss of Brigadier Pennycuick, C.B., and of the gallant officers and men who have honourably fallen in the service of the country.

"It has afforded the Governor-General the highest gratification to observe that the conduct of the troops generally was worthy of all praise.

"The Governor-General, indeed, is concerned to think that any order or misapprehension of an order could have produced the movements of the right brigade of cavalry which His Excellency the Commander-in-Chief reports.

"To the artillery, European and native, to the cavalry on the left, and to the European and native infantry, the Governor-General offers his hearty thanks, especially to those corps, European and native, which His Excellency reports to have acted under trying circumstances with a gallantry worthy of the greatest admiration.

"The Governor-General will have sincere satisfaction in bringing the services of this army under the favourable notice of H.M.'s Government and the Honourable East India Company.

"A salute of 21 guns has been ordered to be fired from every principal station of the Army of India.

"The Governor-General repeats to the Commander-in-Chief and to the army the assurance of his cordial thanks, and expresses his confident belief that the victory which, under Divine providence, they have won, will exercise a most important influence on the successful progress of the war in which they are engaged.

"By order of the Right Honourable the Governor-General of India.

(*Signed*) "H. M. Elliott,
"*Secretary to the Government of India,*
"*with the Governor-General.*"

"FROM HIS EXCELLENCY THE COMMANDER-IN-CHIEF, TO THE RIGHT HONOURABLE THE GOVERNOR-GENERAL OF INDIA.

HEADQUARTERS CAMP, CHILLIANWALLAH, *January* 16*th*, 1849.

"MY LORD,

"Major Mackeson, your Lordship's political agent with my camp, officially communicated to me, on the 10th instant, the fall of Attock and the advance of Sirdar Chutter Sing in order to concentrate his force with the army in my front, under Shere Sing, already amounting to from 30,000 to 40,000 men, with 62 guns, concluding his letter thus: 'I would urge, in the event of your Lordship's finding yourself strong enough with the army under your command to strike an effectual blow at the enemy in our front, that the blow should be struck with the least possible delay.'

"Concurring entirely with Major Mackeson, and feeling that I was perfectly competent effectually to overthrow Shere Sing's army, I moved from Loah Tibba, at daylight on the 12th, to Dingee, about 12 miles. Having learnt from my spies, and from other sources of information, that Shere Sing still held with his right the village of Lukhneewalla and Futtehshaw-ke-Chuck, having the great body of his force at the village of Woolianwalla, with his left at Russool, on the Jhelum, strongly occupying the southern extremity of a low range of hills, intersected by ravines, which extend nearly to that village, I made my arrangements accordingly that evening, and communicated them to the commanders of the several divisions; but to insure correct information as to the nature of the country, which I believed to be excessively difficult and ill-adapted to the advance of a regular army, I determined upon moving on this village with a view to reconnoitre.

"On the morning of the 13th the force advanced. I made a considerable detour to my right, partly in order to distract the enemy's attention, but principally to get as clear as I could of the jungle, on which it would appear that the enemy mainly relied.

"We approached this village about 12 o'clock, and I found, on a mound close to it, a strong picquet of the enemy's cavalry and infantry, which we at once dispersed, obtaining from the mound a very extended view of the country before us, and the enemy drawn out in battle array, he having, either during the night or that morning, moved out of his several positions, and occupied the ground in our front, which, though not dense, was still a difficult jungle, his right in advance of Futtehshaw-ke-Chuck, and his left on the furrowed hills before described.

"The day being so far advanced, I decided upon taking up a position in rear of the village, in order to reconnoitre my front, finding that I could not turn the enemy's flanks, which rested upon a dense jungle, extending nearly to Hillah, which I had previously occupied for some time, and the neighbourhood of which I knew, and upon the raviney hills near Russool, without detaching a force to a distance; this I considered both inexpedient and dangerous.

"The engineer department had been ordered to examine the country before us and the Quartermaster-General was in the act of taking up ground for the encampment, when the enemy advanced some horse artillery, and opened fire on the skirmishers in front of the village.

"I immediately ordered them to be silenced by a few rounds from our heavy guns, which advanced to an open space in front of the village. Their fire was instantly returned by that of nearly the whole of the enemy's field artillery; thus exposing the position of his guns, which the jungle had hitherto concealed.

"It was now evident that the enemy intended to fight, and would probably advance his guns so as to reach the encampment during the night.

"I therefore drew up in order of battle, Sir Walter Gilbert's division on the right, flanked by Brigadier Pope's brigade of cavalry, which I strengthened by the 14th Light Dragoons, well aware that the enemy was strong in cavalry upon his left. To this were attached three troops of horse artillery under Lieutenant-Colonel Grant.

"The heavy guns were in the centre.

"Brigadier-General Campbell's division formed the left, flanked by Brigadier White's brigade of cavalry, and three troops of horse artillery under Lieutenant-Colonel Brind.

"The field batteries were with the infantry divisions.

"Thus formed, the troops were ordered to lie down, whilst the heavy guns, under Major Horsford, ably seconded by Brevet-Majors Ludlow and Sir Richmond Shakespear, opened a well-directed and powerful fire upon the enemy's centre, where his guns appeared principally to be placed; and this fire was ably supported on the flanks by the field batteries of the infantry divisions.

"After about an hour's fire the enemy appeared to be, if not silenced, sufficiently disabled to justify an advance upon his position and guns.

"I then ordered my left division to advance, which had to move over a great extent of ground, and in front of which the enemy seemed not to have many guns. Soon after, I directed Sir Walter Gilbert to advance, and sent orders to Brigadier Pope to protect the flank and support the movement. Brigadier Penny's brigade was held in reserve, while the irregular cavalry under Brigadier Hearsey, with the 20th Native Infantry, was ordered to protect the enormous amount of provision and baggage that so hampers the movement of an Indian army.

"Some time after the advance, I found that Brigadier Pennycuick's brigade had failed in maintaining the position it had carried, and immediately ordered Brigadier Penny's reserve to its support; but Brigadier-General Campbell, with that steady coolness and military decision for which he is so remarkable, having pushed on his left brigade and formed line to his right, carried everything before him, and soon overthrew that portion of the enemy which had obtained a temporary advantage over his right brigade.

"This last brigade, I am informed, mistook for the signal to move in double time the action of their brave leaders, Brigadier Pennycuick and Lieutenant-Colonel Brooks (two officers not surpassed for sound judgment and military daring in this or any other army), who waved their swords over their heads as they cheered on their gallant comrades. This unhappy mistake led to the Europeans outstripping the native corps, which could not keep pace, and arriving, completely blown, at a belt of thick jungle, where they got into some confusion, and Lieutenant-Colonel Brooks, leading the 24th, was killed between the enemy's guns. At this moment a large body of infantry, which supported their guns, opened upon them so destructive a fire that the brigade was forced to retire, having lost their gallant and lamented leader, Brigadier Pennycuick, and the three other field-officers of the 24th, and nearly half the regiment before it gave way, the native regiment, when it came up, also suffering severely. In justice to this brigade, I must be allowed to state that they behaved heroically, and, but for their too hasty, and consequently disorderly advance, would have emulated the conduct of their left brigade, which, left unsupported for a time, had to charge to their front and right wherever an enemy appeared. The brigade of horse artillery on their left, under Lieutenant-Colonel Brind, judiciously and gallantly aiding, maintained an effective fire.

"Major-General Sir J. Thackwell, on the extreme left and rear, charged the enemy's cavalry wherever they showed themselves.

"The right attack of infantry, under that able officer Major-General Sir Walter Gilbert, was most praiseworthy and successful. The left brigade, under Brigadier Mountain, advanced under a heavy fire upon the enemy's guns, in a manner that did credit to the brigadier and his gallant brigade, which came first into action and suffered severely; the right brigade, under Brigadier Godby, ably supported the advance.

"This division nobly maintained the character of the Indian army, taking and spiking the whole of the enemy's guns, in their front, and dispersing the Sikhs wherever they were seen.

"The Major-General reports most favourably of the fire of his field battery.

"The right brigade of cavalry, under Brigadier Pope, was not, I regret to say, so successful. Either by some order, or misapprehension of an order, they got into much confusion, hampered the fine brigade of horse artillery, which, while getting into action, against a body of the enemy's cavalry that was coming down upon them, had their horses separated from their guns by the false movements of our cavalry, and notwithstanding the heroic conduct of the gunners, four of whose guns were disabled to an extent which rendered their withdrawal, at the moment, impossible. The moment the artillery was extricated and the cavalry re-formed, a few rounds put to flight the enemy that had occasioned this confusion.

"With this exception, the conduct of the troops generally was most exemplary, some corps, both European and native, acting under most trying circumstances (from the temporary failure on our left centre and right, and the cover which the jungle afforded to the enemy's movements), and with a gallantry worthy of the highest admiration.

"Although the enemy, who defended not only his guns but his position, with desperation, was driven in much confusion, and with heavy loss, from every part of it and the greater part of his field artillery was actually captured, the march of brigades to their flanks to repel parties that had rallied, and the want of numbers and consequent support to our right flank, aided by the cover of the jungle and the close of the day, enabled him, upon our further advance in pursuit, to return and carry off unobserved the greater portion of the guns we had thus gallantly carried at the point of the bayonet.

"I remained with Brigadier-General Campbell's division, which had been reinforced by Brigadier Mountain's brigade, until near 8 o'clock, in order to effect the bringing in of the captured ordnance and of the wounded, and I hoped to bring in the rest of the guns next morning. But I did not feel justified in remaining longer out. The night was very dark. I knew not how far I had advanced. There were no wells nearer than the line of this village. The troops had been arduously employed all day, and there was every appearance of a wet night; rain did fall before morning.

"I should have felt greater satisfaction if I were enabled to state that my expectations in regard to the guns had been realised; but although a brigade of cavalry, under Brigadier White, with a troop of horse artillery, were on the ground soon after daylight, we found that the enemy, assisted by the neighbouring villagers, had carried off their guns, excepting twelve, which we had brought in the night before. Most of the captured waggons I had caused to be blown up before leaving the ground.

"The victory was complete, as to the total overthrow of the enemy;

and his sense of utter discomfiture and defeat will, I trust, soon be made apparent, unless, indeed, the rumours prevalent this day, of his having been joined by Chutter Sing, prove correct.

"I am informed that the loss of the Sikhs has been very great, and chiefly amongst their old and tried soldiers. In no action do I remember seeing so many of an enemy's slain upon the same space —Sobraon perhaps only excepted.

"I have now, my Lord, stated the general movements of this army previous to and during the action of Chillianwallah, and as that action was characterised by peculiar features, which rendered it impossible for the Commander-in-Chief to witness all the operations of the force, I shall beg leave to bring prominently to your Lordship's notice the names of the several officers and corps particularly mentioned by the divisional commanders.

"I have already stated the obligations I am under to Major-General Sir Joseph Thackwell and Sir Walter Gilbert, and to Brigadier-General Campbell, for their most valuable services. I warmly concur with them in the thanks which they have expressed to the several brigadiers and officers commanding corps, and to the troops generally.

"Sir Joseph Thackwell names, with much satisfaction, Brigadier White's conduct of his brigade; Major Yerbury, commanding 3rd Light Dragoons; the gallant charge of Captain Unett, in command of a squadron of that corps; Major Mackenzie, commanding the 8th, with a squadron detached in support of the artillery. He further notices the assistance he derived from the zeal and activity of Captain Pratt, Assistant Adjutant-General, and Lieutenant Tucker, Deputy Assistant Quartermaster-General of his division; of Captain Cautley, Major of Brigade, of his aide-de-camp, Lieutenant Thackwell, and of Lieutenant Simpson, Sub-Assistant Commissary-General.

"Brigadier-General Campbell speaks in terms of admiration of the Fifth Brigade, led on by that distinguished officer, Brigadier Pennycuick, and particularly of the gallant exertions of H.M.'s 24th Foot, under the command of Lieutenant-Colonel Brooks; and the good and steady advance of the 25th and 45th Native Infantry, under the command of Lieutenant-Colonel Corbett and Major Williams. He particularises the undaunted example set to his brigade by Brigadier Hoggan; the continued steadiness and gallantry of H.M.'s 61st Regiment, commanded by Lieutenant-Colonel Macleod, under the most trying circumstances; the distinguished conduct of Major Fleming and the officers of the 36th Native Infantry; and of the 46th Native Infantry, under Major Tudor; as also the able and zealous exertions of the Brigade Major, Captain Keiller. The Brigadier-General also brings to notice his obligations to Major Tucker, Assistant Adjutant-General of the army; and to Captain Goldie and Lieutenant Irwin, of the Engineers, who were sent to his assistance, and the cordial and able support which he received from Major Ponsonby, his Assistant Adjutant-General; and he particularly mentions the conduct of Ensign Garden, his Deputy Assistant Quartermaster-General; and Captain Haythorne, his aide-de-camp; further naming Lieutenant Grant, of H.M.'s 24th Regiment; Lieutenant Powys, of H.M.'s 61st, who attended him as orderly officers; and of Lieutenant and Adjutant Shadwell, of H.M.'s 98th, who was with him as a volunteer.

"Sir Walter Gilbert speaks warmly of the charge led by Brigadier Mountain, against a large battery of the enemy, and followed up on the right by Brigadier Godby; and of the subsequent conduct of these officers; as also of the conduct of Major Chester, Assistant Adjutant-General; and Lieutenant Galloway, Deputy Assistant Quartermaster-General of the division; of Lieutenant Colt, his aide-de-camp; of Captain Sherwill, and Lieutenant Macdonnell, Majors of Brigades; and of Captain Glasfurd and Lieutenant W. E. Morton, of the Engineers.

"The Major-General further mentions the undaunted bravery on this occasion of H.M.'s 29th Regiment, under Lieutenant-Colonel Congreve; the distinguished conduct of the 2nd European Regiment, under Major Steel; and the manner in which Majors Smith and Way, of the 29th, and Major Talbot, of the 2nd Europeans, seconded their able commanders. He also expresses his thanks to Lieutenant-Colonel Jack, commanding the 30th Native Infantry; Major Banfield, commanding the 56th Native Infantry, who was mortally wounded; Major Corfield, commanding the 31st Native Infantry, and Major McCausland, commanding the 70th Native Infantry, for the manner in which they led their regiments into action; naming likewise Captain Nembhard, of the 56th, who succeeded to the command of that corps; Captain Dawes, commanding the field-battery of the division; and Captain Robbins, of the 15th, who acted as his aide-de-camp.

"The Reserve, consisting of the 15th Native Infantry, and eight companies of the 69th Native Infantry, was ably handled by Brigadier Penny, well seconded by Lieutenant-Colonels Sibbald and Mercer, commanding the corps. The Brigadier particularly mentions the steady conduct of the rifle company of the 69th, under Captain Sissmore; and acknowledges the services of Captain Macpherson, his Major of Brigade, and Brevet-Captain Morris, of the 20th Native Infantry, who attended him as orderly officer.

"Brigadier-General Tennant, commanding the artillery division, rendered me every aid, and presided over the noble arm of which he is the head, most creditably to himself and most beneficially to the service. The Brigadier-General particularly mentions Brigadier J. Brooke, who commanded the whole of the horse artillery; Brigadier Huthwaite, commanding the foot artillery; Lieutenant-Colonels C. Grant and J. Brind; Major R. Horsford and Major Mowatt; all of whom were in important commands. He further brings to notice Captain J. Abercrombie, Deputy Assistant Adjutant-General; Lieutenant Tombs, Deputy Assistant Quartermaster-General, his aide-de-camp; Lieutenant Olpherts; Captain Hogge, Commissary of Ordnance; and Lieutenant de Tassier, who attended him as orderly officer.

"I have, in the beginning of this despatch, noticed the services of Brevet-Major Sir Richmond Shakespear and Brevet-Major Ludlow, in command of the heavy batteries, under the general superintendence of Major Horsford; and it only remains for me to add that the conduct of Major Fordyce, Captains Warner and Duncan, Lieutenants Robinson and Walker, commanding troops and field-batteries, as well as the officers and men of the artillery generally, have been named in terms of praise by the divisional commander.

"Lieutenants C. V. Cox and E. Kaye, Brigade Majors of this arm, have also been named by their respective brigadiers.

"From the engineer department, under Major Tremenhere, I received active assistance, ably aided by Captain Durand, Lieutenants R. Baird-Smith, and Goodwyn.

"To the general staff I am greatly indebted. Lieutenant-Colonel Gough, C.B., Quartermaster-General, and Major Lugard, Acting Adjutant-General, and Captain C. Otter, Acting Assistant Adjutant-General of Her Majesty's forces; Lieutenant-Colonel P. Grant, C.B., Adjutant-General of the army; Major C. Elkins (killed), a valued and much-regretted officer, Deputy Adjutant-General; and Major Tucker, Assistant Adjutant-General of the army; Lieutenant-Colonel W. Garden, C.B., Quartermaster-General of the army; Lieutenant W. F. Tytler, Assistant Quartermaster-General, and Lieutenant Paton, Deputy

Assistant Quartermaster-General of the army; Lieutenant-Colonel Birch, Judge-Advocate-General; and Lieutenant G. B. Johnson, Deputy Judge-Advocate-General; Major G. Thomson, Assistant Commissary-General; Lieutenant-Colonel J. G. W. Curtis, Assistant Commissary-General; Captain C. Campbell, Paymaster to the army; Captain J. Lang, postmaster; and H. Franklin, Esq., Inspector-General of Her Majesty's hospitals.

"To my personal staff I am also much indebted: Captain F. P. Haines, military secretary; Major N. Bates, aide-de-camp; Lieutenant A. Bagot, aide-de-camp; Lieutenant S. J. Hire, aide-de-camp; Captain Gabbett, aide-de-camp; Lieutenant G. N. Hardinge, aide-de-camp; and Lieutenant W. G. Prendergast, Persian interpreter.

"The unwearied exertions of Dr. Penny, superintending surgeon, and of Dr. MacRae, field-surgeon, in the care of the wounded, have been beyond all praise.

"The Earl of Gifford kindly accompanied me throughout the operations, and was most useful in conveying my orders to the several divisions and brigades. I had also the advantage throughout the day of the active services of Lieutenant-Colonel Sir Henry M. Lawrence, Major Mackeson, Mr. Cocks, C.S., Captain Nicholson, and Lieutenant Robinson, as well as of Major Anstruther, of the Madras Artillery, and Lieutenant H. O. Mayne, of the 6th Madras Light Cavalry.

"Captain Ramsay, Joint Deputy Commissary-General, with the several officers of that department, has been most indefatigable, and hitherto kept the army well supplied.

"I have, etc.
(*Signed*) "GOUGH, *General*,
"*Commander-in-Chief*."

"FROM MAJOR-GENERAL SIR JOSEPH THACKWELL, K.C.B., COMMANDING CAVALRY DIVISION.

"CAMP CHILIANWALLA, 16*th January*, 1849.

"SIR,

"I have the honour to report for the information of His Excellency the Right Honourable the Commander-in-Chief, that the cavalry under my command advanced on the flanks of the army on the morning of the 13th inst. from Dinghie, detailed as in the margin.[1] The right brigade in columns of squadrons, left in front. The left brigade in column of the same front, right in front. Both brigades covered by advanced guards and strong flanking parties, those on the right to patrol to the foot of the hills. A squadron of the 8th Light Cavalry under Captain Moore I ordered to form the advance guard to the left column of artillery, and this squadron remained then employed during the operations of the day. I had intended to join the right column of cavalry, but on the enemy's line being discovered in quite a different position from what it was imagined they occupied the previous evening, I thought it likely they had as large a body of cavalry on their right flank as on their left, and particularly as officers sent out to reconnoitre reported that they were making a movement towards our left rear. I remained therefore with the cavalry of this wing, though I did not believe the report. On the infantry deploying into line and advancing, the cavalry did the same on the left of Lieutenant-Colonel Brind's battery, the left regiment being refused, the movement of

[1] 1st Brigade: Brigadier White, C.B.; H.M.'s 3rd Light Dragoons, Major Yerbury; 5th Light Cavalry, Captain Wheatley; 8th Light Cavalry, Major Mackenzie; 2nd Brigade, Brigadier Pope, C.B.; H.M.'s 9th Lancers, Major Grant, C.B.; H.M.'s 14th Light Dragoons, Lieutenant-Colonel King; 1st Light Cavalry, Lieutenant-Colonel Bradford; 6th Light Cavalry, Major Coventry.

the infantry greatly to the left caused Lieutenant-Colonel Brind's guns and the cavalry to make a similar movement, which soon brought the line under the fire of a strong battery on the enemy's right flank, upon which Lieutenant-Colonel Brind's guns soon opened with great effect; and after a cannonade of nearly three-quarters of an hour, judging that many of the enemy's guns were disabled, I ordered the 5th Light Cavalry and a squadron of the 3rd Light Dragoons to charge a body of cavalry which threatened our left, to drive them back and take the enemy's guns in flank, whilst a part of the remaining cavalry was to charge them in front. This intention was abandoned in consequence of the 5th Light Cavalry being driven back. From the gallant charge of the squadron of the 3rd Light Dragoons under Captain Unett, who dispersed the troops opposed to him and the good countenance of the remainder of the cavalry, this cavalry did not dare to advance in pursuit of the 5th Light Cavalry, neither did their cavalry and infantry near their guns dare make any offensive movement. At this time I directed Lieutenant-Colonel Brind to make a movement to his right to support the Third Division of Infantry which had suddenly moved to the same flank. The enemy then directed their fire more rigorously from about six guns which had not been silenced, and the cavalry sustained some casualties in making their flank movement to the right and eventually to the left rear of Chilianwalla.

"I greatly regret to learn of the misconduct of the cavalry of the right wing, but as their movement did not come under my observation I have the honour to transmit the reports of Brigadier Pope, Lieutenant-Colonel Bradford, Lieutenant-Colonel King, and Major Grant, and it would appear that many faults were committed, such as having no reserve in the rear to support and prevent the right flank being turned; secondly, that care had not been taken not to get before the artillery on the left, the denseness of the jungle being no excuse for such a false step; and thirdly, that commanding officers, although they might have heard the words 'Three's about,' did not take upon themselves to charge any of the enemy's cavalry which were pressing upon the line. Could I have anticipated such an untoward circumstance as occurred I should have been on the spot to have given the benefit of my experience to an officer deemed fully competent to have command of a brigade of cavalry, and I feel assured from what I have heard of Brigadier Pope and the conduct of the cavalry of the right that their retrograde movement originated more from mistake than a fear of encountering an insignificant enemy, and I have every confidence on the next occasion the 9th Lancers and the 14th Light Dragoons will remove the stigma now cast upon them and earn similar laurels to those their predecessors gained in the Peninsula. I only regret for myself that I have not the gift of ubiquity. In furtherance of my desire to give every information to His Excellency I have the honour to forward the reports named in the margin,[1] by which, although I cannot advance much in favour of the 2nd Brigade of Cavalry, I have much to say in praise of the 1st Brigade, two regiments of which did their duty entirely to my satisfaction; and although the 5th Light Cavalry met with a check, but soon rallied, the gallant conduct of the squadron of the 3rd Light Dragoons under Captain Unett, who was severely wounded, has contributed more to establish the invincibility of the British cavalry than the accidental mishaps occasioning the loss of a few lives can have aided in lessening a faith in it. Having thus detailed the operations of the Cavalry Division from observation and report, I have now a pleasing duty to perform in bringing to the notice

[1] Brigadier Pope, Lieutenant-Colonel King, Lieutenant-Colonel Bradford,* Major Grant.
* Not yet received.

of the Right Honourable the Commander-in-Chief that Brigadier White conducted his brigade much to my satisfaction, as did also Major Yerbury, Major Mackenzie, and Captain Wheatley their regiments; and I must particularise the gallant charge made by the squadron under Captain Unett and the steady support given by the squadron of the 8th Light Cavalry under Captain Moone to the guns of Lieutenant-Colonel Brind's battery and to Captain Warner's guns when acting with the Third Division of Infantry to the last moment. I am also well satisfied with the conduct of the officers of this brigade and of the men, with few exceptions. I also beg to bring to his Lordship's notice the zeal and activity of my Deputy Assistant Adjutant-General, Captain Pratt, and my Deputy Assistant Quartermaster-General Lieutenant Tucker, and I have derived great assistance from both on all occasions. I also beg to notice the zeal and activity of my aide-de-camp, Lieutenant Thackwell, and the Sub.-Assistant Commissary-General of the Cavalry Division, Lieutenant Simpson, also the zeal and activity of Captain Cautley, Major of Brigade 1st Cavalry Brigade. Lieutenant-Colonel Brind, Major Fordyce, Captains Warner and Duncan attached to this Brigade of Cavalry conducted their operations greatly to my satisfaction and they and the officers and men are deserving of all praise. I beg leave to transmit a return of killed, wounded, and missing.

"I have the honour to be,

"Sir,

"Your most obedient Servant,

"JOSEPH THACKWELL,

"*Major-General Commanding Cavalry Division Army of the Punjab.*

"LIEUTENANT-COLONEL GRANT, C.B.,

"*Adjutant-General of the Army Headquarters.*"

"FROM THE RIGHT HONOURABLE THE COMMANDER-IN-CHIEF, TO THE RIGHT HONOURABLE THE GOVERNOR-GENERAL OF INDIA, ETC., ETC., ETC.

"HEADQUARTERS CAMP, RAMNUGGAR, *December 10th*, 1848.

"MY LORD,

"In continuation of my letter of the 5th instant, I have now the honour to enclose to your Lordship a copy of Major-General Sir Joseph Thackwell's despatch, dated the 6th idem, but only received last night, detailing the operations of the force under his command, after it had been detached from my headquarters.

"I can only repeat the warm approval I have already expressed of the conduct of the Major-General and of every officer and man under his command, and I beg your Lordship's favourable consideration of the services of those named by Sir Joseph Thackwell.

"I beg to enclose a rough sketch of the operations of the 3rd instant.

"I have, etc.,

(*Signed*) "GOUGH."

"FROM MAJOR-GENERAL SIR JOSEPH THACKWELL, K.C.B. AND K.H., TO LIEUTENANT-COLONEL GRANT, C.B., ADJUTANT-GENERAL OF THE ARMY.

"CAMP, HEYLEH, *December 6th*, 1848.

"SIR,

"I have the honour to report, for the information of the Right Honourable the Commander-in-Chief, that, agreeably to His Excellency's orders, I left the camp at Ramnuggar, with the troops named in the margin,[1] at about half-past 3 o'clock on the morning of December 1st, 1848, instead of at 1 o'clock, as I had ordered, some of the troops having lost their way among the intricacies of the rear of the encampment, and proceeded to the vicinity of the ford on the Chenab at Runnee-Khanke-Puttun, distant thirteen miles from Ramnuggar, which, owing to the broken ground and narrow roads, where any existed for the first four miles, I did not reach before 11 o'clock. The enemy had infantry at this ford, which report afterwards magnified to 4,000 men, but the villagers said it was much deeper than the one at Allee-Shere-ke-Chuck, a mile higher up the river. I am much indebted to Lieutenant Paton, Deputy Assistant Quartermaster-General, for his anxious exertions in examining this ford; and from this report I came to the conclusion that this ford of Allee-ke-Chuck could not have artillery on the left bank of the river to cover the passage of the troops, from the insecure bottom of the first ford, neither could the pontoon train be of much use, for the same reason, and the deep sands which lay between the fords. The pontoon train might have been laid over the main stream under cover of a battery, near the enemy's infantry; but beyond the river the sands seemed wet and insecure, and a branch of the river beyond them was said to be deep, with a muddy bottom. Under all these disadvantages, I came to the decision that it was more advisable to try the passage of the river near Wuzeerabad, where Captain Nicholson, assistant to the Resident at Lahore, informed me that at the ferry were seventeen boats, and a ford not more than three feet ten inches deep, with a good bottom, than to run the risk of a severe loss by passing the river near the enemy. This survey of the ford occupied three hours, and at 2 o'clock I put the column in movement to the ford and ferry at Wuzeerabad, which was in possession of Lieutenant Nicholson's pattans, where the leading infantry arrived about 5 o'clock in the afternoon, having made a march of about twenty-five miles. The 6th brigade of infantry and some of the guns were passed over the Chenab immediately, and I am indebted to Brigadier-General Campbell, Lieutenant-Colonel Grant, horse artillery, and Captain Smith of the Engineers, for their great exertions in forwarding this object. Brigadier Eckford I hoped would have crossed the river by the three fords that evening; but as it became too dark and hazy for such an operation, he halted for the night on the dry sands near the last branch of the river. Major Tait, 3rd Irregular Cavalry, was enabled to pass over three of his *risallas*, in doing which, I am sorry to say, three sowars and one horse were drowned. On the morrow the infantry, cavalry, and all the troops were soon over the river by ferry and ford, and all the baggage and commissariat animals passed the same by 12 o'clock, without any further loss.

[1] Major Christie's troop Horse Artillery; Captain Huish's troop Horse Artillery; Captain Warner's troop Horse Artillery; Captain Kinleside, No. 5 Light Field Battery; Captain Austin, No. 10 Light Field Battery; Captain Robinson and two 18-pounders, under the command of Lieutenant-Colonel Grant, Horse Artillery; 2 companies of pioneers; the pontoon train; 1st Brigade of Cavalry, commanded by Brigadier White; 3rd Light Dragoons, commanded by Major Yerbury; 5th Light Cavalry, commanded by Captain Wheatley; 8th Light Cavalry, commanded by Captain Moone; 3rd Irregular Cavalry, commanded by Major Tait; 12th Irregular Cavalry, commanded by Lieutenant Cunningham; 3rd Brigade of Infantry, Brigadier Eckford; 31st Native Infantry, Major Corfield; 56th Native Infantry, Major Bamfield; Third Division of Infantry, Brigadier-General Campbell commanding; Sixth Brigade of Infantry, Brigadier Pennycuick; H.M.'s 24th Foot, Major Harris; 2 flank companies, 2nd Battalion company, 22nd Native Infantry, Major Sampson; Eighth Brigade of Infantry, Brigadier Hoggan; H.M.'s 61st Foot, Lieutenant-Colonel McLeod; 36th Native Infantry, Major Flemyng; 46th Native Infantry, Major Tudor. Of the above detail, the following returned in charge of the two 18-pounders and pontoon train: 2 guns of No. 10 Light Field Battery; 12th Irregular Cavalry; 2 companies 22nd Native Infantry.

"At 2 p.m., after the troops had dined, I marched, in order of battle, three brigade columns of companies, at half distance left in front, at deploying interval; the First Brigade of Cavalry, in the same order on the right, with strong flanking parties and rearguard, and the 3rd Irregular Cavalry on the left, with orders to patrol to the river and clear the right bank, aided by infantry, if necessary. In this order I arrived at Doorawal at dusk, about twelve miles from the ferry, and halted for the night. On Sunday, December 3rd, at daylight, the troops proceeded in the same order towards the Sikh position, and I intended to have reconnoitred and commenced an attack upon it by 11 o'clock; hearing, however, when within about four miles of it or less, that reinforcements were expected to pass over the Chenab at the ford near Ghurree-ke-Puttun, it became necessary to secure that post, and which had been found without an enemy an hour before, but to which it now seemed that a body of about 600 of the enemy were seen approaching, and I detached a wing of the 56th Native Infantry, and two *risallas* of the 3rd Irregular Cavalry under Major Tait, who secured the post and frustrated the attempt of the enemy. This caused so much delay, that enough of daylight would not be left for the advance and attack on the left and rear of the enemy's position. About 2 p.m. some of the enemy's guns opened on a patrol of the 5th Light Cavalry, and he was seen advancing in large bodies of cavalry and infantry, and the picquets, which occupied three villages with large plantations of sugar-cane, being too much in advance to be supported, fell back without any loss, and the enemy occupied these villages with cavalry on the right, guns and bodies of infantry, and the main body of their cavalry with horse artillery were on their left. When the enemy's guns opened, I ordered Brigadier-General Campbell to deploy the infantry into line in front of the village of Sudoolapoor, Brigadier Eckford and part of Brigadier Hoggan's brigade being extended, in order not to be outflanked. It was not until the enemy came well within range of our guns that I caused them to open their fire, which they then did with great effect. The enemy tried to turn both our flanks, which having foreseen, I had caused Captain Warner to move his troop of artillery to the left of the infantry, and had sent the 5th Light Cavalry to the left to support these guns, and to act in conjunction with the two *risallas* of the 3rd Irregular Cavalry under Captain Biddulph, who were posted on open ground, and these soon drove the enemy back. The attempt to turn our right was met by extending the 8th Light Cavalry and H. M.'s 3rd Light Dragoons, supported by Major Christie's troop of artillery. As the cavalry on the right advanced, the enemy's sowars gave way, and they fell back on their infantry, having lost some men by the skirmishers of the 3rd Light Dragoons. After a cannonade of about two hours the fire of the enemy slackened, and I sent Lieutenant Paton to desire the cavalry on the right to charge and take the enemy's guns, if possible, intending to support them by moving the brigades in échelon from the right at intervals according to circumstances; but as no opportunity offered for the cavalry to charge, and so little of the daylight remained, I deemed it safer to remain in my position than attempt to drive back an enemy so strongly posted on their right and centre, with the prospect of having to attack their entrenched position afterwards. From this position the Sikhs began to retire at about 12 o'clock at night, as was afterwards ascertained, and as was conjectured by the barking of the dogs in their rear. I have every reason to believe that Shere Singh attacked with twenty guns, and nearly the whole of the Sikh army were employed against my position, which was by no means what I could have wished it; but the fire of our artillery was so effective, that he did not dare to bring his masses to the front, and my brave, steady, and ardent infantry, whom I had caused to lie down to avoid the heavy fire, had no chance of firing a shot, except a few companies on the left of the line. The enemy's loss has been severe ; ours comparatively small. I regret not being able to capture the enemy's guns ; but with the small force of cavalry—two regiments on the right only—it would have been a matter of difficulty for tired cavalry to overtake horse artillery, fresh and well mounted. In these operations the conduct of all has merited my warmest praise, and the patient endurance of the artillery, cavalry, European infantry and sepoys, under privations of no ordinary nature, has been most praiseworthy.

"To Brigadier-General Campbell I am much indebted for his able assistance during these movements, and to Lieutenant-Colonel Grant, commanding the artillery, Major Christie, Captains Huish, Warner, Austin, and Kinleside, and the officers and men under their command, I cannot bestow too much praise for their skill and gallantry in overcoming the fire of a numerous artillery, some of which were of heavy calibre. I am also greatly indebted to Captain Smith of the Engineers, for his exertions in passing over guns at the Wuzeerabad ferry, and for his assistance in conveying my orders on various occasions. And my thanks are due to Lieutenants Yule and Crommelin of the same corps, and Lieutenant Bacon of the Sappers ; to Lieutenant Paton, Deputy Assistant Quartermaster-General, my best thanks are due for his exertions and assistance in the advance of the troops, and during the action ; and to Captain Nicholson, assistant to the Resident at Lahore, I beg to offer my best thanks for his endeavours to procure intelligence of the enemy's movements, for his endeavours to procure supplies for the troops, and his able assistance on all occasions. Captain Pratt, my Deputy Assistant Adjutant-General; Lieutenant Tucker, Deputy Assistant Quartermaster-General, and Lieutenant Thackwell, aide-de-camp, have been most zealous in performing their respective duties, and have rendered me every assistance ; and I feel assured that if the cavalry and infantry had been brought into close action, I should have had the great satisfaction of thanking brigadiers commanding, officers of corps, and the officers and men, for their gallantry and noble bearing in action, as I now do for their steadiness and good conduct. To Major Mainwaring, Captains Gerrard, Simpson, Faddy, and James, I am much indebted for their exertions in their respective departments.

"I beg further to state that on the morning of the 4th I put the troops in motion to pursue the enemy, who had retreated during the preceding night, and encamped about eleven miles from the Chenab, on the road to Jullalpore, the 9th Lancers having been pushed to the front, but without seeing anything of the enemy, who had retreated by the Jhelum, Jullalpore, and Pind Dadun Khan roads. On the following day I arrived at this place, and sent two regiments of cavalry on the road to Dingee, one of them, the 14th Light Dragoons, and two regiments of cavalry, and a troop of horse artillery on the road to Jullalpore. The latter party observed two bodies of the enemy of about 800 and 400 men each, imagined to be a strong rearguard, about eight miles from this, and behind a thick jungle which reaches to the river ; and the former went to Dingee, which place the enemy had left, and the villagers said had gone over to the Jhelum. Both parties returned to this camp without, I am sorry to say, having overtaken any of the enemy's troops or guns.

"I beg leave to enclose a return of the killed and wounded.

"I have, etc.,

(*Signed*) "Jos. Thackwell, *Major-General,*

"*Commanding the advanced post of the Army.*"

"THE SECRETARY WITH THE GOVERNOR-GENERAL TO THE ADJUTANT-GENERAL OF THE ARMY.

"FEROZEPORE, *January* 31*st*, 1849.

"I am directed to acknowledge the receipt of His Excellency the Commander-in-Chief's despatches, dated the 5th, 10th, and 16th ultimo, reporting the particulars of an action with the enemy at Sadoolapore, and the passage of the Chenab by Major-General Sir Joseph Thackwell, K.C.B.

"The Governor-General regrets to find that he inadvertently omitted to issue instructions founded on a minute which he had recorded on the subject of the despatches under acknowledgment. His Lordship begs to congratulate the Commander-in-Chief on the success of the measures which he adopted for effecting the passage of the Chenab, and to convey to him the assurance of his satisfaction with, and his best thanks for, the judicious arrangements by which he was enabled, with comparatively little loss, to carry into execution his plans for the passage of that difficult river, and for compelling the retreat of the Sikh army from the formidable position which they occupied on its further bank, after they had been engaged, and beaten back by the forces under Major-General Sir Joseph Thackwell. The result of His Excellency's movements, in driving the Sikh army from their entrenchments, and forcing them to retire on the other extremity of the Dooab, was of much importance.

"The Governor-General offers his best thanks to Major-General Sir Joseph Thackwell for his successful direction of the force under his command, and for the dispositions by which he compelled the enemy to retire, and ultimately to quit the ground he had occupied. The Governor-General tenders his best thanks to Brigadier-General Campbell for the able assistance which he rendered to Major-General Sir Joseph Thackwell, and to Lieutenant-Colonel Grant for the powerful and effective use which he made of the artillery under his command.

"The Governor-General has had much gratification in observing the terms in which the Commander-in-Chief has spoken of the army under his command in the field; and he concurs with His Excellency in bestowing upon them the praise which is their due."

GENERAL ORDERS BY THE RIGHT HONOURABLE THE GOVERNOR-GENERAL OF INDIA.

CAMP FEROZEPORE, *February* 24*th*, 1849.

The following notifications from the Foreign Department are published for the information of the army:

"*Notification.*

"FOREIGN DEPARTMENT, CAMP, FEROZEPORE, *February* 23*rd*, 1849.

"The Governor-General has the gratification of intimating to the President in Council, and notifying for public information, that he has this day received a despatch from Major Mackeson, C.B., agent to the Governor-General with the Commander-in-Chief, conveying the intelligence that the forces under His Excellency the Commander-in-Chief on the 21st instant attacked and routed the Sikh army in the neighbourhood of Goojerat.

"The enemy was beaten in every point and retreated in disorder, leaving in the hands of the British troops, by whom he was pursued, a great portion of his artillery, his ammunition, and the whole of his standing camp.

"The official despatches of His Excellency the Commander-in-Chief will be published as soon as they are received.

"The Governor-General directs that a salute of 21 guns shall be fired, at every principal station of the army, on the receipt of this notification.

"By order of the Right Honourable the Governor-General of India.

(*Signed*) "H. M. ELLIOTT,
"*Secretary to the Governor of India,*
"*with the Governor-General.*"

Notification.

FOREIGN DEPARTMENT, CAMP, FEROZEPORE, *February 24th*, 1849.

The Right Honourable the Governor-General directs the publication of the following letter from His Excellency the Commander-in-Chief, reporting the complete defeat of the Sikh army on the 21st instant. The detailed despatches will be published hereafter.

"FROM HIS EXCELLENCY THE COMMANDER-IN-CHIEF IN INDIA, TO THE RIGHT HONOURABLE THE GOVERNOR-GENERAL.

"CAMP IN FRONT OF GOOJERAT, *February 21st*, 1849.

"MY LORD,

"I have the honour to report to your Lordship that I have this day obtained a victory of no common order, either in its character, or, I trust, in its effects.

"I was joined yesterday by Brigadier Markham's brigade, Brigadier-General Dundas having joined late the preceding night. I moved on in the afternoon of yesterday, as soon as these troops were refreshed, from Trikur, to the village of Shadiwal, and at 7 this morning I moved to the attack, which commenced at half-past 8 o'clock, and by 1 o'clock I was in possession of the whole of the Sikh position, with all his camp equipag , baggage, magazines, and, I hope, a large proportion of his guns; the exact number I cannot at present state, from the great extent of his position and length of pursuit, as I followed up the enemy from four to five miles on the Bimber road, and pushed on Sir Joseph Thackwell with the cavalry. The rout has been most complete; the whole road for twelve miles is strewed with guns, ammunition waggons, arms, and baggage.

"My loss was comparatively small (I hope within 300 killed and wounded) when it is considered I had to attack 60,000 Sikhs, in a very strong position, armed with upwards of 60 guns. The loss of the enemy must have been very severe.

"The conduct of the whole army, in every arm, was conspicuous for steadiness in movement and gallantry in action. The details I shall furnish hereafter.

"I have, etc.,
(*Signed*) "GOUGH, *General,*
"*Commander-in-Chief in India.*

"By order of the Right Honourable the Governor-General of India.
(*Signed*) "H. M. ELLIOT.
"*Secretary to the Government of India, with the Governor-General.*
(*Signed*) "J. STUART, *Colonel,*
"*Secretary to the Government of India, Military Department,*
"*with the Governor-General.*"

FORT WILLIAM, FOREIGN DEPARTMENT, *March 9th*, 1849.

The President in Council is pleased to direct the publication of the following notification issued by the Right Honourable the Governor-General at his Lordship's headquarters, with a despatch from His Excellency the Commander-in-Chief, reporting the details of the complete victory which was gained over the Sikh force at Goojerat, on the 21st ultimo, by the army under His Excellency's command.

By order of the President of the Council of India in Council.

(*Signed*) FRED. JAS. HALLIDAY,
Officiating Secretary to the Government of India.

"GENERAL ORDER OF THE RIGHT HONOURABLE THE GOVERNOR-GENERAL OF INDIA.

"FOREIGN DEPARTMENT, CAMP, FEROZEPORE, *March 1st*, 1849.

"The Governor-General, having received from His Excellency the Commander-in-Chief a despatch, reporting the details of the brilliant victory which was gained by the British army at Goojerat, on the 21st ultimo, directs that it be published for the information of the army and of the people of India.

"The Sikh army, under the command of Sirdar Chutter Sing and of Rajah Shere Sing, combined with the Afghan troops in the service of the Ameer of Cabool, were posted in great strength near to the town of Goojerat.

"Their numbers were estimated at 60,000 men, and 59 guns were brought by them into action.

"On the morning of the 21st they were attacked by the forces under the personal command of His Excellency the Commander-in-Chief. A powerful and sustained cannonade by the British artillery compelled them, after some time, to retire from the positions they had well and resolutely maintained.

"The subsequent advance of the British army drove them back at once from every point, and retreat having been speedily converted into rout, they fled in the utmost disorder, and, abandoning their guns, and throwing away their arms, were pursued by the artillery and cavalry till the evening, for many miles beyond the town.

"Fifty-three pieces of the enemy's artillery, his camp, his baggage, his magazines, and vast stores of ammunition left in the hands of the British troops, bear testimony to the completeness and to the importance of the victory that has been won.

"The Governor-General, in the name and on behalf of the Government of India, most cordially congratulates his Excellency the Commander-in-Chief and the whole army on the glorious success which, under the blessing of Divine providence, their skill and gallantry have achieved; and he offers to his Excellency, to the generals, the officers, non-commissioned officers and soldiers of the force, his grateful acknowledgments of the services they have thus rendered to the Government and to their country.

"The Governor-General begs especially to thank Major-General Sir Joseph Thackwell, K.C.B. and K.H.; Major-General Sir W. Gilbert, K.C.B.; Major-General Whish, C.B.; Brigadier-General Campbell, C.B.; and Brigadier-General the Honourable H. Dundas, C.B., for the ability and judgment with which they directed the operations of the divisions respectively under their command.

"To the chief engineer, Brigadier Cheape, C.B.; to the officers commanding brigades, Brigadier Brooke, C.B.; Brigadier Huthwaite, C.B.; Brigadier Leeson; to Brigadier White, C.B., Brigadier Hearsey and Brigadier Lockwood, C.B.; to Brigadier Hervey and Brigadier Markham; to Brigadier Mountain, C.B.; Brigadier Capon and Brigadier Hoggan; Brigadier Carnegy and Brigadier McLeod, the best thanks of the Governor-General are due.

"The services of Brigadier-General Tennant and of the artillery of the force have been recorded in the despatch of the Commander-in-Chief in terms of which they may justly be proud.

"The Governor-General cordially joins with his Excellency in acknowledging their merit, and in bestowing upon them the praise they have earned so well.

"To Major Lugard, to Lieutenant-Colonel Gough, C.B., and to the officers of the general staff of Her Majesty's army; to Lieutenant-Colonel Grant, C.B.; to Lieutenant-Colonel Garden, C.B.; and to the officers of the general staff of the army; to Captain Ramsay and the officers of the commissariat department; to Mr. Franklin, inspector-general of Her Majesty's hospitals; to Dr. Renny and the officers of the medical department, and to the officers of his Excellency's personal staff, the Governor-General offers his best thanks, and assures them of his full appreciation of their services.

"And to all the troops of every arm, European and native, the Governor-General desires to convey his entire approbation of their steady and gallant conduct throughout the day; particularly to a portion of the 9th Lancers and the Scinde Horse for their charge against the Afghan cavalry; to the Third Brigade of Infantry, under Brigadier Penny, C.B., for their attack on the village of Kalra; and to a portion of Brigadier Hervey's brigade for their charge led by Lieutenant-Colonel Franks, C.B., all of which have been specially reported by his Excellency the Commander-in-Chief.

"The Governor-General estimates highly the important results which the battle gained on the 21st ultimo is calculated to produce. He entertains a hope that the conviction, which the events of that day must force upon all, of the vast superiority which the British army derived from the possession of science and military resource, will induce the enemy shortly to abandon a contest which is a hopeless one.

"The war in which we are engaged must be prosecuted with vigour and determination, to the entire defeat and dispersion of all who are in arms against us, whether Sikhs or Afghans.

"The Governor-General has ever felt, and feels, unbounded confidence in the army which serves in India. He relies fully on the conviction that their services will be given cheerfully and gallantly, as heretofore, whatever may be the obstacles opposed to them; and he does not doubt that, with the blessing of Heaven, such full success will continue to follow their efforts as shall speedily give to the Government of India the victory over its enemies, and restore the country to the enjoyment of peace.

"The Governor-General will not fail earnestly to commend the past services of this army to the favourable consideration of Her Majesty's Government and of the Honourable East India Company.

"A salute of 21 guns has been ordered at every principal station of the army in India.

"By order of the Right Honourable the Governor-General of India.

(*Signed*) "H. M. Elliot,
"*Secretary to the Government of India,*
"*with the Governor-General.*"

"FROM THE RIGHT HONOURABLE THE COMMANDER-IN-CHIEF IN INDIA, TO THE RIGHT HONOURABLE THE GOVERNOR-GENERAL OF INDIA.

"HEADQUARTERS CAMP, GOOJERAT, *February 26th*, 1849.

"MY LORD,

"By my letter of the 21st instant, written on the field of battle immediately after the action, your Lordship will have been made acquainted with the glorious result of my operations on that day against the Sikh army, calculated from all credible reports at 60,000 men of all arms and 59 pieces of artillery, under the command of Sirdar Chutter Sing and Rajah Shere Sing, with a body of 1,500 Afghan Horse led by Akram Khan, son of the Ameer Dost Mahomed Khan; a result, my Lord, glorious indeed for the ever-victorious army of India! the ranks of the enemy broken, their position carried, their guns, ammunition, camp equipage, and baggage captured, their flying masses driven before the victorious pursuers from mid-day to dusk, receiving a most severe punishment in their flight; and, my Lord, with gratitude to a merciful Providence, I have the satisfaction of adding that, notwithstanding the obstinate resistance of the enemy, this triumphant success, this brilliant victory has been achieved with comparatively little loss on our side.

"The number of guns taken in action and captured in the line of pursuit, I now find to be 53.

"The official report made by the Adjutant-General of the army on the 20th instant will have informed your Lordship that I had directed Brigadier-General the Honourable H. Dundas to join me by forced marches, and that I had closed up to so short a distance of the Sikh army that they could not possibly attempt the passage of the Chenab, in order to put into execution their avowed determination of moving upon Lahore, make a retrograde movement by the Kooree Pass (the only practicable one for guns), or indeed quit their position, without my being able to attack them and defeat their movement.

"On the 18th instant Brigadier Markham had proceeded from Ramnuggar up the left bank of the river to Kanokee, to which I had directed 47 boats to be sent up. On the morning of the 20th this officer crossed the Chenab, by my instructions, and joined me at 11 o'clock a.m. At the same time Lieutenant-Colonel Byrne was directed to move down the left bank, from the position he held in front of Wuzeerabad, with two corps of infantry and four guns, leaving two regiments of irregular cavalry to watch the fords, and to prevent any marauding parties or bodies of the routed enemy from effecting a passage.

"On the same day a reconnaissance was made of the enemy's position, and it was ascertained that their camp nearly encircled the town of Goojerat, their regular troops being placed immediately fronting us between the town and a deep watercourse, the dry bed of the River Dwara—this nullah, which is very tortuous, passing round nearly two sides of the town of Goojerat, diverging to a considerable distance on the north and west faces, and then taking a southerly direction, running through the centre of the ground I occupied at Shadiwal. Thus the enemy's position on the right was generally strengthened, the nullah giving cover to his infantry in front of his guns, whilst another deep, though narrow wet nullah running from the east of the town and falling into the Chenab, in the direction of Wuzeerabad, covered his left.

"The ground between these nullahs, for a space of nearly three miles, being well calculated for the operations of all arms, and presenting no obstacle to the movement of my heavy guns, I determined to make my principal attack in that direction and dispose my force accordingly.

"On the extreme left I placed the Bombay column, commanded by Brigadier the Honourable H. Dundas, supported by Brigadier White's brigade of cavalry and the Scinde Horse, under Sir Joseph Thackwell, to protect the left and to prevent large bodies of Sikh and Afghan cavalry from turning that flank; with this cavalry I placed Captains Duncan and Huish's troops of horse artillery, whilst the infantry was covered by the Bombay troop of horse artillery under Major Blood.

"On the right of the Bombay column, and with its right resting on the nullah, I placed Brigadier-General Campbell's division of infantry, covered by No. 5 and No. 10 Light Field Batteries, under Major Ludlow and Lieutenant Robertson, having Brigadier Hoggan's brigade of infantry in reserve.

"Upon the right of the nullah I placed the infantry division of Major-General Sir Walter Gilbert, the heavy guns, 18 in number, under Majors Day and Horsford, with Captain Shakspear and Brevet-Major Sir Richmond Shakspear, commanding batteries, being disposed in two divisions upon the flanks of his left brigade.

"This line was prolonged by Major-General Whish's division of infantry, under Brigadier Markham, in support of second line, and the whole covered by three troops of horse artillery, Major Fordyce's, Captains Mackenzie's, and Anderson's; No. 17 Light Field Battery, under Captain Dawes, with Lieutenant-Colonel Lane's and Captain Kinleside's troops of horse artillery, in a second line in reserve, under Lieutenant-Colonel Brind.

"My right flank was protected by Brigadiers Hearsey's and Lockwood's brigades of cavalry, with Captain Warner's troop of horse artillery.

"The 5th and 6th Light Cavalry, with the Bombay Light Field Battery, and the 45th and 69th Regiments, under the command of Lieutenant-Colonel Mercer, most effectually protected my rear and baggage.

"With my right wing I proposed penetrating the centre of the enemy's line, so as to turn the position of their force in rear of the nullah, and thus enable my left wing to cross it with little loss, and in co-operation with the right to double upon the centre of the wing of the enemy's force opposed to them.

"At half-past 7 o'clock the army advanced in the order described, with the precision of a parade movement. The enemy opened their fire at a very long distance, which exposed to my artillery both the position and range of their guns. I halted the infantry just out of fire, and advanced the whole of my artillery, covered by skirmishers.

"The cannonade now opened upon the enemy was the most magnificent I ever witnessed, and as terrible in its effect.

"The Sikh guns were served with their accustomed rapidity, and the enemy well and resolutely maintained his position; but the terrific force of our fire obliged them, after an obstinate resistance, to fall back. I then deployed the infantry, and directed a general advance, covering the movement by my artillery as before.

"The village of Burra-Kalra, the left one of those of that name, in which the enemy had concealed a large body of infantry, and which was apparently the key of their position, lay immediately in the line of Major-General Sir Walter Gilbert's advance, and was carried in the most brilliant style, by a spirited attack of the Third Brigade, under Brigadier Penny, consisting of the 2nd Europeans, 31st and 70th Regiments of Native Infantry, which drove the enemy from their cover with great slaughter.

"A very spirited and successful movement was also made about the same time against a heavy body of the enemy's troops, in and about Chota-Kalra, by part of Brigadier Hervey's brigade, most gallantly led by Lieutenant-Colonel Franks of H.M.'s 10th Foot.

"The heavy artillery continued to advance with extraordinary celerity, taking up successive forward positions, driving the enemy from those they had retired to, whilst the rapid advance and beautiful fire of the horse artillery and light field batteries, which I strengthened by bringing to the front the two reserved troops of horse artillery under Lieutenant-Colonel Brind—Brigadier Brooke having the general superintendence of the whole of the horse artillery—broke the ranks of the enemy at all points. The whole infantry line now rapidly advanced and drove the enemy before it; the nullah was cleared, several villages stormed, the guns that were in position carried, the camp captured, and the enemy routed in every direction!—the right wing and Brigadier-General Campbell's division passing in pursuit to the eastward, the Bombay column to the westward of the town.

"The retreat of the Sikh army, thus hotly pressed, soon became a perfect flight, all arms dispersing over the country, rapidly pursued by our troops for a distance of twelve miles, their track strewed with their wounded, their arms, and military equipments, which they threw away to conceal that they were soldiers.

"Throughout the operations thus detailed, the cavalry brigades on the flanks were threatened, and occasionally attacked by vast masses of the enemy's cavalry, which were, in every instance, put to flight by the steady movements and spirited manœuvres of our cavalry, most zealously and judiciously supported by the troops of horse artillery attached to them, from whom the enemy received the severest punishment.

"On the left, a most successful and gallant charge was made upon the Afghan cavalry and a large body of Goorchurras, by the Scinde Horse and a party of the 9th Lancers, when some standards were captured.

"The determined front shown by the 14th Light Dragoons and the other cavalry regiments on the right, both regular and irregular, completely overawed the enemy, and contributed much to the success of the day; the conduct of all in following up the fugitive enemy was beyond all praise.

"A competent force, under the command of Major-General Sir Walter Gilbert, resumed the pursuit towards the Jhelum on the following morning, with a view of cutting off the enemy from the only practicable gun road to the Jhelum. Another division of infantry, under Brigadier-General Campbell, advanced on the road to Bimber, scouring the country in that direction, to prevent their carrying off the guns by that route, and a body of cavalry, under Lieutenant-Colonel Bradford, successfully pushed on several miles into the hills, and twenty-four from Goojerat, accompanied by that most energetic political officer, Captain Nicholson, for the same purpose, whilst I remained in possession of the field for the purpose of supporting these operations, covering the fords of the Chenab, and destroying the vast magazine of ammunition left scattered about in all directions. I am happy to add that these combinations have been entirely successful, the detached parties coming at every step on the wreck of the dispersed and flying foe.

"Having thus endeavoured to convey to your Lordship the particulars of the operations of the battle of Goojerat, I beg now to offer my heartfelt congratulations to your Lordship, and to the Government of India, upon this signal victory achieved under the blessing of Divine providence by the united efforts and indomitable gallantry of the noble army under my command, a victory, my Lord, as glorious to the army that gained it as it must be satisfactory to yourself and the Government of India, from the very important and decisive results to be expected from it.

"It is quite impossible for me sufficiently to express my admiration of the gallant and steady conduct of the officers and men, as well native as European, upon this occasion.

"The brilliant service they have performed in so signally defeating so vastly a superior force, amongst whom were the *élite* of the old Khalsa army, making a last, united, and desperate struggle, will speak for itself, and will, I am confident, be justly estimated by your Lordship.

"I cannot too strongly express to your Lordship my deep sense of obligation to the general officers and brigadier-generals in command of divisions, who so ably carried out my views and directed the operations of their troops on this day.

"I beg to annex for your Lordship's information the reports I have received from them, and to bring most prominently to your Lordship's notice the brigadiers commanding brigades; the commanding officers of regiments and of troops of horse artillery and light field batteries; and the several officers of the divisional and brigade staff enumerated in these reports, in terms of such just commendation.

"I feel much indebted to Major-General Sir Joseph Thackwell for the able and judicious manner he manœuvred the cavalry, with horse artillery attached, on the left, keeping in check the immensely superior force of the enemy, whose main object was to turn my flanks. I am also greatly indebted to this tried and gallant officer for his valuable assistance and untiring exertions throughout the present and previous operations as second in command with this force.

"To Major-General Sir Walter Gilbert, whose services upon this, as on all former occasions, were invaluable, and ever marked by energy, zeal, and devotion; as well as to Major-General Whish, Brigadier-Generals Campbell and Dundas, for their able assistance, I am deeply indebted.

"To Brigadier-General Tennant, commanding that splendid arm, the artillery, to whose irresistible power I am mainly indebted for the glorious victory of Goojerat, I am indeed most grateful. Conspicuous as the artillery has ever proved itself, never was its superiority over that of the enemy, its irresistible and annihilating power, more truthfully shown than in this battle. The heavy batteries manœuvred with the celerity of light guns, and the rapid advance, the scientific and judicious selection of points of attack, the effective and well-directed fire of the troops of horse artillery and light field-batteries, merit my warmest praise; and I beg most earnestly to recommend their brave and gallant commanders, with the several officers named in Brigadier-General Tennant's report, to your Lordship's most favourable notice.

"From Brigadier Cheape, the chief engineer, and the talented officers in that department as named in the Brigadier's report, I have received the most valuable assistance in reconnoitring the enemy's position and on the field of battle. The Sappers and Pioneers, under that most able officer, Captain Siddons, did excellent service, and were ever in front to overcome any obstacle to the advance of the artillery.

"To the officers of the general staff of Her Majesty's service, Major Lugard, Acting Adjutant-General, and Lieutenant-Colonel Gough, Quartermaster-General of Her Majesty's troops in India, my best thanks are due; their exertions upon the present occasion and throughout the recent operations were most valuable, and I beg to bring them under your Lordship's most favourable notice. I am equally indebted to

Captain Otter, Acting Assistant Adjutant-General of Her Majesty's forces, for his valuable services.

"To the officers of the general staff of the army, Lieutenant-Colonel Grant, Adjutant-General, and Lieutenant-Colonel Garden, Quarter-master-General, whose most onerous and very important duties have invariably been conducted to my entire satisfaction, I am under the greatest obligation. Their valuable assistance in the field, and their indefatigable exertions throughout operations of no ordinary character, deserve my warmest thanks and your Lordship's approbation.

"To Lieutenant-Colonel Birch, Judge-Advocate-General, I am much indebted for his assistance upon every occasion.

"To Major Tucker, Deputy Adjutant-General, a most gallant, energetic, and valuable officer; to Lieutenant-Colonel Drummond, Deputy Quartermaster-General, whose services have been most praiseworthy; to Major Chester, Assistant Adjutant-General, and Lieutenant Tytler, Assistant Quartermaster-General; Lieutenant Johnston, Deputy Judge-Advocate-General; Major G. Thompson, and Lieutenant-Colonel Curtis, Assistant Commissary-General; Captain C. Campbell, paymaster to the army, I offer my best thanks for their services whilst attending me in the field, and the efficient manner they have performed their several duties.

"Mr. Franklin, inspector-general of Her Majesty's hospitals, has been unceasing in his exertions in rendering every aid to the sick and wounded of the royal service, and giving the benefit of his long professional experience in such duties; as has Doctor Renny, superintending surgeon of this army, who has been indefatigable in his professional exertions and well-organised medical arrangements.

"I feel I cannot too prominently bring to notice the valuable exertions of Doctor MacRae, field-surgeon, and of the medical officers of the army generally; they have been most unwearied and praiseworthy.

"To Captain Ramsay, Deputy Commissary-General, and to the officers of his department, I am much indebted, and feel grateful for their unceasing and successful exertions, amidst all difficulties, to supply the troops, and thus preserve the efficiency of the army.

"The officers of my personal staff have well merited my best thanks and your Lordship's favourable notice. Captain Haines, military secretary, who has rendered me most valuable aid; Brevet-Major Bates, aide-de-camp; Lieutenant A. Bagot, aide-de-camp; Lieutenant S. J. Hire, aide-de-camp; Captain Gabbett, aide-de-camp; Lieutenant G. Hardinge, aide-de-camp; and Lieutenant W. G. Prendergast, my Persian interpreter.

"I beg also to acknowledge the valuable assistance I have received from the political officers, Major Mackeson, Mr. Cocks, Captain Nicholson, and Lieutenant Robinson, both in the field and throughout the operations. I regret to add that Mr. Cocks was seriously wounded during the action in a rencontre with a Sikh horseman.

"I would also bring to your Lordship's notice the name of Lieutenant Stannus, of the 5th Light Cavalry. This officer has commanded the cavalry party attached to my escort throughout the operations to my entire satisfaction. He was severely wounded on the 21st, when gallantly charging a party of the enemy's horsemen.

"Major Anstruther, of the Madras Artillery; Lieutenant Mayne, of the Madras Cavalry, and Captain Showers, of the 14th Native Infantry, attended me in the field.

"I have most unwillingly been delayed from sooner forwarding this despatch, from the circumstance of having only this day received Brigadier-General the Honourable H. Dundas's report, and some of the casualty returns have not even yet reached us. As soon as the whole come in, a full, amended, general return shall be transmitted without loss of time for your Lordship's information.

"I have the pleasure to enclose a plan of the battle of Goojerat, also a return of the captured ordnance.

"I have, etc.
(*Signed*) "GOUGH, *General*,
"*Commander-in-Chief in India*.

"P.S.—The casualty lists having arrived, I have the honour to enclose the return of killed and wounded, which, I am sorry to see, is so much heavier than I at first anticipated. Several of these were occasioned by accidental explosions of the enemy's tumbrils and magazines after the action. "G."

"FROM MAJOR-GENERAL SIR J. THACKWELL, K.C.B. AND K.H., COMMANDING CAVALRY DIVISION, TO LIEUTENANT-COLONEL GRANT, C.B., ADJUTANT-GENERAL OF THE ARMY.

"HEADQUARTERS CAMP, GOOJERAT, *February 25th*, 1849.

"SIR,

I have the honour to report, for the information of His Excellency the Right Honourable the Commander-in-Chief, the operations of the division of cavalry under my command, in the battle fought on the 21st instant, near the town of Goojerat.

"The left column of cavalry, under the command of Brigadier White, C.B., consisting of the troops named in the margin,[1] was assembled in column of troops at half distance, right in front, at deploying interval, on the left of the Bombay column of infantry, at 7 o'clock in the morning.

"The right column, composed of troops named in the margin,[2] under the command of Brigadiers Hearsey and Lockwood, C.B., were formed in column left in front at the same hour.

"The Third Division of Infantry and the Bombay brigade, all on the left of the nullah, leading towards Goojerat, being under my immediate superintendence, I remained on the left flank of the army; and I make no doubt Brigadier-Generals Dundas and Campbell have made you fully acquainted with the operations of the troops under their command.

"On approaching the village of Nurrawalla, just without the range of the enemy's batteries, the infantry deployed into line, and Brigadier White formed his cavalry in front of that village with his left back, and parallel to a gentle rising of the ground on which was posted the enemy's right, consisting of a large body of Afghans and Goorchurra Horse. From this position a fire of round shot was opened, and the enemy's cavalry extended to the right, so as to threaten to turn our left flank. To oppose the enemy's guns, I ordered Captain Duncan to move his troop of horse artillery to the front, which he did in good style, and opened his fire within five hundred or six hundred yards.

[1] Left Column. First Brigade of Cavalry, Brigadier White, C.B., commanding: H.M.'s 3rd Dragoons, Major Yerbury; H.M.'s 9th Lancers Lieutenant-Colonel Fullerton; 8th Light Cavalry, Major Mackenzie; Scinde Horse, Captain Malcolm; Captain Duncan's troop Horse Artillery; Captain Huish's troop Horse Artillery.

[2] Right Column. Second Brigade of Cavalry, Brigadier Lockwood, C.B., commanding: H.M.'s 14th Light Dragoons, Lieutenant-Colonel King; 1st Light Cavalry, Lieutenant-Colonel Bradford; 2 *risallas* 11th Irregular Cavalry, Captain Masters; 2 *risallas* 14th Irregular Cavalry, Lieutenant Robarts; Captain Warner's troop Horse Artillery.

Fourth Brigade Cavalry, Brigadier Hearsey commanding: 3rd Irregular Cavalry, Major Tait, C.B.; 9th Irregular Cavalry, Major Christie. The 5th and 6th Light Cavalry were left in the rear to protect the baggage.

This movement was followed by the advance of Captain Huish's troop, and both did considerable execution upon the enemy, but did not prevent the attempt of the Afghans to outflank our left. The Scinde Horse were on the left of my line, and I ordered them to advance with a squadron of the 9th Lancers, under Captain Campbell, a part of the former to be in reserve, and supported by a squadron of the 9th Lancers, under Major Grant, C.B., in échelon on the right. These troops made a most brilliant charge upon the enemy; at the same time I advanced the guns and cavalry towards the enemy's line. The fire of the guns soon put the Goorchurras in retreat, and the glorious charge of the troops on the left caused their whole force to seek safety in retreat by the Burradurree. A gun was captured during these proceedings; but as we were then considerably in advance of the left of the infantry —although Captain Duncan was enabled to enfilade a battery opposed to them—and ignorant of the force the enemy might have between the Burradurree and the town, a space covered with trees, it became necessary to proceed with caution; yet I soon was enabled to open a fire upon the enemy, both on the right and left of the former place which caused them considerable loss, and hastened their retreat.

"I may here observe that all the enemy's tents were left standing near the Burradurree, and on the Sikh right of the town, with probably much baggage in them, all of which were probably plundered by the camp followers.

"The enemy being now in full retreat, I moved Brigadier White's brigade well to the left front, and soon forced the enemy from the Jhelum road, and eventually from that of Bimber, also cutting off large bodies of the enemy, much baggage, and many guns, which were secured by this brigade, as well as the troops of the Second and Fourth Brigades, which had been ordered to join in the pursuit. At 4.20 p.m., none of the enemy being in sight, and being, as was said by the villagers, nine or ten miles from Goojerat, I discontinued the pursuit and returned to camp at this place. In this pursuit Captains Duncan and Huish's troops of artillery, latterly joined by Major Leeson and Major Blood's troop of the same arm, brought their guns to bear upon the enemy with good effect on several occasions, and their advance was as rapid as the intersected nature of the ground (by nullahs) would admit, and the 9th Lancers and 8th Light Cavalry made gallant attempts to close with the enemy's cavalry, which, however, were frustrated by the rapid retreat of the latter, yet a great number of the enemy were slain by this brigade in the pursuit. I witnessed the activity of Captain Unett, and part of his squadron of the 3rd Light Dragoons, and Brigadier White mentions that the whole of that regiment was actively engaged in this work of retribution.

"Being an eye-witness to all the movements of the First Brigade, I have great satisfaction in stating that Brigadier White conducted them very much to my satisfaction. I am also well satisfied with the manner in which Lieutenant-Colonel Fullerton, Majors Yerbury and Mackenzie commanded their respective regiments, and in which Major Grant supported the charge of cavalry on the left. The charge of the Scinde Horse reflects the highest credit on Captain Malcolm, and I have great pleasure in having witnessed the gallant bearing of all the officers and men of this brigade during the operations of the day; and I feel sure that their only regret was that the enemy's cavalry so often declined the attack.

"To Captains Duncan and Huish and Majors Leeson and Blood I am much indebted for the manner in which they brought their guns into action whenever an opportunity occurred, and the steadiness and good conduct of both officers and men were very conspicuous.

"I have now the pleasing duty to state, that I have received every assistance and support from my Deputy Assistant Adjutant-General, Captain Pratt, on the present occasion, as well as during the campaign. To my Deputy Assistant Quartermaster-General, Lieutenant Tucker, I am greatly indebted for his zeal, activity, intelligence, and successful endeavours to procure intelligence of the movements of the enemy during the operations. He, as well as my aide-de-camp, Lieutenant Thackwell, Lieutenant Young, of the Engineers, Lieutenant Carter, of the Pioneers, and Cornet Beatson, of the 6th Light Cavalry, accompanied me during the battle, and afforded me essential service in carrying my orders on various occasions during the operations of the day.

"Brigadier White states how greatly he was satisfied with the conduct of his Brigade Major, Captain Cautley, and the whole of the officers and men of his brigade.

"As the operations of the Second and Fourth Brigades of cavalry did not come under my observation, except towards the latter end of the pursuit, I have the honour to forward Brigadier Lockwood's report, and it would appear therefrom that he conducted his brigade judiciously; and I am gratified to learn that both officers and men behaved greatly to his satisfaction, and that the 14th Light Dragoons and 1st Light Cavalry conducted themselves gallantly, and evinced every anxiety to close with the enemy. I am happy to observe that the Brigadier has mentioned with great approbation the conduct of Lieutenant-Colonels Bradford and King, in command of their regiments; and I cannot avoid here stating, for the information of his Lordship, that I observed with much satisfaction the zeal and judgment evinced by both officers, when in command of considerable bodies of cavalry detached from the camp at Chillianwallah on important duties.

"I regret that I have not yet received any report from Brigadier Hearsey, or return of casualties from his brigade or the Scinde Horse; these will be forwarded when they arrive.

"I have, etc.

(*Signed*) "J. THACKWELL.

"*Major-General commanding Cavalry Division.*"

INDEX

www.ingramcontent.com/pod-product-compliance
Ingram Content Group UK Ltd.
Pitfield, Milton Keynes, MK11 3LW, UK
UKHW021837270726
14058UKWH00002B/204

9 781843 425489